International Series in Operations Research & Management Science

Volume 203

Series Editor

Frederick S. Hillier
Stanford University, CA, USA

Special Editorial Consultant

Camille C. Price
Stephen F. Austin State University, TX, USA

For further volumes:
http://www.springer.com/series/6161

Masatoshi Sakawa • Hitoshi Yano • Ichiro Nishizaki

Linear and Multiobjective Programming with Fuzzy Stochastic Extensions

Masatoshi Sakawa
Department of System Cybernetics
Graduate School of Engineering
Hiroshima University
Higashi-Hiroshima, Japan

Ichiro Nishizaki
Department of System Cybernetics
Graduate School of Engineering
Hiroshima University
Higashi-Hiroshima, Japan

Hitoshi Yano
Department of Social Sciences
Graduate School of Humanities
 and Social Sciences
Nagoya City University
Nagoya, Japan

ISSN 0884 8289 ISSN 2214-7934 (electronic)
ISBN 978-1-4614-9398-3 ISBN 978-1-4614-9399-0 (eBook)
DOI 10.1007/978-1-4614-9399-0
Springer New York Heidelberg Dordrecht London

Library of Congress Control Number: 2013953319

© Springer Science+Business Media New York 2013
This work is subject to copyright. All rights are reserved by the Publisher, whether the whole or part of the material is concerned, specifically the rights of translation, reprinting, reuse of illustrations, recitation, broadcasting, reproduction on microfilms or in any other physical way, and transmission or information storage and retrieval, electronic adaptation, computer software, or by similar or dissimilar methodology now known or hereafter developed. Exempted from this legal reservation are brief excerpts in connection with reviews or scholarly analysis or material supplied specifically for the purpose of being entered and executed on a computer system, for exclusive use by the purchaser of the work. Duplication of this publication or parts thereof is permitted only under the provisions of the Copyright Law of the Publisher's location, in its current version, and permission for use must always be obtained from Springer. Permissions for use may be obtained through RightsLink at the Copyright Clearance Center. Violations are liable to prosecution under the respective Copyright Law.
The use of general descriptive names, registered names, trademarks, service marks, etc. in this publication does not imply, even in the absence of a specific statement, that such names are exempt from the relevant protective laws and regulations and therefore free for general use.
While the advice and information in this book are believed to be true and accurate at the date of publication, neither the authors nor the editors nor the publisher can accept any legal responsibility for any errors or omissions that may be made. The publisher makes no warranty, express or implied, with respect to the material contained herein.

Printed on acid-free paper

Springer is part of Springer Science+Business Media (www.springer.com)

To our parents and families

Preface

Since G.B. Dantzig invented the celebrated simplex method around 1947, linear programming, an optimization method for maximizing or minimizing a linear objective function subject to linear constraints, attracted an immense amount of interest from both practitioners and academicians. Nowadays, with the significant advances in computer technology, linear programming, together with its extensions, has been widely used in the fields of operations research, industrial engineering, systems science, management science, and computer science.

From a probabilistic point of view, between 1955 and 1960, linear programming problems with random variable coefficients, called stochastic programming problems, were introduced. They are two-stage models and chance constrained programming. In two-stage models including simple recourse models, a shortage or an excess arising from the violation of the constraints is penalized, and then the expectation of the amount of the penalties for the constraint violation is minimized. Considering the stochastic constraints are not always satisfied, it is natural to permit constraint violations up to specified probability levels, which leads to the idea of chance constraints meaning that the constraints involving random variables need to be satisfied with a certain probability or over. For reflecting the diversity of criteria for optimizing the stochastic objective functions, optimization criteria different from expectation and variance are also provided to maximize the probability of the objective functions being smaller than or equal to target values as well as to minimize the target values under a given probability.

However, the consideration of several criteria in the actual decision-making process requires multiobjective approaches rather than that of a single objective. One of the major systematic approaches to multicriteria decision making under constraints is multiobjective programming as a generalization of traditional single-objective programming. For such multiobjective programming problems, it is significant to realize that multiple objectives are often incommensurable and conflict with each other. With this observation, in multiobjective programming problems, the notion of Pareto optimality or efficiency has been introduced instead of the optimality concept for single-objective problems. However, decisions with Pareto optimality or efficiency are not uniquely determined; the final decision must be selected by a decision maker, which well represents the subjective judgments, from the set of Pareto optimal or efficient solutions. For deriving a compromise or satisficing solution through interactions with the decision maker, interactive methods for multiobjective programming have been developed.

Recalling the imprecision or fuzziness inherent in human judgments, however, two types of inaccuracies of human judgments should be incorporated in multiobjective optimization problems. One is the fuzzy goal of the decision maker for each of the objective functions, and the other is the experts' ambiguous understanding of the nature of the parameters in the problem-formulation process. For handling and tackling such kinds of imprecision or vagueness in human judgments, it is not hard to imagine that the conventional multiobjective optimization approaches, such as a deterministic or even a probabilistic approach, cannot always be applied. The motivation for multiobjective optimization under imprecision or fuzziness comes from this observation.

In most practical situations, however, it is natural to consider that the uncertainty in real-world decision-making problems is often expressed by a fusion of fuzziness and randomness rather than either fuzziness or randomness. Through the use of stochastic models including the expectation model, the variance model, the probability model, the fractile model, and the simple recourse model together with chance constrained programming techniques, several multiobjective stochastic programming problems were formulated. By considering the imprecision of a decision maker's judgments for stochastic objective functions and/or constraints in multiobjective problems, fuzzy multiobjective stochastic programming problems were introduced and several interactive fuzzy satisficing methods to derive a satisficing solution for the decision maker have been developed.

Such five major topics, linear programming, multiobjective programming, fuzzy programming, stochastic programming, and fuzzy stochastic programming, are presented in a comprehensive manner in this book. In particular, the last four topics together comprise the main characteristics of this book, and special stress is placed on interactive decision-making aspects of multiobjective programming for human-centered systems in most realistic situations under fuzziness and/or randomness. Chapter 2 is a concise and condensed description of the theory of linear programming and its algorithms. Chapter 3 discusses fundamental notions and methods of multiobjective linear programming and concludes with interactive multiobjective linear programming. In Chap. 4, starting with clear explanations of fuzzy linear programming and fuzzy multiobjective linear programming, interactive fuzzy multiobjective linear programming is presented. Multiobjective linear programming problems involving fuzzy parameters are then formulated and linear programming-based interactive fuzzy programming is also discussed. Chapter 5 gives detailed explanations of fundamental notions and methods of stochastic programming including two-stage programming and chance constrained programming. As a natural extension of Chaps. 5 and 6 develops several interactive fuzzy programming approaches to multiobjective stochastic programming problems. Applications of linear programming, multiobjective programming, fuzzy programming, stochastic programming, and fuzzy stochastic programming to purchase and transportation planning for food retailing are considered in Chap. 7. Throughout this book, as well as comparing a number of solution methods including interactive ones, simple examples with two decision variables of production planning are provided to illustrate their main ideas. At the end of each chapter, an adequate number of problems can be found. Some of these problems test basic understanding and give routine exercise, while others concentrate on theoretical aspects. The readers could test and develop their understandings of the material. The book is self-contained because of the three appendices and solutions to problems. Appendix A contains a brief summary of the topics from linear algebra. Pertinent results from nonlinear programming are summarized in Appendix B. Appendix C is a clear explanation of the Excel solver, one of the easiest ways to solve optimization problems, through the use of simple examples of linear and nonlinear programming.

The intended readers of this book, which can be used both as a reference and as a textbook, are undergraduate students, graduate students, and researchers and practitioners in the fields of operations research, industrial engineering, systems science, management science, computer science, and other engineering disciplines that deal with the subjects of linear programming, multiobjective programming, fuzzy programming, stochastic programming, and fuzzy stochastic programming. The book can be used in several ways. Chapters 1, 2 and Appendix C with selected applications from Chap. 7 comprise the material for a basic undergraduate course on linear programming. As time permits, material from Chap. 3 can be included. It can be used in a one semester advanced undergraduate course on linear and multiobjective programming and then in a one semester graduate course emphasizing fuzzy and stochastic programming. The book can also be utilized in a one semester course on linear and multiobjective programming with fuzzy stochastic extensions by omitting some topics.

Higashi-Hiroshima, Japan	Masatoshi Sakawa
Nagoya, Japan	Hitoshi Yano
Higashi-Hiroshima, Japan	Ichiro Nishizaki

Contents

1 Introduction 1
  1.1 Linear Programming in Two Dimensions 1
  1.2 Extensions of Linear Programming................................. 3
  Problems 5

2 Linear Programming 7
  2.1 Algebraic Approach to Two-Dimensional Linear Programming..... 7
  2.2 Typical Examples of Linear Programming Problems................ 10
  2.3 Standard Form of Linear Programming 12
  2.4 Simplex Method 18
  2.5 Two-Phase Method 28
  2.6 Revised Simplex Method 39
  2.7 Duality 49
  2.8 Dual Simplex Method 55
  Problems 67

3 Multiobjective Linear Programming 73
  3.1 Problem Formulation and Solution Concepts 73
  3.2 Scalarization Methods... 77
    3.2.1 Weighting Method 78
    3.2.2 Constraint Method 81
    3.2.3 Weighted Minimax Method 83
  3.3 Linear Goal Programming 87
  3.4 Compromise Programming 92
  3.5 Interactive Multiobjective Linear Programming 96
  Problems 102

4 Fuzzy Linear Programming 105
  4.1 Fuzzy Sets and Fuzzy Decision.................................... 105
  4.2 Fuzzy Linear Programming 115
  4.3 Fuzzy Multiobjective Linear Programming 119

	4.4 Interactive Fuzzy Multiobjective Linear Programming	123
	4.5 Interactive Fuzzy Linear Programming with Fuzzy Parameters	131
	Problems	145
5	**Stochastic Linear Programming**	**149**
	5.1 Elementary Probability	149
	5.2 Two-Stage Programming	160
	5.3 Chance Constrained Programming	164
	5.3.1 Expectation Model	168
	5.3.2 Variance Model	171
	5.3.3 Probability Model	174
	5.3.4 Fractile Model	181
	Problems	191
6	**Interactive Fuzzy Multiobjective Stochastic Linear Programming**	**197**
	6.1 Multiobjective Chance Constrained Programming	197
	6.1.1 Expectation Model	199
	6.1.2 Variance Model	205
	6.1.3 Probability Model	210
	6.1.4 Fractile Model	217
	6.2 Multiobjective Simple Recourse Optimization	222
	Problems	229
7	**Purchase and Transportation Planning for Food Retailing**	**233**
	7.1 Linear Programming Formulation	233
	7.2 Multiobjective Linear Programming Formulation	244
	7.3 Fuzzy Multiobjective Linear Programming Formulation	251
	7.4 Fuzzy Multiobjective Linear Stochastic Programming Formulation	257
	Problems	269
A	**Linear Algebra**	**271**
	A.1 Vector	271
	A.2 Matrix	274
B	**Nonlinear Programming**	**281**
	B.1 Problem Formulation	281
	B.2 Basic Notions and Optimality Conditions	284
C	**Usage of Excel Solver**	**289**
	C.1 Setup for Solver	289
	C.2 Solving a Production Planning Problem	289
	C.3 Solving a Diet Problem	298
	C.4 Solving a Nonlinear Programming Problem	301

Solutions 305

References 323

Index 337

Chapter 1
Introduction

In this chapter, as an introductory numerical example, a simple production planning problem is considered. A production planning problem having two decision variables is formulated as a linear programming problem, and a graphical method for obtaining an optimal solution is illustrated. Moreover, by considering environmental quality, a two-objective linear programming problem is formulated, and the notion of Pareto optimality is outlined.

1.1 Linear Programming in Two Dimensions

First, consider the following simple production planning problem as an example of a problem that can be solved using linear programming.

Example 1.1 (Production planning problem). A manufacturing company desires to maximize the total profit from producing two products P_1 and P_2 utilizing three different materials M_1, M_2, and M_3. The company knows that to produce 1 ton of product P_1 requires 2 tons of material M_1, 3 tons of material M_2, and 4 tons of material M_3, while to produce 1 ton of product P_2 requires 6 tons of material M_1, 2 tons of material M_2, and 1 ton of material M_3. The total amounts of available materials are limited to 27, 16, and 18 tons for M_1, M_2, and M_3, respectively. It also knows that product P_1 yields a profit of 3 million yen per ton, while P_2 yields 8 million yen (see Table 1.1). Given these limited materials, the company is trying to figure out how many units of products P_1 and P_2 should be produced to maximize the total profit. ◊

Let x_1 and x_2 denote decision variables representing the numbers of tons produced of products P_1 and P_2, respectively. Using these decision variables, this production planning problem can be formulated as the following linear programming problem:

Table 1.1 Production conditions and profit

	Product P_1	Product P_2	Amounts available
Material M_1 (ton)	2	6	27
Material M_2 (ton)	3	2	16
Material M_3 (ton)	4	1	18
Profit (million yen)	3	8	

Maximize the linear profit function

$$3x_1 + 8x_2$$

subject to the linear constraints

$$2x_1 + 6x_2 \leq 27$$
$$3x_1 + 2x_2 \leq 16$$
$$4x_1 + x_2 \leq 18$$

and nonnegativity conditions for these decision variables

$$x_1 \geq 0,\ x_2 \geq 0.$$

For convenience in our subsequent discussion, let the opposite of the total profit be

$$z = -3x_1 - 8x_2,$$

and convert the profit maximization problem to the problem to minimize z under the above constraints, i.e.,

$$\begin{aligned}
\text{minimize } z &= -3x_1 - 8x_2 \\
\text{subject to} \quad & 2x_1 + 6x_2 \leq 27 \\
& 3x_1 + 2x_2 \leq 16 \\
& 4x_1 + x_2 \leq 18 \\
& x_1 \geq 0,\quad x_2 \geq 0.
\end{aligned}$$

It is easy to see that, in the x_1-x_2 plane, the linearly constrained set of points (x_1, x_2) satisfying the above constraints is the boundary lines and interior points of the convex pentagon $ABCDE$ shown in Fig. 1.1.

The set of points satisfying $-x_1 - 2x_2 = z$ for a fixed value of z is a line in the x_1-x_2 plane. As z is varied, the line is moved parallel to itself. The optimal value of this problem is the smallest value of z for which the corresponding line has at least one point in common with the linearly constrained set $ABCDE$. As can been seen from Fig. 1.1, this occurs at point D. Hence, the optimal solution to this problem is

$$x_1 = 3,\quad x_2 = 3.5,\quad z = -37.$$

1.2 Extensions of Linear Programming

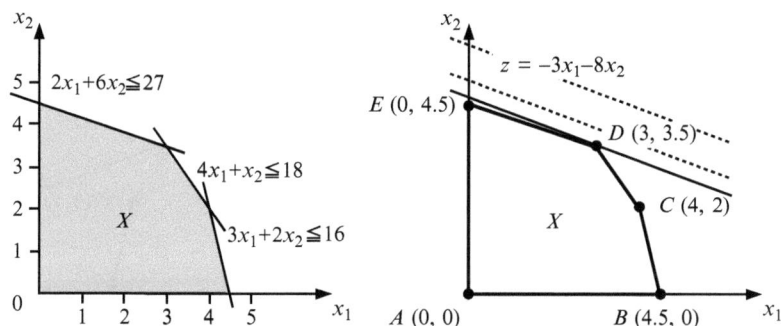

Fig. 1.1 Feasible region and optimal solution for production planning problem

It is significant to realize here that the optimal solution is located at a vertex of the linearly constrained set since the constrained set has a finite number of vertices and the contours of constant value of the objective function are linear. Note that vertices are usually called extreme points in linear programming.

The values of the objective function corresponding to the extreme points A, B, C, D, and E are 0, -5, -7.5, -10, and -9, respectively. Therefore, for example, starting from the extreme point A, if we move $A \to B \to C \to D$ or $A \to E \to D$ such that the values of the objective function are decreasing, it seems to be possible to reach the extreme point which gives the minimum value of z.

Obviously, for more than two or three variables, such a graphical method cannot be applied and it becomes necessary to characterize extreme points algebraically. The simplex method for linear programming originated by Dantzig (1963) is well known and widely used as a powerful computational procedure for solving linear programming problems. The simplex method consists of two phases. Phase I finds an initial extreme point of the feasible region or gives the information that none exists due to the inconsistency of the constraints. In Phase II, starting from an initial extreme point, it determines whether it is optimal or not. If not, it finds an adjacent extreme point at which the value of z is less than or equal to the previous value. The process is repeated until it finds an optimal solution or gives the information that the optimal value is unbounded. The details of linear programming can be found in standard texts including Dantzig (1963), Dantzig and Thapa (1997), Gass (1958), Hadley (1962), Hillier and Lieberman (1990), Ingnizio and Cavalier (1994), Luenberger (1973, 1984, 2008), Nering and Tucker (1993), and Thie (1988).

1.2 Extensions of Linear Programming

Recall the production planning problem discussed in Example 1.1.

Example 1.2 (Production planning with environmental considerations). Unfortunately, however, in the production process, it is pointed that out producing 1 ton of

Fig. 1.2 Feasible region and solutions maximizing the total profit and minimizing the pollution

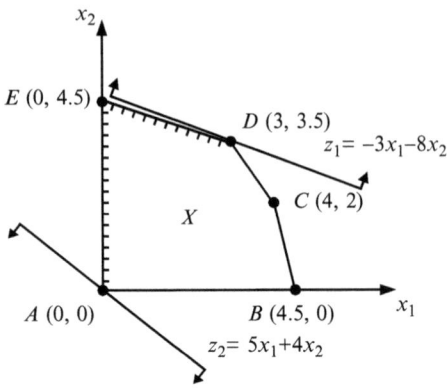

P1 and P2 yields 5 and 4 units of pollution, respectively. Thus, the manager should not only maximize the total profit but also minimize the amount of pollution.

For simplicity, assume that the amount of pollution is a linear function of two decision variables x_1 and x_2 such as

$$5x_1 + 4x_2$$

where x_1 and x_2 denote the numbers of tons produced of products P_1 and P_2, respectively.

Considering environmental quality, the production planning problem can be reformulated as the following two-objective linear programming problem:

$$\begin{aligned}
\text{minimize } z_1 &= -3x_1 - 8x_2 \\
\text{minimize } z_2 &= 5x_1 + 4x_2 \\
\text{subject to } \quad & 2x_1 + 6x_2 \leq 27 \\
& 3x_1 + 2x_2 \leq 16 \\
& 4x_1 + x_2 \leq 18 \\
& x_1 \geq 0, \quad x_2 \geq 0.
\end{aligned}$$

◊

Before discussing the solution concepts of multiobjective linear programming problems, it is instructive to consider the geometric interpretation of the two-objective linear programming problem in Example 1.2. The feasible region X for this problem in the x_1-x_2 plane is composed of the boundary lines and interior points of the convex pentagon $ABCDE$ in Fig. 1.2. Among the five extreme points A, B, C, D, and E, observe that z_1 is minimized at the extreme point $D(3, 3.5)$ while z_2 is minimized at the extreme point $A(0, 0)$.

As will be discussed in Chap. 3, these two extreme points A and D are obviously Pareto optimal solutions since they cannot improve respective objective functions z_1 and z_2 anymore. In addition to the extreme points A and D, the extreme point E and all of the points of the segments AE and ED are Pareto optimal solutions since they can be improved only at the expense of either z_1 or z_2. However, all of the remaining feasible points are not Pareto optimal since there always exist other feasible points which improve at least one of the objective functions.

In another perspective, recalling the imprecision or fuzziness inherent in human judgments, two types of inaccuracies of human judgments should be incorporated in multiobjective optimization problems. One is the fuzzy goals of the decision maker for each of the objective functions, and the other is the experts' ambiguous understanding of the nature of the parameters in the problem-formulation process. The motivation for multiobjective optimization under imprecision or fuzziness comes from this observation. In Chap. 4, we will deal with fuzzy linear programming and fuzzy multiobjective linear programming. Multiobjective linear programming problems involving fuzzy parameters will be also discussed.

From an uncertain viewpoint different with fuzziness, linear programming problems with random variable coefficients, called stochastic programming problems, are developed. They are two-stage models and constrained programming. In two-stage models, a shortage or an excess arising from the violation of the constraints is penalized, and then the expectation of the amount of the penalties for the constraint violation is minimized. In a model of the constrained programming, from the observation that the stochastic constraints are not always satisfied, the problem is formulated so as to permit constraint violations up to specified probability levels. Chapter 5 will discuss such stochastic programming techniques.

Problems

1.1 A manufacturing company desires to maximize the total profit from producing two products P_1 and P_2 utilizing three different materials M_1, M_2, and M_3. The company knows that to produce 1 ton of product P_1 requires 2 tons of material M_1, 8 tons of material M_2, and 3 tons of material M_3, while to produce 1 ton of product P_2 requires 6 tons of material M_1, 6 tons of material M_2, and 1 ton of material M_3. The total amounts of available materials are limited to 27, 45, and 15 tons for M_1, M_2, and M_3, respectively. It also knows that product P_1 yields a profit of 2 million yen per ton, while P_2 yields 5 million yen. Given these limited materials, the company is trying to figure out how many units of products P_1 and P_2 should be produced to maximize the total profit.

(1) Let x_1 and x_2 denote decision variables for the numbers of tons produced of products P_1 and P_2, respectively. Formulate the problem as a linear programming problem.
(2) Graph the problem in the $x_1 - x_2$ plane, find an optimal solution.

1.2 (Transportation problem)
Consider a planning problem of transporting goods from m warehouses to n retail stores. Assume that a quantity a_i is available at warehouse i, a quantity

b_j is required at store j, and the cost of transportation of one unit of goods from warehouses i to store j is c_{ij}. Also it is assumed that the total amount available is equal to the total required, i.e., $\sum a_i = \sum b_j$. Let x_{ij} be a decision variable for the amount shipped from warehouse i to store j. Formulate a linear programming problem so as to satisfy the shipping requirements and minimize the total transportation cost.

1.3 (Assignment problem)

Suppose that each of n candidates to be assigned to one of n jobs and the number c_{ij} which measures the effectiveness of candidate i in job j is known. Introducing n^2 decision variables x_{ij}, $i = 1,\ldots,n$; $j = 1,\ldots,n$ with the interpretation that $x_{ij} = 1$ if candidate i is assigned to job j, and $x_{ij} = 0$ otherwise, formulate a linear programming problem so as to maximize the overall effectiveness.

Chapter 2
Linear Programming

Since G.B. Dantzig first proposed the simplex method around 1947, linear programming, as an optimization method of maximizing or minimizing a linear objective function subject to linear constraints, has been extensively studied and, with the significant advances in computer technology, widely used in the fields of operations research, industrial engineering, systems science, management science, and computer science.

In this chapter, after an overview of the basic concepts of linear programming via a simple numerical example, the standard form of linear programming and fundamental concepts and definitions are introduced. The simplex method and the two-phase method are presented with the details of the computational procedures. By reviewing the procedure of the simplex method, the revised simplex method, which provides a computationally efficient implementation, is also discussed. Associated with linear programming problems, dual problems are formulated, and duality theory is discussed which also leads to the dual simplex method.

2.1 Algebraic Approach to Two-Dimensional Linear Programming

In Sect. 1.1, we have presented a graphical method for solving the two-dimensional production planning problem of Example 1.1.

Minimize the opposite of the linear total profit

$$z = -3x_1 - 8x_2$$

subject to the linear inequality constraints

$$2x_1 + 6x_2 \leq 27$$
$$3x_1 + 2x_2 \leq 16$$
$$4x_1 + x_2 \leq 18$$

and nonnegativity conditions for all decision variables

$$x_1 \geq 0, \ x_2 \geq 0.$$

Since in multiple dimensions more than two the graphical method used in Sect. 1.1 cannot be applied, it becomes necessary to develop an algebraic method. In this section, as a prelude to the development of the general theory, consider an algebraic approach to two-dimensional linear programming problems for understanding the basic ideas of linear programming. To do so, by introducing the amounts, x_3 (≥ 0), x_4 (≥ 0), and x_5 (≥ 0), of unused (idle) materials for M_1, M_2, and M_3, respectively, and converting the inequalities into the equalities, the problem with the equation $-3x_1 - 8x_2 - z = 0$ for the objective function can then be stated as follows:

Find values of $x_j \geq 0$, $j = 1, 2, 3, 4, 5$ so as to minimize z, satisfying the augmented system of linear equations

$$\left.\begin{array}{rl} 2x_1 + 6x_2 + x_3 & = 27 \\ 3x_1 + 2x_2 + x_4 & = 16 \\ 4x_1 + x_2 + x_5 & = 18 \\ -3x_1 - 8x_2 -z & = 0. \end{array}\right\} \quad (2.1)$$

In (2.1), setting $x_1 = x_2 = 0$ yields $x_3 = 27$, $x_4 = 16$, $x_5 = 18$, and $z = 0$, which corresponds to the extreme point A in Fig. 1.1. Now, from the fourth equation of (2.1) for the objective function, we see that any increase in the values of x_1 and x_2 from 0 to positive would decrease the value of the objective function z. Considering that the profit of P_2 is larger than that of P_1 (in the above formulation, the opposite of the profit is smaller), choose to increase x_2 from 0 to a positive value, while keeping $x_1 = 0$. In Fig. 1.1, this corresponds to the movement from the extreme point A to E along the edge AE. From (2.1), if x_2 can be made positive, the values of x_3, x_4, and x_5 decrease. However, since x_3, x_4, and x_5 cannot become negative, the increase amount of x_2 is restricted by the first three equations of (2.1). In the first three equations of (2.1), remaining $x_1 = 0$, the values of x_2 to be increased are restricted to at most $27/6 = 4.5$, $16/2 = 8$, and $18/1 = 18$, respectively. Hence, the largest permissible value of x_2 not yielding the negative values of x_3, x_4, and x_5 is the smallest of 4.5, 8, and 18, that is, 4.5. Increasing the values of x_2 from 0 to 4.5 yields $x_3 = 0$, which implies that the available amount of material M_1 is used up.

Dividing the first equation of (2.1) by the coefficient 6 of x_2 and eliminating x_2 from the second, third, and fourth equations yields

2.1 Algebraic Approach to Two-Dimensional Linear Programming

$$\left.\begin{aligned}
\frac{1}{3}x_1 + x_2 + \frac{1}{6}x_3 & & & = 4.5 \\
\frac{7}{3}x_1 & - \frac{1}{3}x_3 + x_4 & & = 7 \\
\frac{11}{3}x_1 & - \frac{1}{6}x_3 & + x_5 & = 13.5 \\
-\frac{1}{3}x_1 & + \frac{4}{3}x_3 & & -z = 36.
\end{aligned}\right\} \quad (2.2)$$

In (2.2), setting $x_1 = x_3 = 0$ yields $x_2 = 4.5$, $x_4 = 7$, $x_5 = 13.5$, and $z = -36$. This implies the resulting point $(x_1, x_2) = (0, 4.5)$ corresponds to the extreme point E and the value of the objective function z is decreased from 0 to -36.

Next, from the fourth equation of (2.2), keeping $x_3 = 0$, by increasing the value of x_1 from 0 to positive, the value of z can be decreased. This corresponds to the movement from the extreme point E to D along the edge ED in Fig. 1.1. From the first three equations of (2.2), to keep the values of x_2, x_4, and x_5 nonnegative, the values of x_1 to be increased are restricted to at most $4.5/(1/3) = 13.5$, $7/(7/3) = 3$, and $13.5/(11/3) \simeq 3.682$, respectively. Hence, increasing the values of x_2 from 0 to 3, the smallest among them, yields $x_4 = 0$, which implies that the available amount of material M_2 is used up.

Dividing the second equation of (2.2) by the coefficient 7/3 of x_1 and eliminating x_1 from the first, third, and fourth equations yields

$$\left.\begin{aligned}
x_2 + \frac{3}{14}x_3 & - \frac{1}{7}x_4 & & = 3.5 \\
x_1 & - \frac{1}{7}x_3 + \frac{3}{7}x_4 & & = 3 \\
& \frac{5}{14}x_3 - \frac{11}{7}x_4 + x_5 & & = 2.5 \\
& \frac{9}{7}x_3 + \frac{1}{7}x_4 & -z & = 37.
\end{aligned}\right\} \quad (2.3)$$

In (2.3), setting $x_3 = x_4 = 0$ yields $x_1 = 3$, $x_2 = 3.5$, $x_5 = 2.5$, and $z = -37$, which corresponds to the extreme point D in Fig. 1.1, and the value of z is decreased from -36 to -37.

From the fourth equation of (2.3), both coefficients of x_3 and x_4 are positive. This means that increasing the value of x_3 or x_4 increases the value of z. Therefore, the minimum of z is -37, that is, the maximum of the total profit is 37 million yen, and the production numbers of products P_1 and P_2 are 3 and 3.5 tons, respectively.

2.2 Typical Examples of Linear Programming Problems

Thus far, we have outlined linear programming through the two-dimensional production planning problem, which can be generalized as the following production planning problem with n decision variables.

Example 2.1 (Production planning problem). A manufacturing company has fixed amounts of m different resources at its disposal. These resources are used to produce n different commodities. The company knows that to produce one unit of commodity j, a_{ij} units of resource i are required. The total number of units of resource i available is b_i. It also knows that a profit per unit of commodity j is c_j. It desires to produce a combination of commodities which will maximize the total profit.

Let x_j denote a decision variable for the production amount of commodity j. Since the amount of resource i that is used must be less than or equal to the available number b_i of units of resource i, we have, for each $i = 1, 2, \ldots, m$, a linear inequality

$$a_{i1}x_1 + a_{i2}x_2 + \cdots + a_{in}x_n \leq b_i.$$

As a negative x_j has no appropriate interpretation, it is required that $x_j \geq 0$, $j = 1, 2, \ldots, n$. The profit arising from producing x_j units of commodity j is calculated as $c_j x_j$. Our formulation is represented as a linear programming problem where the linear profit function

$$c_1 x_1 + c_2 x_2 + \cdots + c_n x_n \qquad (2.4)$$

is maximized subject to the linear inequality constraints

$$\left.\begin{array}{r} a_{11}x_1 + a_{12}x_2 + \cdots + a_{1n}x_n \leq b_1 \\ a_{21}x_1 + a_{22}x_2 + \cdots + a_{2n}x_n \leq b_2 \\ \cdots\cdots\cdots\cdots\cdots\cdots\cdots \\ a_{m1}x_1 + a_{m2}x_2 + \cdots + a_{mn}x_n \leq b_m \end{array}\right\} \qquad (2.5)$$

and nonnegativity conditions for all decision variables

$$x_j \geq 0, \quad j = 1, 2, \ldots, n. \qquad (2.6)$$

$\diamond$

Compared with such a production planning problem maximizing the linear objective function of the total profit subject to the linear inequality constraints in a direction of the less than or equal to symbol $\leq$, the following diet problem minimizing the linear objective function of the total cost subject to the linear inequality constraints in a direction of the greater than or equal to symbol $\geq$ is well

2.2 Typical Examples of Linear Programming Problems

known as a nearly symmetric one. It should be noted here that both of the problems have the nonnegativity conditions for all decision variables $x_j \geq 0$, $j = 1, \ldots, n$.

Example 2.2 (Diet problem). How can we determine the most economical diet that satisfies the basic minimum nutritional requirements for good health? Assume n different foods are available at the market and the selling price for food j is c_j per unit. Moreover, there are m basic nutritional ingredients for the human body, and at least b_i units of nutrient i are required everyday to achieve a balanced diet for good health. In addition, assume that each unit of food j contains a_{ij} units of nutrient i. The problem is to determine the most economical diet that satisfies the basic minimum nutritional requirements.

For this problem, let x_j, $j = 1, \ldots, n$ denote a decision variable for the number of units of food j in the diet, and then it is required that $x_j \geq 0$, $j = 1, \ldots, n$. The total amount of nutrient i

$$a_{i1}x_1 + a_{i2}x_2 + \cdots + a_{in}x_n$$

contained in the purchased foods must be greater than or equal to the daily requirement b_i of nutrient i. Thus, the economic diet can be represented as a linear programming problem where the linear cost function

$$c_1 x_1 + c_2 x_2 + \cdots + c_n x_n \tag{2.7}$$

is minimized subject to the linear constraints

$$\left. \begin{array}{l} a_{11}x_1 + a_{12}x_2 + \cdots + a_{1n}x_n \geq b_1 \\ a_{21}x_1 + a_{22}x_2 + \cdots + a_{2n}x_n \geq b_2 \\ \cdots\cdots\cdots\cdots\cdots\cdots\cdots\cdots\cdots \\ a_{m1}x_1 + a_{m2}x_2 + \cdots + a_{mn}x_n \geq b_m \end{array} \right\} \tag{2.8}$$

and nonnegativity conditions for all decision variables

$$x_j \geq 0, \quad j = 1, 2, \ldots, n. \tag{2.9}$$

$\Diamond$

To develop a better understanding, as a simple numerical example of the diet problem, we present the following diet problem with two decision variables and three constraints.

Example 2.3 (Diet problem with 2 decision variables and 3 constraints). A housewife is planning a menu by utilizing two foods F_1 and F_2 containing three nutrients N_1, N_2, and N_3 in order to meet the nutritional requirements at a minimum cost.

Each 1 g (gram) of the food F_1 contains 1 mg (milligram) of N_1, 1 mg of N_2, and 2 mg of N_3; and each 1 g of the food F_2 contains 3 mg of N_1, 2 mg of N_2, and 1 mg of N_3. The recommended amounts of the nutrients N_1, N_2, and N_3 are known

Table 2.1 Data for two foods diet problem

	Food F_1 (g)	Food F_2 (g)	Minimum requirement
Nutrient N_1 (mg)	1	3	12
Nutrient N_2 (mg)	1	2	10
Nutrient N_3 (mg)	2	1	15
Price (thousand yen)	4	3	

to be at least 12 mg, 10 mg, and 15 mg, respectively. Also, it is known that the costs per gram of the foods F_1 and F_2 are, respectively, 4 and 3 thousand yen. These data concerning the nutrients and foods are summarized in Table 2.1.

The housewife's problem is to determine the purchase volumes of foods F_1 and F_2 which minimize the total cost satisfying the nutritional requirements for the nutrients N_1, N_2, and N_3.

Let x_j denote a decision variable for the number of units of food F_j to be purchased, and then we can formulate the corresponding linear programming problem minimizing the linear cost function

$$4x_1 + 3x_2 \qquad (2.10)$$

subject to the linear constraints

$$\left. \begin{array}{l} x_1 + 3x_2 \geq 12 \\ x_1 + 2x_2 \geq 10 \\ 2x_1 + x_2 \geq 15 \end{array} \right\} \qquad (2.11)$$

and nonnegativity conditions for all variables

$$x_1 \geq 0, \ x_2 \geq 0. \qquad (2.12)$$

2.3 Standard Form of Linear Programming

In order to deal with such nearly symmetrical production planning problems and diet problems in a unified way, the standard form of linear programming is defined as follows:

The standard form of linear programming is to minimize the linear objective function

$$z = c_1 x_1 + c_2 x_2 + \cdots + c_n x_n \qquad (2.13)$$

2.3 Standard Form of Linear Programming

subject to the linear equality constraints

$$\left.\begin{array}{c} a_{11}x_1 + a_{12}x_2 + \cdots + a_{1n}x_n = b_1 \\ a_{21}x_1 + a_{22}x_2 + \cdots + a_{2n}x_n = b_2 \\ \cdots\cdots\cdots\cdots\cdots\cdots \\ a_{m1}x_1 + a_{m2}x_2 + \cdots + a_{mn}x_n = b_m \end{array}\right\} \quad (2.14)$$

and nonnegativity conditions for all decision variables

$$x_j \geq 0, \quad j = 1, 2, \ldots, n, \qquad (2.15)$$

where the a_{ij}, b_i, and c_j are fixed real constants. In particular, b_i is called a right-hand side constant, and c_j is sometimes called a cost coefficient in a minimization problem, while called a profit coefficient in a maximization one.

In this book, the standard form of linear programming is written in the following form:

$$\left.\begin{array}{ll} \text{minimize} & z = c_1 x_1 + c_2 x_2 + \cdots + c_n x_n \\ \text{subject to} & a_{11}x_1 + a_{12}x_2 + \cdots + a_{1n}x_n = b_1 \\ & a_{21}x_1 + a_{22}x_2 + \cdots + a_{2n}x_n = b_2 \\ & \cdots\cdots\cdots\cdots\cdots\cdots \\ & a_{m1}x_1 + a_{m2}x_2 + \cdots + a_{mn}x_n = b_m \\ & x_j \geq 0, \quad j = 1, 2, \ldots, n, \end{array}\right\} \quad (2.16)$$

or using summation notation, it is compactly rewritten as

$$\left.\begin{array}{ll} \text{minimize } z = \sum_{j=1}^{n} c_j x_j \\ \text{subject to} \quad \sum_{j=1}^{n} a_{ij} x_j = b_i, \quad i = 1, \ldots, m \\ \quad x_j \geq 0, \quad j = 1, \ldots, n. \end{array}\right\} \quad (2.17)$$

By introducing an n dimensional row vector c, an $m \times n$ matrix A, an n dimensional column vector x, and an m dimensional column vector b, the standard form of linear programming can be then written in a more compact vector–matrix form as follows:

$$\left.\begin{array}{ll} \text{minimize } z = cx \\ \text{subject to} \quad Ax = b \\ \quad x \geq 0, \end{array}\right\} \quad (2.18)$$

where
$$c = (c_1, c_2, \ldots, c_n), \tag{2.19}$$

$$A = \begin{bmatrix} a_{11} & a_{12} & \cdots & a_{1n} \\ a_{21} & a_{22} & \cdots & a_{2n} \\ \vdots & \vdots & \ddots & \vdots \\ a_{m1} & a_{m2} & \cdots & a_{mn} \end{bmatrix}, \quad x = \begin{pmatrix} x_1 \\ x_2 \\ \vdots \\ x_n \end{pmatrix}, \quad b = \begin{pmatrix} b_1 \\ b_2 \\ \vdots \\ b_m \end{pmatrix}, \tag{2.20}$$

and $\mathbf{0}$ is an n dimensional column vector with zero components.

Moreover, by denoting the jth column of an $m \times n$ matrix A by

$$p_j = \begin{pmatrix} a_{1j} \\ a_{2j} \\ \vdots \\ a_{mj} \end{pmatrix}, \quad j = 1, 2, \ldots, n \tag{2.21}$$

and writing $A = [p_1 \ p_2 \ \cdots \ p_n]$, the standard form linear programming (2.16) can also be represented in column form:

$$\left. \begin{array}{rl} \text{minimize} & z = c_1 x_1 + c_2 x_2 + \cdots + c_n x_n \\ \text{subject to} & p_1 x_1 + p_2 x_2 + \cdots + p_n x_n - b \\ & x_j \geq 0, \quad j = 1, 2, \ldots, n. \end{array} \right\} \tag{2.22}$$

In the standard form of linear programming (2.16), the objective function

$$z = c_1 x_1 + c_2 x_2 + \cdots + c_n x_n$$

can be treated as just another equation, i.e.,

$$-z + c_1 x_1 + c_2 x_2 + \cdots + c_n x_n = 0, \tag{2.23}$$

and by including it in an augmented system of equations, the problem can then be stated as follows:

Find values of the nonnegative decision variables $x_1 \geq 0, x_2 \geq 0, \ldots, x_n \geq 0$ so as to minimize z, satisfying the augmented system of linear equations

$$\left. \begin{array}{r} a_{11} x_1 + a_{12} x_2 + \cdots + a_{1n} x_n = b_1 \\ a_{21} x_1 + a_{22} x_2 + \cdots + a_{2n} x_n = b_2 \\ \cdots\cdots\cdots\cdots\cdots\cdots\cdots \\ a_{m1} x_1 + a_{m2} x_2 + \cdots + a_{mn} x_n = b_m \\ -z + c_1 x_1 + c_2 x_2 + \cdots + c_n x_n = 0. \end{array} \right\} \tag{2.24}$$

It should be noted here that the standard form of linear programming deals with a linear minimization problem with nonnegative decision variables and linear equality

2.3 Standard Form of Linear Programming

constraints. We introduce a mechanism to convert any general linear programming problem into the standard form. A linear inequality can be easily converted into an equality. When the ith constraint is represented as

$$\sum_{j=1}^{n} a_{ij} x_j \leq b_i, \quad i = 1, 2, \ldots, m, \tag{2.25}$$

by adding a nonnegative slack variable $x_{n+i} \geq 0$ such that

$$\sum_{j=1}^{n} a_{ij} x_j + x_{n+i} = b_i, \quad i = 1, 2, \ldots, m, \tag{2.26}$$

the inequality (2.25) becomes the equality (2.26).

Similarly, if the ith constraint is

$$\sum_{j=1}^{n} a_{ij} x_j \geq b_i, \quad i = 1, 2, \ldots, m, \tag{2.27}$$

by subtracting a nonnegative surplus variable $x_{n+i} \geq 0$ such that

$$\sum_{j=1}^{n} a_{ij} x_j - x_{n+i} = b_i, \quad i = 1, 2, \ldots, m, \tag{2.28}$$

we can also transform the inequality (2.27) into the equality (2.28). It should be noted here that both the slack variables and the surplus variables must be nonnegative in order that the inequalities (2.25) and (2.27) are satisfied for all $i = 1, 2, \ldots, m$.

If, in the original formulation of the problem, some decision variable x_k is not restricted to be nonnegative, it can be replaced with the difference of two nonnegative variables, i.e.,

$$x_k = x_k^+ - x_k^-, \quad x_k^+ \geq 0, \, x_k^- \geq 0. \tag{2.29}$$

If an objective function is to be maximized, we simply multiply the objective function by -1 to convert a maximization problem into a minimization problem.

Recall that, in the algebraic method for the production planning problem of Example 1.1, multiplying the objective function by -1 and introducing the three nonnegative slack variables x_3, x_4, and x_5 yields the following standard form of linear programming:

$$\left. \begin{aligned} \text{minimize } z &= -3x_1 - 8x_2 \\ \text{subject to} \quad & 2x_1 + 6x_2 + x_3 = 27 \\ & 3x_1 + 2x_2 + x_4 = 16 \\ & 4x_1 + x_2 + x_5 = 18 \\ & x_j \geq 0, \quad j = 1, 2, 3, 4, 5. \end{aligned} \right\} \tag{2.30}$$

For the general production planning problem with n decision variables, by introducing the m nonnegative slack variables x_{n+i} (≥ 0), $i = 1, \ldots, m$, it can be converted into the following standard form of linear programming:

$$\left.\begin{aligned}
\text{minimize} \quad & c_1 x_1 + c_2 x_2 + \cdots + c_n x_n \\
\text{subject to} \quad & a_{11} x_1 + a_{12} x_2 + \cdots + a_{1n} x_n + x_{n+1} = b_1 \\
& a_{21} x_1 + a_{22} x_2 + \cdots + a_{2n} x_n \quad\quad + x_{n+2} = b_2 \\
& \quad\quad \cdots\cdots\cdots\cdots\cdots \\
& a_{m1} x_1 + a_{m2} x_2 + \cdots + a_{mn} x_n \quad\quad\quad + x_{n+m} = b_m \\
& x_j \geq 0, \quad j = 1, 2, \ldots, n, n+1, \ldots, n+m.
\end{aligned}\right\} \quad (2.31)$$

Similarly, for the diet problem with n decision variables, introducing the m nonnegative surplus variables x_{n+i} (≥ 0), $i = 1, \ldots, m$ yields the following standard form of linear programming:

$$\left.\begin{aligned}
\text{minimize} \quad & c_1 x_1 + c_2 x_2 + \cdots + c_n x_n \\
\text{subject to} \quad & a_{11} x_1 + a_{12} x_2 + \cdots + a_{1n} x_n - x_{n+1} = b_1 \\
& a_{21} x_1 + a_{22} x_2 + \cdots + a_{2n} x_n \quad\quad - x_{n+2} = b_2 \\
& \quad\quad \cdots\cdots\cdots\cdots\cdots \\
& a_{m1} x_1 + a_{m2} x_2 + \cdots + a_{mn} x_n \quad\quad\quad - x_{n+m} = b_m \\
& x_j \geq 0, \quad j = 1, 2, \ldots, n, n+1, \ldots, n+m.
\end{aligned}\right\} \quad (2.32)$$

The basic ideas of linear programming are to first detect whether solutions satisfying equality constraints and nonnegativity conditions exist and, if so, to find a solution yielding the minimum value of z.

However, in the standard form of linear programming (2.16) or (2.18), if there is no solution satisfying the equality constraint, or if there exists only one, we do not need optimization. Also, if any of the equality constraints is redundant, i.e., a linear combination of the others, it could be deleted without changing any solutions of the system. Therefore, we are mostly interested in the case where the system of linear equations (2.16) is nonredundant and has an infinite number of solutions.

For that purpose, assume that the number of variables exceeds the number of equality constraints, i.e.,

$$n > m \quad\quad (2.33)$$

and the system of linear equations is linearly independent, i.e.,

$$\text{rank}(A) = m. \quad\quad (2.34)$$

Under these assumptions, we introduce a number of definitions for the standard form of linear programming (2.16) or (2.18).[1]

---

[1] These assumptions, introduced to establish the principle theoretical results, will be relaxed in Sect. 2.5 and are no longer necessary when solving general linear programming problems.

2.3 Standard Form of Linear Programming

Definition 2.1 (Feasible solution). A feasible solution to the linear programming problem (2.16) is a vector $x = (x_1, x_2, \ldots, x_n)^T$ which satisfies the linear equalities and the nonnegativity conditions of (2.16).[2]

Definition 2.2 (Basis matrix). A basis matrix is an $m \times m$ nonsingular submatrix formed by choosing some m columns of the rectangular matrix A. Observe that A contains at least one basis matrix due to rank$(A) = m$.

Definition 2.3 (Basic solution). A basic solution to the linear programming problem (2.16) is a solution obtained by setting $n - m$ variables (called nonbasic variables) equal to zeros and solving for the remaining m variables (called basic variables). A basic solution is also a unique vector determined by choosing a basis matrix from the $m \times n$ matrix A and solving the resulting square, nonsingular system of equations for the m variables. The set of all basic variables is called the basis.

Definition 2.4 (Basic feasible solution). A basic feasible solution to the linear programming problem (2.16) is a basic solution which satisfies not only the linear equations but also the nonnegativity conditions of (2.16), that is, all basic variables are nonnegative. Observe that at most m variables can be positive by Definition 2.3.

Definition 2.5 (Nondegenerate basic feasible solution). A nondegenerate basic feasible solution to the linear programming problem (2.16) is a basic solution with exactly m positive x_j, that is, all basic variables are positive.

Definition 2.6 (Optimal solution). An optimal solution to the linear programming problem (2.16) is a feasible solution which also minimizes z in (2.16). The corresponding value of z is called the optimal value.

The number of basic solutions is the number of ways that m variables are selected from a group of n variables, i.e.,

$$_nC_m = \frac{n!}{(n-m)!m!}.$$

Example 2.4 (Basic solutions). Consider the basic solutions of the standard form of the linear programming (2.30) discussed in Example 1.1.

Choosing x_3, x_4, and x_5 as basic variables, we have the corresponding basic solution $(x_1, x_2, x_3, x_4, x_5) = (0, 0, 27, 16, 18)$ which is a nondegenerate basic feasible solution and corresponds to the extreme point A in Fig. 1.1. After making another choice of x_1, x_2, and x_4 as basic variables, solving

$$\begin{aligned} 2x_1 + 6x_2 &= 27 \\ 3x_1 + 2x_2 + x_4 &= 16 \\ 4x_1 + x_2 &= 18 \end{aligned}$$

---

[2] In this book, the superscript T denotes the transpose operation for a vector or a matrix.

yields $x_1 = 81/22$, $x_2 = 36/11$, and $x_4 = -35/22$. The resulting basic solution $(x_1, x_2, x_3, x_4, x_5) = (81/22, 36/11, 0, -35/22, 0)$ is not feasible.

Choosing x_1, x_2, and x_5 as basic variables, we solve

$$\begin{aligned} 2x_1 + 6x_2 &= 27 \\ 3x_1 + 2x_2 &= 16 \\ 4x_1 + x_2 + x_5 &= 18, \end{aligned}$$

and then we have a basic feasible solution $(x_1, x_2, x_3, x_4, x_5) = (3, 3.5, 0, 0, 2.5)$. It corresponds to the extreme point D in Fig. 1.1 which is an optimal solution. ◊

2.4 Simplex Method

For generalizing the basic ideas of linear programming grasped in the algebraic approach to the two-dimensional production planning problem of Example 1.1, consider the following linear programming problem with basic variables $x_1, x_2, \ldots, x_m$:

Find values of $x_1 \geq 0, x_2 \geq 0, \ldots, x_n \geq 0$ so as to minimize z, satisfying the augmented system of linear equations

$$\left.\begin{aligned} x_1 & + \bar{a}_{1,m+1}x_{m+1} + \bar{a}_{1,m+2}x_{m+2} + \cdots + \bar{a}_{1n}x_n = \bar{b}_1 \\ x_2 & + \bar{a}_{2,m+1}x_{m+1} + \bar{a}_{2,m+2}x_{m+2} + \cdots + \bar{a}_{2n}x_n = \bar{b}_2 \\ & \quad \cdots\cdots\cdots\cdots\cdots\cdots\cdots\cdots\cdots\cdots\cdots\cdots \\ x_m & + \bar{a}_{m,m+1}x_{m+1} + \bar{a}_{m,m+2}x_{m+2} + \cdots + \bar{a}_{mn}x_n = \bar{b}_m \\ -z & + \bar{c}_{m+1}x_{m+1} + \bar{c}_{m+2}x_{m+2} + \cdots + \bar{c}_n x_n = -\bar{z}. \end{aligned}\right\} \quad (2.35)$$

As in the previous section, here it is assumed that $n > m$ and the system of m equality constrains is nonredundant. As with the augmented system of equations (2.35), a system of linear equations in which each of the variables $x_1, x_2, \ldots, x_m$ has a coefficient of unity in one equation and zeros elsewhere is called a canonical form or a basic form. In a canonical form, the variables $x_1, x_2, \ldots, x_m$ and $(-z)$ are called basic variables, and the remaining variables $x_{m+1}, x_{m+2}, \ldots, x_n$ are called nonbasic variables. In such a canonical form, observing that $(-z)$ always is a basic variable, with no further notice, only $x_1, x_2, \ldots, x_m$ are called basic variables.

It is useful to set up such a canonical form (2.35) in tableau form as shown in Table 2.2. This table is called a simplex tableau, in which only the coefficients of the algebraic representation in (2.35) are given.

From the canonical form (2.35) or the simplex tableau given in Table 2.2, it follows directly that a basic solution with basic variables $x_1, x_2, \ldots, x_m$ becomes

$$x_1 = \bar{b}_1, \ x_2 = \bar{b}_2, \ldots, \ x_m = \bar{b}_m, \ x_{m+1} = x_{m+2} = \cdots = x_n = 0 \quad (2.36)$$

2.4 Simplex Method

Table 2.2 Simplex tableau

Basis	x_1	x_2	$\cdots$	x_m	x_{m+1}	x_{m+2}	$\cdots$	x_n	Constants
x_1	1				$\bar{a}_{1,m+1}$	$\bar{a}_{1,m+2}$	$\cdots$	$\bar{a}_{1n}$	$\bar{b}_1$
x_2		1			$\bar{a}_{2,m+1}$	$\bar{a}_{2,m+2}$	$\cdots$	$\bar{a}_{2n}$	$\bar{b}_2$
$\vdots$			$\ddots$		$\vdots$	$\vdots$	$\cdots$	$\vdots$	$\vdots$
x_m				1	$\bar{a}_{m,m+1}$	$\bar{a}_{m,m+2}$	$\cdots$	$\bar{a}_{mn}$	$\bar{b}_m$
$-z$					$\bar{c}_{m+1}$	$\bar{c}_{m+2}$	$\cdots$	$\bar{c}_n$	$-\bar{z}$

and the value of the objective function is

$$z = \bar{z}. \tag{2.37}$$

If

$$\bar{b}_1 \geq 0, \bar{b}_2 \geq 0, \ldots, \bar{b}_m \geq 0, \tag{2.38}$$

then the solution $(x_1, \ldots, x_m, x_{m+1}, \ldots, x_n) = (\bar{b}_1, \ldots, \bar{b}_m, 0, \ldots, 0)$ is a basic feasible solution. In this case, the corresponding canonical form (tableau) is called a feasible canonical form (tableau). If, for one or more i, $\bar{b}_i = 0$ holds, then it is said that the basic feasible solution is degenerate.

As an example that we can directly formulate a feasible canonical form, consider the production planning problem of Example 2.1. For this problem, by introducing m slack variables $x_{n+i} \geq 0, i = 1, 2, \ldots, m$ and multiplying the objective function by -1 to convert the maximization problem into a minimization problem, the following canonical form is obtained:

$$\left.\begin{aligned} a_{11}x_1 + a_{12}x_2 + \cdots + a_{1n}x_n + x_{n+1} \phantom{+ x_{n+2}} \phantom{+ x_{n+m}} &= b_1 \\ a_{21}x_1 + a_{22}x_2 + \cdots + a_{2n}x_n \phantom{+ x_{n+1}} + x_{n+2} \phantom{+ x_{n+m}} &= b_2 \\ \cdots\cdots\cdots\cdots\cdots\cdots\cdots\cdots\cdots & \\ a_{m1}x_1 + a_{m,2}x_2 + \cdots + a_{mn}x_n \phantom{+ x_{n+1}} \phantom{+ x_{n+2}} + x_{n+m} &= b_m \\ c_1x_1 + c_2x_2 + \cdots + c_nx_n \phantom{+ x_{n+1}} \phantom{+ x_{n+2}} \phantom{+ x_{n+m}} -z &= 0. \end{aligned}\right\} \tag{2.39}$$

In this formulation, by using the m slack variables $x_{n+1}, x_{n+2}, \ldots, x_{n+m}$ as basic variables, it is evident that (2.39) is a canonical form, and then the corresponding basic solution is

$$x_1 = x_2 = \cdots = x_n = 0, x_{n+1} = b_1, \ldots, x_{n+m} = b_m. \tag{2.40}$$

From the fact that the right-hand side constant b_i means the available amount of resource i, it should be nonnegative, i.e., $b_i \geq 0, i = 1, 2, \ldots, m$, and therefore this canonical form is feasible.

In contrast, for the diet problem of Example 2.2, introducing m surplus variables $x_{n+i} \geq 0$, $i = 1, 2, \ldots, m$ and then multiplying both sides of the resulting constraints by -1 yields a basic solution

$$x_1 = x_2 = \cdots = x_n = 0, x_{n+1} = -b_1, \ldots, x_{n+m} = -b_m. \quad (2.41)$$

Unfortunately, however, since $b_i \geq 0$, $i = 1, 2, \ldots, m$, this operation cannot lead a feasible canonical form.

In the following discussions of this section, assume that the canonical form (2.35) is feasible. That is, starting with the canonical form (2.35) with the basic solution

$$x_1 = \bar{b}_1, \; x_2 = \bar{b}_2, \ldots, \; x_m = \bar{b}_m, \; x_{m+1} = x_{m+2} = \cdots = x_n = 0,$$

we assume that this basic solution is feasible, i.e.,

$$\bar{b}_1 \geq 0, \; \bar{b}_2 \geq 0, \ldots, \bar{b}_m \geq 0.$$

From the last equation in (2.35), we have

$$z = \bar{z} + \bar{c}_{m+1} x_{m+1} + \bar{c}_{m+2} x_{m+2} + \cdots + \bar{c}_n x_n.$$

Since $x_{m+1} = x_{m+2} = \cdots = x_n = 0$, one finds $z = \bar{z}$. This equation provides even more valuable information than this. By merely glancing at the numbers $\bar{c}_j$, $j = m+1, m+2, \ldots, n$, one can tell if this basic feasible solution is optimal or not. Furthermore, one can find a better basic feasible solution if it is not optimal. Consider first the optimality of the canonical form, given by the following theorem.

Theorem 2.1 (Optimality test). *In the feasible canonical form (2.35), if all coefficients $\bar{c}_{m+1}, \bar{c}_{m+2}, \ldots, \bar{c}_n$ of the last equation are nonnegative, i.e.,*

$$\bar{c}_j \geq 0, \quad j = m+1, m+2, \ldots, n, \quad (2.42)$$

then the basic feasible solution is optimal.

Proof. The last equation of (2.35) can be rewritten as

$$z = \bar{z} + \bar{c}_{m+1} x_{m+1} + \bar{c}_{m+2} x_{m+2} + \cdots + \bar{c}_n x_n.$$

The nonbasic variables $x_{m+1}, x_{m+2}, \ldots, x_n$ are presently zeros, and they are restricted to be nonnegative. If $\bar{c}_j \geq 0$ for $j = m+1, m+2, \ldots, n$, then from $\bar{c}_j x_j \geq 0$, $j = m+1, m+2, \ldots, n$, increasing any x_j cannot decrease the objective function z. Thus, since any change in the nonbasic variables cannot decrease z, the present solution must be optimal. □

The coefficient $\bar{c}_j$ of x_j in (2.35) represents the rate of change of z with respect to the nonbasic variable x_j. From this observation, the coefficient $\bar{c}_j$ is called the relative cost coefficient or, alternatively, the reduced cost coefficient.

2.4 Simplex Method

The optimality condition (2.42) is sometimes referred to as the optimality criterion or the simplex criterion. The feasible canonical form satisfying the optimality criterion is called the optimal canonical form or the optimal basic form, and the simplex tableau satisfying the optimality criterion is also called the optimal tableau.

Note that since $\bar{c}_j = 0$ for all basic variables, the optimality criterion (2.42) could also be stated simply as $\bar{c}_j \geq 0$ for all $j = 1, 2, \ldots, n$ in place of $\bar{c}_j \geq 0$ for all $j = m+1, \ldots, n$.

In addition to the optimality, the relative cost coefficients can also tell if there are multiple optima. Assume that for all nonbasic variables x_j, $\bar{c}_j \geq 0$, and for some nonbasic variable x_k, $\bar{c}_k = 0$. In that case, if the increase of x_k does not violate the constraints, there are multiple optima because no change in z results. Hence, the following theorem can be derived.

Theorem 2.2 (Unique optimal solution). *In the feasible canonical form (2.35), if $\bar{c}_j > 0$ for all nonbasic variables, then the basic feasible solution is the unique optimal solution.*

Of course, if, for some nonbasic variable x_j, $\bar{c}_j < 0$, then z can be decreased by increasing x_j. Consider a method for finding better solutions than the current nonoptimal solution.

If there is at least one negative coefficient, say $\bar{c}_j < 0$, then, under the assumption of nondegeneracy, i.e., $\bar{b}_i > 0$ for all i, it is always possible to generate another basic feasible solution with an improved value of the objective function. If there are two or more negative coefficients, we choose a variable x_s with the smallest relative cost coefficient

$$\bar{c}_s = \min_{\bar{c}_j < 0} \bar{c}_j \tag{2.43}$$

and increase the value of x_s.

Although this choice may not lead to the greatest possible decrease in z (since only a limited extent of increase of x_s may be allowed), it is at least intuitively a good rule for choosing a variable to be made a basic one. It is the one used in practice today because (i) it is simple and (ii) it generally leads to an optimal solution in fewer iterations than just choosing any $\bar{c}_s < 0$.

After a nonbasic variable x_s is selected to be a basic one, we increase the value of x_s from zero, holding the other nonbasic variables zeros. Observe the effect of this operation on the current basic variables. From (2.35), each of the current basic variables can be represented as a function of x_s:

$$\left.\begin{aligned} x_1 &= \bar{b}_1 - \bar{a}_{1s} x_s \\ x_2 &= \bar{b}_2 - \bar{a}_{2s} x_s \\ &\cdots\cdots\cdots \\ x_m &= \bar{b}_m - \bar{a}_{ms} x_s \\ z &= \bar{z} + \bar{c}_s x_s. \end{aligned}\right\} \tag{2.44}$$

Since the coefficient $\bar{c}_s$ of the last equation in (2.44) is negative, i.e., $\bar{c}_s < 0$, increasing the value of x_s decreases the value of z. The only factor limiting the increase of x_s is that all of the variables $x_1, x_2, \ldots, x_m$ must be nonnegative. In other words, keeping the feasibility of the solution requires

$$x_i = \bar{b}_i - \bar{a}_{is}x_s \geq 0, \quad i = 1, 2, \ldots, m. \tag{2.45}$$

However, if all the coefficients $\bar{a}_{is}$, $i = 1, 2, \ldots, m$ are nonpositive, i.e.,

$$\bar{a}_{is} \leq 0, \quad i = 1, 2, \ldots, m, \tag{2.46}$$

then x_s can increase infinitely. Hence since $\bar{c}_s < 0$, from the last equation of (2.44), it follows that

$$z = \bar{z} + \bar{c}_s x_s \to -\infty.$$

Thus, we have the following theorem.

Theorem 2.3 (Unboundedness). *If in the feasible canonical form (2.35), for some nonbasic variable x_s, the coefficients $\bar{a}_{is}$, $i = 1, 2, \ldots, m$ are nonpositive and the coefficient $\bar{c}_s$ is negative, i.e.,*

$$\bar{a}_{is} \leq 0, \, i = 1, 2, \ldots, m, \quad \text{and} \quad \bar{c}_s < 0, \tag{2.47}$$

then the optimal value is unbounded.

If, however, at least one $\bar{a}_{is}$ is positive, then x_s cannot be increased indefinitely since eventually some basic variable, say x_i, will decrease beyond zero and become negative. From (2.44), x_i becomes zero when the coefficient $\bar{a}_{is}$ is positive and x_s raises to $\bar{b}_i/\bar{a}_{is}$, i.e.,

$$x_s = \frac{\bar{b}_i}{\bar{a}_{is}}, \quad \bar{a}_{is} > 0. \tag{2.48}$$

The value of x_s is maximized under the condition of the nonnegativity of the basic variables x_i, $i = 1, 2, \ldots, m$, and it is given by

$$\min_{\bar{a}_{is} > 0} \frac{\bar{b}_i}{\bar{a}_{is}} = \frac{\bar{b}_r}{\bar{a}_{rs}} = \theta. \tag{2.49}$$

The basic variable x_r determined by (2.49) then becomes nonbasic, and instead, the nonbasic variable x_s becomes basic. That is, x_r becomes zero while x_s increases from zero to $\bar{b}_r/\bar{a}_{rs} = \theta \, (\geq 0)$. Also, from the last equation of (2.44), the value of objective function z decreases by $|\bar{c}_s x_s| = |\bar{c}_s \theta|$.

A new canonical form in which x_s is selected as a basic variable in place of x_r can be easily obtained by pivoting on $\bar{a}_{rs}$, which is called the pivot element determined by (2.43) and (2.49). That is, finding $\bar{c}_s = \min_{\bar{c}_j < 0} \bar{c}_j$ tells us that the pivot term is in column s, and finding the minimum $\bar{b}_r/\bar{a}_{rs}$ of all the ratios $\bar{b}_i/\bar{a}_{is}$ such that $\bar{a}_{is} > 0$ tells us that it is in row r.

Fundamental to linear programming is a pivot operation defined as follows.

2.4 Simplex Method

Definition 2.7 (Pivot operation). A pivot operation consists of m elementary steps for replacing a linear system with an equivalent system in which a specified variable has a coefficient of unity in one equation and zeros elsewhere. The detailed steps are as follows:

(i) Select the nonzero element a_{rs} in row (equation) r and column s, which is called the pivot element.
(ii) Replace the rth equation with the rth equation multiplied by $1/a_{rs}$.
(iii) For each $i = 1, 2, \ldots, m$ except $i = r$, replace the ith equation with the sum of the ith equation and the replaced rth equation multiplied by $-a_{is}$.

In linear programming, pivot operations are sometimes counted by the term "cycle." Now, a pivot operation on $\bar{a}_{rs} (\neq 0)$ is performed to the feasible canonical form

$$
\begin{aligned}
x_1 &+ \bar{a}_{1,m+1}x_{m+1} + \cdots + \bar{a}_{1s}x_s + \cdots + \bar{a}_{1n}x_n = \bar{b}_1 \\
x_2 &+ \bar{a}_{2,m+1}x_{m+1} + \cdots + \bar{a}_{2s}x_s + \cdots + \bar{a}_{2n}x_n = \bar{b}_2 \\
&\cdots\cdots\cdots\cdots \\
x_r &+ \bar{a}_{r,m+1}x_{m+1} + \cdots + \bar{a}_{rs}x_s + \cdots + \bar{a}_{rn}x_n = \bar{b}_r \\
&\cdots\cdots\cdots\cdots \\
x_m &+ \bar{a}_{m,m+1}x_{m+1} + \cdots + \bar{a}_{ms}x_s + \cdots + \bar{a}_{mn}x_n = \bar{b}_m \\
-z &+ \bar{c}_{m+1}x_{m+1} + \cdots + \bar{c}_s x_s + \cdots + \bar{c}_n x_n = -\bar{z},
\end{aligned}
\tag{2.50}
$$

where $\bar{b}_i \geq 0, i = 1, 2, \ldots, m$, and then we have the new canonical form

$$
\begin{aligned}
x_1 &+ \bar{a}^*_{1r}x_r + \bar{a}^*_{1,m+1}x_{m+1} + \cdots + 0 + \cdots + \bar{a}^*_{1n}x_n = \bar{b}^*_1 \\
x_2 &+ \bar{a}^*_{2r}x_r + \bar{a}^*_{2,m+1}x_{m+1} + \cdots + 0 + \cdots + \bar{a}^*_{2n}x_n = \bar{b}^*_2 \\
&\cdots\cdots\cdots\cdots \\
&\bar{a}^*_{rr}x_r + \bar{a}^*_{r,m+1}x_{m+1} + \cdots + x_s + \cdots + \bar{a}^*_{rn}x_n = \bar{b}^*_r \\
&\cdots\cdots\cdots\cdots \\
&\bar{a}^*_{mr}x_r + x_m + \bar{a}^*_{m,m+1}x_{m+1} + \cdots + 0 + \cdots + \bar{a}^*_{mn}x_n = \bar{b}^*_m \\
&\bar{c}^*_r x_r -z + \bar{c}^*_{m+1}x_{m+1} + \cdots + 0 + \cdots + \bar{c}^*_n x_n = -\bar{z}^*,
\end{aligned}
\tag{2.51}
$$

where the superscript $*$ is added to a revised coefficient, and the revised coefficients for $j = r, m+1, m+2, \ldots, n$ are calculated as follows:

$$\bar{a}^*_{rj} = \frac{\bar{a}_{rj}}{\bar{a}_{rs}}, \quad \bar{b}^*_r = \frac{\bar{b}_r}{\bar{a}_{rs}}, \tag{2.52}$$

$$\bar{a}^*_{ij} = \bar{a}_{ij} - \bar{a}_{is}\frac{\bar{a}_{rj}}{\bar{a}_{rs}}, \quad \bar{b}^*_i = \bar{b}_i - \bar{a}_{is}\frac{\bar{b}_r}{\bar{a}_{rs}}, \quad i = 1, 2, \ldots, m; \; i \neq r, \tag{2.53}$$

$$\bar{c}^*_j = \bar{c}_j - \bar{c}_s\frac{\bar{a}_{rj}}{\bar{a}_{rs}}, \quad -\bar{z}^* = -\bar{z} - \bar{c}_s\frac{\bar{b}_r}{\bar{a}_{rs}}. \tag{2.54}$$

Table 2.3 Pivot operation on $\bar{a}_{rs}$

Cycle	Basis	x_1	$\cdots$	x_r	$\cdots$	x_m	x_{m+1}	$\cdots$	x_s	$\cdots$	x_n	Constants
ℓ	x_1	1					$\bar{a}_{1,m+1}$	$\cdots$	$\bar{a}_{1s}$	$\cdots$	$\bar{a}_{1n}$	$\bar{b}_1$
	$\vdots$		$\ddots$				$\vdots$		$\vdots$		$\vdots$	$\vdots$
	x_r			1			$\bar{a}_{r,m+1}$	$\cdots$	$[\bar{a}_{rs}]$	$\cdots$	$\bar{a}_{rn}$	$\bar{b}_r$
	$\vdots$				$\ddots$		$\vdots$		$\vdots$		$\vdots$	$\vdots$
	x_m					1	$\bar{a}_{m,m+1}$	$\cdots$	$\bar{a}_{ms}$	$\cdots$	$\bar{a}_{mn}$	$\bar{b}_m$
	$-z$						$\bar{c}_{m+1}$	$\cdots$	$\bar{c}_s$	$\cdots$	$\bar{c}_n$	$-\bar{z}$
$\ell+1$	x_1	1		$\bar{a}^*_{1r}$			$\bar{a}^*_{1,m+1}$	$\cdots$	0	$\cdots$	$\bar{a}^*_{1n}$	$\bar{b}^*_1$
	$\vdots$		$\ddots$	$\vdots$			$\vdots$		$\vdots$		$\vdots$	$\vdots$
	x_s			$\bar{a}^*_{rr}$			$\bar{a}^*_{r,m+1}$	$\cdots$	1	$\cdots$	$\bar{a}^*_{rn}$	$\bar{b}^*_r$
	$\vdots$			$\vdots$	$\ddots$		$\vdots$		$\vdots$		$\vdots$	$\vdots$
	x_m			$\bar{a}^*_{mr}$		1	$\bar{a}^*_{m,m+1}$	$\cdots$	0	$\cdots$	$\bar{a}^*_{mn}$	$\bar{b}^*_m$
	$-z$			$\bar{c}^*_r$	$\cdots$		$\bar{c}^*_{m+1}$	$\cdots$	0	$\cdots$	$\bar{c}^*_n$	$-\bar{z}^*$

$$\bar{a}^*_{rj} = \frac{\bar{a}_{rj}}{\bar{a}_{rs}}, \quad \bar{b}^*_r = \frac{\bar{b}_r}{\bar{a}_{rs}}$$

$$\bar{a}^*_{ij} = \bar{a}_{ij} - \bar{a}_{is}\frac{\bar{a}_{rj}}{\bar{a}_{rs}} = \bar{a}_{ij} - \bar{a}_{is}\bar{a}^*_{rj}, \quad \bar{b}^*_i = \bar{b}_i - \bar{a}_{is}\frac{\bar{b}_r}{\bar{a}_{rs}} = \bar{b}_i - \bar{a}_{is}\bar{b}^*_r \quad (i \neq r)$$

$$\bar{c}^*_j = \bar{c}_j - \bar{c}_s\frac{\bar{a}_{rj}}{\bar{a}_{rs}} = \bar{c}_j - \bar{c}_s\bar{a}^*_{rj}, \quad -\bar{z}^* = -\bar{z} - \bar{c}_s\frac{\bar{b}_r}{\bar{a}_{rs}} = -\bar{z} - \bar{c}_s\bar{b}^*_r$$

Since the pivot element $\bar{a}_{rs}$ is determined by (2.43) and (2.49), it is expected that the new canonical form (2.51) with basic variables $x_1, x_2, \ldots, x_{r-1}, x_s, x_{r+1}, \ldots, x_m$ also becomes feasible. This fact can be formally verified as follows.

It is obvious that $\bar{b}^*_r = \bar{b}_r/\bar{a}_{rs} \geq 0$. For i ($i \neq r$) such that $\bar{a}_{is} > 0$, from (2.49), it follows that

$$\bar{b}^*_i = \bar{b}_i - \frac{\bar{a}_{is}}{\bar{a}_{rs}}\bar{b}_r = \bar{a}_{is}\left(\frac{\bar{b}_i}{\bar{a}_{is}} - \frac{\bar{b}_r}{\bar{a}_{rs}}\right) \geq 0,$$

and for i ($i \neq r$) such that $\bar{a}_{is} \leq 0$, one finds that

$$\bar{b}^*_i = \bar{b}_i - \frac{\bar{a}_{is}}{\bar{a}_{rs}}\bar{b}_r \geq 0.$$

Hence, it holds that $\bar{b}^*_i \geq 0$ for all i, and then (2.51) is a feasible canonical form.

The pivot operation on $\bar{a}_{rs}$ replacing x_r with x_s as a new basic variable can be summarized in Table 2.3.

As described so far, starting with a feasible canonical form and updating it through a series of pivot operations, the simplex method seeks for an optimal solution satisfying the optimality criterion or the unboundedness information. The procedure of the simplex method, starting with a feasible canonical form, can be summarized as follows.

2.4 Simplex Method

Procedure of the Simplex Method

Start with a feasible canonical form (simplex tableau).

Step 1 If all of the relative cost coefficients are nonnegative, i.e., $\bar{c}_j \geq 0$ for all indices j of the nonbasic variables, then the current solution is optimal, and stop. Otherwise, by using the relative cost coefficients $\bar{c}_j$, find the index s such that

$$\min_{\bar{c}_j < 0} \bar{c}_j = \bar{c}_s.$$

Step 2 If all of the coefficients in column s are nonpositive, i.e., $\bar{a}_{is} \leq 0$ for all indices i of the basic variables, then the optimal value is unbounded, and stop.

Step 3 If some of $\bar{a}_{is}$ are positive, find the index r such that

$$\min_{\bar{a}_{is}>0} \frac{\bar{b}_i}{\bar{a}_{is}} = \frac{\bar{b}_r}{\bar{a}_{rs}} = \theta.$$

Step 4 Perform the pivot operation on $\bar{a}_{rs}$ for obtaining a new feasible canonical form (simplex tableau) with x_s replacing x_r as a new basic variable. The coefficients of the new feasible canonical form after pivoting on $\bar{a}_{rs} \neq 0$ are calculated as follows:

(i) Replace row r (the rth equation) with row r multiplied by $1/\bar{a}_{rs}$ (divide row r by $\bar{a}_{rs}$), i.e.,

$$\bar{a}^*_{rj} = \frac{\bar{a}_{rj}}{\bar{a}_{rs}}, \quad \bar{b}^*_r = \frac{\bar{b}_r}{\bar{a}_{rs}}.$$

(ii) For each $i = 1, 2, \ldots, m$ except $i = r$, replace row i (the ith equation) with the sum of row i and the revised row r multiplied by $-\bar{a}_{is}$, i.e.,

$$\bar{a}^*_{ij} = \bar{a}_{ij} - \bar{a}_{is}\bar{a}^*_{rj}, \quad \bar{b}^*_i = \bar{b}_i - \bar{a}_{is}\bar{b}^*_r.$$

(iii) Replace row $m+1$ (the $(m+1)$th equation for the objective function) with the sum of row $m+1$ and the revised row r multiplied by $-\bar{c}_s$, i.e.,

$$\bar{c}^*_j = \bar{c}_j - \bar{c}_s\bar{a}^*_{rj}, \quad -\bar{z}^* = -\bar{z} - \bar{c}_s\bar{b}^*_r.$$

Return to step 1.

It should be noted here that when multiple candidates exist for the index s of the variable entering the basis in step 1 or the index r of the variable leaving the basis in step 3, for the sake of convenience, we choose the smallest index.

Table 2.4 Simplex tableau for Example 1.1

Cycle	Basis	x_1	x_2	x_3	x_4	x_5	Constants
0	x_3	2	[6]	1			27
	x_4	3	2		1		16
	x_5	4	1			1	18
	$-z$	-3	-8				0
1	x_2	1/3	1	1/6			4.5
	x_4	[7/3]		$-1/3$	1		7
	x_5	11/3		$-1/6$		1	13.5
	$-z$	$-1/3$		4/3			36
2	x_2		1	3/14	$-1/7$		3.5
	x_1	1		$-1/7$	3/7		3
	x_5			5/14	$-11/7$	1	2.5
	$-z$			9/7	1/7		37

Example 2.5 (Simplex method for the production planning problem of Example 1.1).
Using the simplex method, solve the production planning problem in the standard form given in Example 1.1:

$$\begin{aligned}
\text{minimize } z &= -3x_1 - 8x_2 \\
\text{subject to } \quad 2x_1 + 6x_2 + x_3 &= 27 \\
3x_1 + 2x_2 \quad\quad + x_4 &= 16 \\
4x_1 + x_2 \quad\quad\quad\quad + x_5 &= 18 \\
x_j \geq 0, \quad j &= 1, 2, 3, 4, 5.
\end{aligned}$$

Introducing the slack variables x_3, x_4, and x_5 and using them as the basic variables, we have the initial basic feasible solution

$$x_1 = x_2 = 0, \quad x_3 = 27, \quad x_4 = 16, \quad x_5 = 18,$$

which is shown at cycle 0 of the simplex tableau given in Table 2.4.

At cycle 0, since the minimum of $\bar{c}_1$ and $\bar{c}_2$ is

$$\min(-3, -8) = -8 < 0,$$

x_2 becomes a new basic variable. The minimum ratio, $\min_{\bar{a}_{i2}>0} \bar{b}_i/\bar{a}_{i2}$, is calculated as

$$\min\left(\frac{27}{6}, \frac{16}{2}, \frac{18}{1}\right) = \frac{27}{6} = 4.5,$$

and then x_3 becomes a nonbasic variable. From $s = 2$ and $r = 1$, the pivot element is 6 bracketed by [ ] in Table 2.4. After the pivot operation on 6, the result at cycle 1 is obtained.

2.4 Simplex Method

Table 2.5 Simplex tableau with multiple optima

Cycle	Basis	x_1	x_2	x_3	x_4	x_5	Constants
0	x_3	2	[6]	1			27
	x_4	3	2		1		16
	x_5	4	1			1	18
	$-z$	-1	-3				
1	x_2	1/3	1	1/6			4.5
	x_4	[7/3]		$-1/3$	1		7
	x_5	11/3		$-1/6$		1	13.5
	$-z$	0		1/2			13.5
2	x_2		1	3/14	$-1/7$		3.5
	x_1	1		$-1/7$	3/7		3
	x_5			5/14	$-11/7$	1	2.5
	$-z$			1/2	0		13.5

At cycle 1, since the negative relative cost coefficient is only $-1/3$, x_1 becomes a basic variable. Since

$$\min\left(\frac{4.5}{1/3}, \frac{7}{7/3}, \frac{13.5}{11/3}\right) = \frac{7}{7/3} = 3,$$

7/3 bracketed by [ ] becomes the pivot element. After the pivot operation on 7/3, the result at cycle 2 is obtained. At cycle 2, all of the relative cost coefficients become positive, and then the following optimal solution is obtained:

$$x_1 = 3, \ x_2 = 3.5 \ (x_3 = x_4 = 0, x_5 = 2.5), \quad z = -37$$

The above optimal solution corresponds to the extreme point D in Fig. 1.1. ◊

Example 2.6 (Example with multiple optima). To show a simple linear programming problem having multiple optima, consider the following modified production planning problem in which the coefficients of x_1 and x_2 in the original objective function given in Example 1.1 are changed to 1 and 3, respectively:

$$\begin{aligned}
\text{minimize } z = &\ -x_1 - 3x_2 \\
\text{subject to} \quad & 2x_1 + 6x_2 + x_3 = 27 \\
& 3x_1 + 2x_2 + x_4 = 16. \\
& 4x_1 + x_2 + x_5 = 18 \\
& x_j \geq 0, \quad j = 1, 2, 3, 4, 5
\end{aligned}$$

By using the simplex method, at cycle 1 in Table 2.5, an optimal solution

$$x_1 = 0, \ x_2 = 4.5 \ (x_3 = 0, x_4 = 7, x_5 = 13.5), \quad z = -13.5$$

is obtained, observing the relative cost coefficient of x_1 is zero, which means that the value of the objective function is unchanged even if x_1 becomes positive, provided that it is not violating the constraints. Replacing x_1 with x_4 as a basic variable yields an alternative optimal solution

$$x_1 = 3, x_2 = 3.5 \ (x_3 = x_4 = 0, x_5 = 2.5), \quad z = -13.5$$

giving the same value of the objective function. It should be noted here that the optimal solutions obtained in cycles 1 and 2, respectively, correspond to the extreme points E and D in Fig. 1.1, and all of the points on the line segment ED are also optimal. ◊

2.5 Two-Phase Method

The simplex method requires a basic feasible solution as a starting point. Such a starting point is not always easy to find, and in fact none will exist if the constraints are inconsistent. Phase I of the simplex method finds an initial basic feasible solution or derives the information that no feasible solution exists. Phase II then proceeds from this starting point to an optimal solution or derives the information that the optimal value is unbounded. Both phases use the procedure of the simplex method given in the previous section.

Phase I starts with a linear programming problem in the standard form (2.24), where all the constants b_i are nonnegative. For this purpose, if some b_i is negative, multiply the corresponding equation by -1. In order to set up an initial feasible solution for phase I, the linear programming problem in the standard form is augmented with a set of nonnegative variables $x_{n+1} \geq 0, x_{n+2} \geq 0, \ldots, x_{n+m} \geq 0$, so that the problem becomes as follows:

Find values of $x_j \geq 0$, $j = 1, 2, \ldots, n, n+1, \ldots, n+m$ so as to minimize z, satisfying the augmented system of linear equations

$$\left.\begin{aligned}
a_{11}x_1 + a_{12}x_2 + \cdots + a_{1n}x_n + x_{n+1} \phantom{+x_{n+2}\ \ } &= b_1 \ (\geq 0) \\
a_{21}x_1 + a_{22}x_2 + \cdots + a_{2n}x_n \phantom{+x_{n+1}} + x_{n+2} &= b_2 \ (\geq 0) \\
\cdots\cdots\cdots\cdots\cdots\cdots\cdots & \\
a_{m1}x_1 + a_{m2}x_2 + \cdots + a_{mn}x_n + x_{n+m} &= b_m \ (\geq 0) \\
c_1x_1 + c_2x_2 + \cdots + c_nx_n -z &= 0.
\end{aligned}\right\}$$
(2.55)

The newly introduced nonnegative variables $x_{n+1} \geq 0, x_{n+2} \geq 0, \ldots, x_{n+m} \geq 0$ are called artificial variables.

In the canonical form (2.55), using the artificial variables $x_{n+1}, x_{n+2}, \ldots, x_{n+m}$ as basic variables, the following initial basic feasible solution is directly obtained:

$$x_1 = x_2 = \cdots = x_n = 0, \quad x_{n+1} = b_1 \geq 0, \ldots, x_{n+m} = b_m \geq 0. \quad (2.56)$$

2.5 Two-Phase Method

Although a basic feasible solution to (2.55) such as (2.56) is not always feasible to the original system, basic feasible solutions to (2.55) such that all the artificial variables $x_{n+1}, x_{n+2}, \ldots, x_{n+m}$ are equal to zeros are also feasible to the original system. Thus, one way to find a basic feasible solution to the original system is to start from the initial basic solution (2.56) and use the simplex method to drive a basic feasible solution such that all the artificial variables are equal to zeros. This can be done by minimizing a function of the artificial variables

$$w = x_{n+1} + x_{n+2} + \cdots + x_{n+m} \tag{2.57}$$

subject to the equality constraints (2.55) and the nonnegativity conditions for all variables. By its very nature, the function (2.57) is sometimes called the infeasibility form.

That is, the phase I problem is to find values of $x_1 \geq 0, x_2 \geq 0, \ldots, x_n \geq 0$, $x_{n+1} \geq 0, \ldots, and\, x_{n+m} \geq 0$ so as to minimize w, satisfying the augmented system of linear equations

$$\left.\begin{aligned}
a_{11}x_1 + a_{12}x_2 + \cdots + a_{1n}x_n + x_{n+1} \phantom{+x_{n+m}} &= b_1 \ (\geq 0) \\
a_{21}x_1 + a_{22}x_2 + \cdots + a_{2n}x_n \phantom{+x_{n+1}} + x_{n+2} \phantom{+x_{n+m}} &= b_2 \ (\geq 0) \\
\cdots\cdots\cdots\cdots\cdots\cdots\cdots & \\
a_{m1}x_1 + a_{m2}x_2 + \cdots + a_{mn}x_n \phantom{+x_{n+1}} + x_{n+m} &= b_m \ (\geq 0) \\
c_1 x_1 + c_2 x_2 + \cdots + c_n x_n \phantom{+x_{n+m}} - z \phantom{+x_{n+m}} &= 0 \\
x_{n+1} + x_{n+2} + \cdots + x_{n+m} - w &= 0.
\end{aligned}\right\} \tag{2.58}$$

Since the artificial variables are nonnegative, the function w which is the sum of the artificial variables is obviously larger than or equal to zero. In particular, if the optimal value of w is zero, i.e., $w = 0$, then all the artificial variables are zeros, i.e., $x_{n+i} = 0$ for all $i = 1, 2, \ldots, m$. In contrast, if it is positive, i.e., $w > 0$, then no feasible solution to the original system exists because some artificial variables are not zeros and then the corresponding original constraints are not satisfied. Given an initial basic feasible solution, the simplex method generates other basic feasible solutions in turn, and then the end product of phase I must be a basic feasible solution to the original system if such a solution exists.

It should be mentioned that a full set of m artificial variables may not be necessary. If the original system has some variables that can be used as initial basic variables, then they should be chosen in preference to artificial variables. The result is less work in phase I.

For obtaining an initial basic feasible solution through the minimization of w with the simplex method, it is necessary to convert the augmented system (2.58) into the canonical form with the row of $-w$, where w must be expressed by the current nonbasic variables $x_1, x_2, \ldots, x_n$. Since from (2.58) the artificial variables are represented by using the nonbasic variables, i.e.,

$$x_{n+i} = b_i - a_{i1}x_1 - a_{i2}x_2 - \cdots - a_{in}x_n, \quad i = 1, 2, \ldots, m,$$

(2.57) can be rewritten as

$$w = \sum_{i=1}^{m} x_{n+i} = \sum_{i=1}^{m}\left(b_i - \sum_{j=1}^{n} a_{ij} x_j\right) = \sum_{i=1}^{m} b_i - \sum_{j=1}^{n}\left(\sum_{i=1}^{m} a_{ij}\right) x_j, \quad (2.59)$$

which is now expressed by the nonbasic variables $x_1, x_2, \ldots, x_n$. Defining

$$w_0 = \sum_{i=1}^{m} b_i \ (\geq 0), \quad d_j = -\sum_{i=1}^{m} a_{ij}, \quad j = 1, 2, \ldots, n, \quad (2.60)$$

the row of $-w$ is compactly expressed as

$$-w + d_1 x_1 + d_2 x_2 + \cdots + d_n x_n = -w_0. \quad (2.61)$$

In this way, the augmented system (2.58) is converted into the following initial feasible canonical form for phase I with the row of $-w$ in which the artificial variables $x_{n+1}, x_{n+2}, \ldots, x_{n+m}$ are selected as basic variables:

$$\left.\begin{array}{l}
a_{11}x_1 + a_{12}x_2 + \cdots + a_{1n}x_n + x_{n+1} \hspace{3.2cm} = b_1 \ (\geq 0) \\
a_{21}x_1 + a_{22}x_2 + \cdots + a_{2n}x_n \hspace{1.1cm} + x_{n+2} \hspace{2.1cm} = b_2 \ (\geq 0) \\
\hspace{3cm} \cdots\cdots\cdots\cdots\cdots \\
a_{m1}x_1 + a_{m2}x_2 + \cdots + a_{mn}x_n \hspace{2.2cm} + x_{n+m} \hspace{1cm} = b_m \ (\geq 0) \\
c_1 x_1 + c_2 x_2 + \cdots + c_n x_n \hspace{3.2cm} -z \hspace{1cm} = 0 \\
d_1 x_1 + d_2 x_2 + \cdots + d_n x_n \hspace{4.5cm} -w = -w_0.
\end{array}\right\} \quad (2.62)$$

Now it becomes possible to solve the phase I problem as given by (2.62) using the simplex method. Finding the pivot element $\bar{a}_{rs}$ by using the rule

$$\bar{d}_s = \min_{\bar{d}_j < 0} \bar{d}_j \quad (2.63)$$

and

$$\frac{\bar{b}_r}{\bar{a}_{rs}} = \min_{\bar{a}_{is} > 0} \frac{\bar{b}_i}{\bar{a}_{is}} \quad (2.64)$$

and performing the pivot operation on it, we minimize the objective function w in phase I. When

$$\bar{d}_j \geq 0, \quad j = 1, \ldots, n, n+1, \ldots, n+m; \quad w = 0, \quad (2.65)$$

all the artificial variables become zeros. In this case, if all the artificial variables become nonbasic ones, an initial basic feasible solution to the original problem is

2.5 Two-Phase Method

obtained. Hence, after eliminating all the artificial variables together with the row of $-w$, initiate phase II of the simplex method for minimizing the original objective function z.

In the two-phase method, phase I finds an initial basic feasible solution or derives the information that no feasible solution exists, and phase II then proceeds from this starting point to an optimal solution or derives the information that the optimal value is unbounded.

Whenever the original system contains redundancies and often when degenerate solutions occur, artificial variables will remain in the basis at the end of phase I. Thus, it is necessary to prevent their values from becoming positive in phase II. One possible way is to drop all nonartificial variables whose relative cost coefficients for w are positive and all nonbasic artificial variables before starting phase II. To see this, we note that the equation for w at the end of phase I satisfies

$$\sum_{j=1}^{n+m} \bar{d}_j x_j = w - w_0, \qquad (2.66)$$

where $\bar{d}_j \geq 0$ and $w_0 = 0$ since feasible solutions to the original problem exist. For feasibility, w must remain zero in phase II, which means that every x_j corresponding to $\bar{d}_j > 0$ must be zero; hence, all such x_j can be set equal to zero and eliminated from further consideration in phase II. We can also drop any nonbasic artificial variables because we no longer need to consider them. That is, eliminate the columns of the artificial variables leaving from the basis and those of nonbasic variable x_j with $d_j > 0$ in the optimal simplex tableau of phase I. Due to this operation, the objective function w of phase I will not become positive again, and also the values of the artificial variables remaining in the basis will not become positive in phase II. This means that basic solutions generated in phase II are always feasible.

$$d_j = -\sum_{i=1}^{m} a_{ij}, \quad -w_0 = -\sum_{i=1}^{m} b_i$$

Before summarizing the procedure of the two-phase method, the following useful remarks are given. In the simplex tableau, it is customary to omit the artificial variable columns because these, once dropped from the basis, can be eliminated from further consideration. Moreover, if the pivot operations for minimizing w in phase I are also simultaneously performed on the row of $-z$, the original objective function z will be expressed in terms of nonbasic variables at each cycle. Thus, if an initial basic feasible solution is found for the original problem, the simplex method can be initiated immediately on z. Therefore, the row of $-z$ is incorporated into the pivot operations in phase I.

Following the above discussions, the procedure of the two-phase method can be summarized as follows.

Table 2.6 Initial tableau of two-phase method

Basis	x_1	x_2	$\cdots$	x_j	$\cdots$	x_n	Constants
x_{n+1}	a_{11}	a_{12}	$\cdots$	a_{1j}	$\cdots$	a_{1n}	b_1
x_{n+2}	a_{21}	a_{22}	$\cdots$	a_{2j}	$\cdots$	a_{2n}	b_2
$\vdots$	$\vdots$	$\vdots$		$\vdots$		$\vdots$	$\vdots$
x_{n+i}	a_{i1}	a_{i2}	$\cdots$	a_{ij}	$\cdots$	a_{in}	b_i
$\vdots$	$\vdots$	$\vdots$		$\vdots$		$\vdots$	$\vdots$
x_{n+m}	a_{m1}	a_{m2}	$\cdots$	a_{mj}	$\cdots$	a_{mn}	b_m
$-z$	c_1	c_2	$\cdots$	c_j	$\cdots$	c_n	0
$-w$	d_1	d_2	$\cdots$	d_j	$\cdots$	d_n	$-w_0$

Procedure of Two-Phase Method

Phase I Starting with the simplex tableau in Table 2.6, perform the simplex method with the row of $-w$ as an objective function in phase I, where the pivot element is not selected from the row of $-z$, but the pivot operation is performed to the row of $-z$. When an optimal tableau is obtained, if $w > 0$, no feasible solution exists to the original problem. Otherwise, i.e., if $w = 0$, proceed to phase II.

Phase II After dropping all columns of x_j such that $\bar{d}_j > 0$ and the row of $-w$, perform the simplex method with the row of $-z$ as the objective function in phase II.

Example 2.7 (Two-phase method for diet problem with two decision variables). Using the two-phase method, solve the diet problem in the standard form given in Example 2.3.

$$\begin{aligned} \text{minimize } z = {} & 4x_1 + 3x_2 \\ \text{subject to } & x_1 + 3x_2 - x_3 = 12 \\ & x_1 + 2x_2 \quad\quad - x_4 = 10 \\ & 2x_1 + x_2 \quad\quad\quad - x_5 = 15 \\ & x_j \geq 0, \quad j = 1, 2, 3, 4, 5. \end{aligned}$$

After introducing artificial variables x_6, x_7, and x_8 as basic variables in phase I, the two-phase method starts from cycle 0 as shown in Table 2.7, and then the value of w becomes zero, i.e., $w = 0$ at cycle 3. In this example, when phase I has finished at cycle 3, since all of the relative cost coefficients of the row of $-z$ are positive, phase II also finishes. Thus, an optimal solution

$$x_1 = 6.6, \quad x_2 = 1.8 \quad (x_3 = 0, x_4 = 0.2, x_5 = 0) \quad z = 31.8$$

is obtained.

2.5 Two-Phase Method

Table 2.7 Simplex tableau of two-phase method for Example 2.3

Cycle	Basis	x_1	x_2	x_3	x_4	x_5	Constants
0	x_6	1	[3]	−1			12
	x_7	1	2		−1		10
	x_8	2	1			−1	15
	−z	4	3				0
	−w	−4	−6	1	1	1	−37
1	x_2	1/3	1	−1/3			4
	x_7	[1/3]		2/3	−1		2
	x_8	5/3		1/3		−1	11
	−z	3		1			−12
	−w	−2		−1	1	1	−13
2	x_2		1	−1	1		2
	x_1	1		2	−3		6
	x_8			−3	[5]	−1	1
	−z			−5	9		−30
	−w			3	−5	1	−1
3	x_2		1	−0.4		0.2	1.8
	x_1	1		0.2		−0.6	6.6
	x_4			−0.6	1	−0.2	0.2
	−z			0.4		1.8	−31.8
	−w			0		0	0

Example 2.8 (Example of infeasible problem with two decision variables and four constraints). As an example of simple infeasible problem, consider a linear programming problem in the standard form for the diet problem of Example 2.7 including the additional inequality constraint

$$4x_1 + 5x_2 \leq 8.$$

Introducing a slack variable x_6, the problem is converted into the following standard form of linear programming:

$$\begin{aligned}
\text{minimize } z = &\ 4x_1 + 3x_2 \\
\text{subject to } &\ x_1 + 3x_2 - x_3 & = 12 \\
&\ x_1 + 2x_2 \quad\quad - x_4 & = 10 \\
&\ 2x_1 + x_2 \quad\quad\quad\quad - x_5 & = 15 \\
&\ 4x_1 + 5x_2 \quad\quad\quad\quad\quad\quad + x_6 & = 8 \\
&\ x_j \geq 0, \quad j = 1, 2, 3, 4, 5, 6.
\end{aligned}$$

Using the slack variable x_6 and the artificial variables x_7, x_8, and x_9 as initial basic variables, phase I of the simplex method is performed. As shown in Table 2.8, phase I is terminated at cycle 1 because of $d_1 > 0$, $d_3 > 0$, $d_4 > 0$, $d_5 > 0$, and $d_6 > 0$. However, from $w = 27.4 > 0$, no feasible solution exists to this problem.

Table 2.8 Infeasible simplex tableau

Cycle	Basis	x_1	x_2	x_3	x_4	x_5	x_6	Constants
0	x_7	1	3	−1				12
	x_8	1	2		−1			10
	x_9	2	1			−1		15
	x_6	4	[5]				1	8
	$-z$	4	3					0
	$-w$	−4	−6	1	1	1		−37
1	x_7	−1.4		−1			−0.6	7.2
	x_8	−0.6			−1		−0.4	6.8
	x_9	1.2				−1	−0.2	13.4
	x_2	0.8	1				0.2	1.6
	$-z$	1.6					−0.6	−4.8
	$-w$	0.8		1	1	1	1.2	−27.4

It should be noted here that since the slack variable x_6 is used as a basic variable, the row of $-w$ is calculated only from the rows of x_7, x_8, and x_9 in cycle 0. For example, $d_1 = -(1+1+2) = -4$. ◊

Example 2.9 (Example of artificial variables left in the basis). As an example where artificial variables remain as a part of basic variables, consider the following problem:

$$\begin{aligned} \text{minimize } z =\ & 3x_1 + x_2 + 2x_3 \\ \text{subject to}\quad & x_1 + x_2 + x_3 = 10 \\ & 3x_1 + x_2 + 4x_3 - x_4 = 30 \\ & 4x_1 + 3x_2 + 3x_3 + x_4 = 40 \\ & x_j \geq 0, \quad j = 1, 2, 3, 4. \end{aligned}$$

Using the artificial variables x_5, x_6, and x_7 as basic variables, phase I of the simplex method is performed. As shown in Table 2.9, phase I is terminated with $w = 0$ at cycle 1. However, the artificial variables x_6 and x_7 still remain in the basis as a part of basic variables. Since $\bar{d}_2 = 3 > 0$, after dropping the columns of x_2 and the row of $-w$, phase II of the simplex method is performed. At cycle 3, an optimal solution

$$x_1 = 0, \ x_2 = 0, \ x_3 = 10, \ x_4 = 10, \ (x_5 = 0, x_6 = 0, x_7 = 0) \quad z = 20$$

is obtained. ◊

The procedure of the simplex method considered thus far provides a means of going from one basic feasible solution to another one such that the objective function z is lower than the previous value of z if there is no degeneracy or at least equal to it

2.5 Two-Phase Method

Table 2.9 Example of artificial variables left in the basis

Cycle	Basis	x_1	x_2	x_3	x_4	Constants
0	x_5	[1]	1	1		10
	x_6	3	1	4	−1	30
	x_7	4	3	3	1	40
	−z	3	1	2	0	0
	−w	−8	−5	−8	0	−80
1	x_1	1	1	1		10
	x_6		−2	[1]	−1	0
	x_7		−1	−1	1	0
	−z	0	−2	−1	0	−30
	−w	0	3	0	0	0
2	x_1	1			[1]	10
	x_3			1	−1	0
	x_7					0
	−z	0		0	−1	−30
3	x_4	1			1	10
	x_3	1		1		10
	x_7					0
	−z	1		0	0	−20

(as can occur in the degenerate case). It continues until (i) the condition of optimality test (2.42) is satisfied, or (ii) the information of unboundedness on the optimal value is provided. Therefore, in case of no degeneracy, the following convergence theorem can be easily understood.

Theorem 2.4 (Finite convergence of simplex method (nondegenerate case)). *Assuming nondegeneracy at each iteration, the simplex method will terminate in a finite number of iterations.*

Proof. Since the number of basic feasible solutions is at most $_nC_m$ and it is finite, the algorithm of the simplex method fails to finitely terminate only if the same basic feasible solution repeatedly appears. Such repetition implies that the value of the objective function z is the same. Under nondegeneracy, however, since each value of z is lower than the previous, no repetition can occur and therefore the algorithm finitely terminates. □

Recall that there is at least one basic variable whose value is zero in a degenerate basic feasible solution. Such degeneracy may occur in an initial feasible canonical form, and it is also possible that after some pivot operations in the procedure of the simplex method, degenerate basic feasible solution may occur.

For example, in step 3 of the procedure of the simplex method, if the minimum of $\{\bar{b}_i/\bar{a}_{is}$ for all $i \mid \bar{a}_{is} > 0\}$ is attained by two or more basic variables, i.e.,

$$\min_{\bar{a}_{is}>0} \frac{\bar{b}_i}{\bar{a}_{is}} = \theta = \frac{\bar{b}_{r_1}}{\bar{a}_{r_1s}} = \frac{\bar{b}_{r_2}}{\bar{a}_{r_2s}}, \tag{2.67}$$

either x_{r_1} or x_{r_2} can be removed from the basis and the other remains in the basis. In either case, both x_{r_1} and x_{r_2} become zeros, i.e.,

$$\left.\begin{aligned} x_{r_1} &= \bar{b}_{r_1} - \bar{a}_{r_1s}\theta = 0 \\ x_{r_2} &= \bar{b}_{r_2} - \bar{a}_{r_2s}\theta = 0. \end{aligned}\right\} \tag{2.68}$$

Thus, since there is at least one basic variable whose value is zero, the new basic feasible solution is degenerate.

This in itself does not undermine the feasibility of the solution. However, if at some iteration a basic feasible solution is degenerate, the value of objective function z could remain the same for some number of subsequent iterations. Moreover, there is a possibility that after a series of pivot operations without decrease of z, the same basis appears, and then the simplex method may be trapped into an endless loop without termination. This phenomenon is called cycling or circling.[3]

The following example given by H.W. Kuhn shows that the simplex method could be trapped into the cycling problem if the smallest index is used as tie breaker.

Example 2.10 (Kuhn's example of cycling). As an example of cycling, consider the following problem given by H.W. Kuhn:

$$\begin{aligned} \text{minimize } z = & & -2x_4 - 3x_5 + x_6 + 12x_7 & \\ \text{subject to} \quad & x_1 & -2x_4 - 9x_5 + x_6 + 9x_7 &= 0 \\ & x_2 & +\tfrac{1}{3}x_4 + x_5 - \tfrac{1}{3}x_6 - 2x_7 &= 0 \\ & x_3 & + 2x_4 + 3x_5 - x_6 - 12x_7 &= 2 \\ & x_j \geq 0, \quad j = 1, 2, \ldots, 7 \end{aligned}$$

Using x_1, x_2, and x_3 as the initial basic variables and performing the simplex method, we have the result shown in Table 2.10. Observing that the tableau of cycle 6 is completely identical to that of cycle 0 in Table 2.10, one finds that cycling occurs. ◊

To avoid the trap of cycling, some means to prevent the procedure from cycling is required. Observe that in the absence of degeneracy the objective function values in a series of iterations of the simplex method form a strictly decreasing monotone sequence that guarantees the same basis does not repeatedly appear. With a degenerate basic solution, the sequence is no longer strictly decreasing. To prevent the procedure from revisiting the same basis, we need to incorporate another rule to keep a strictly monotone decreasing sequence.

---

[3]In his famous 1963 book, G.B. Dantzig adopted the term "circling" for avoiding possible confusion with the term "cycle," which was used synonymously with iteration.

2.5 Two-Phase Method

Table 2.10 Simplex tableau for Kuhn's example of cycling

Cycle	Basis	x_1	x_2	x_3	x_4	x_5	x_6	x_7	Constants
0	x_1	1			−2	−9	1	9	0
	x_2		1		1/3	[1]	−1/3	−2	0
	x_3			1	2	3	−1	−12	2
	−z				−2	−3	1	12	0
1	x_1	1	9		[1]		−2	−9	0
	x_5		1		1/3	1	−1/3	−2	0
	x_3		−3	1	1		0	−6	2
	−z		3		−1		0	6	0
2	x_4	1	9		1		−2	−9	0
	x_5	−1/3	−2			1	1/3	[1]	0
	x_3	−1	−12	1			2	3	2
	−z	1	12				−2	−3	0
3	x_4	−2	−9		1	9	[1]		0
	x_7	−1/3	−2			1	1/3	1	0
	x_3	0	−6	1		−3	1		2
	−z	0	6			3	−1		0
4	x_6	−2	−9		1	9	1		0
	x_7	1/3	[1]		−1/3	−2		1	0
	x_3	2	3	1	−1	−12			2
	−z	−2	−3		1	12			0
5	x_6	[1]			−2	−9	1	9	0
	x_2	1/3	1		−1/3	−2		1	0
	x_3	1		1	0	−6		-3	2
	−z	−1			0	6		3	0
6	x_1	1			−2	−9	1	9	0
	x_2		1		1/3	[1]	−1/3	−2	0
	x_3			1	2	3	−1	−12	2
	−z				−2	−3	1	12	0

Several methods besides the random choice rule exist for avoiding cycling in the simplex method. Among them, a very simple and elegant (but not necessarily efficient) rule due to Bland (1977) is theoretically interesting. Bland's rule is summarized as follows:

(i) Among all candidates to enter the basis, choose the one with the smallest index.
(ii) Among all candidates to leave the basis, choose the one with the smallest index.

The procedure of the simplex method incorporating Bland's anticycling rule, just specifying the choice of both the entering and leaving variables, can now be given in the following.

Procedure of Simplex Method Incorporating Bland's Rule

Step 1B If all of the relative cost coefficients are nonnegative, i.e., $\bar{c}_j \geq 0$ for all indices j of the nonbasic variables, then the current solution is optimal, and stop. Otherwise, by using the relative cost coefficients $\bar{c}_j$, find the index s such that

$$\min\{j \mid \bar{c}_j < 0\} = s. \quad (j : \text{nonbasic})$$

That is, if there are two or more indices j such that $\bar{c}_j < 0$ for all indices of the nonbasic variables, choose the smallest index s as the index of a nonbasic variable newly entering the basis.

Step 2 If all of the coefficients in column s are nonpositive, i.e., $\bar{a}_{is} \leq 0$ for all indices i of the basic variables, then the optimal value is unbounded, and stop.

Step 3B If some of $\bar{a}_{is}$ are positive, find the index r such that

$$\min_{\bar{a}_{is} > 0} \frac{\bar{b}_i}{\bar{a}_{is}} = \frac{\bar{b}_r}{\bar{a}_{rs}} = \theta.$$

If there is a tie in the minimum ratio test, choose the smallest index r as the index of a basic variable leaving the basis.

Step 4 Perform the pivot operation on $\bar{a}_{rs}$ for obtaining a new feasible canonical form with x_s replacing x_r as a basic variable. Return to step 1B.

It is interesting to note here that the use of Bland's rule for cycling prevention can be proven by contradiction on the basis of the following observation.

In a degenerate pivot operation, if some variable x_q enters the basis, then x_q cannot leave the basis until some other variable with a higher index than q, which was nonbasic when x_q entered, also enters the basis. If this holds, then cycling cannot occur because in a cycle any variable that enters must also leave the basis, which means that there exists some highest indexed variable that enters and leaves the basis. This contradicts the foregoing monotone feature.[4]

In practice, however, such a procedure is found to be unnecessary because the simplex procedure generally does not enter a cycle even if degenerate solutions are encountered. However, an anticycling procedure is simple, and therefore many codes incorporate such a procedure for the sake of safety.

Example 2.11 (Simplex method incorporating Bland's rule for Kuhn's example). Apply the simplex method incorporating Bland's rule to the example given by Kuhn. After only two pivot operations, the algorithm stops, and the result is shown in Table 2.11. At cycle 2 in Table 2.11, degeneracy is ended, and an optimal solution

$$x_1 = 2, x_2 = 0, x_3 = 0, x_4 = 2, x_5 = 0, x_6 = 2, x_7 = 0, \quad z = -2$$

is obtained. ◊

---

[4] The interested reader should refer to the solution of Problem 2.8 for a full discussion of the proof.

2.6 Revised Simplex Method

Table 2.11 Simplex tableau incorporating Bland's rule

Cycle	Basis	x_1	x_2	x_3	x_4	x_5	x_6	x_7	Constants
0	x_1	1			-2	-9	1	9	0
	x_2		1		[1/3]	1	$-1/3$	-2	0
	x_3			1	2	3	-1	-12	2
	$-z$				-2	-3	1	12	0
1	x_1	1	6			-3	-1	-3	0
	x_4		3		1	3	-1	-6	0
	x_3		-6	1		-3	[1]	0	2
	$-z$		6			3	-1	0	0
2	x_1	1	0	1		-6		-3	2
	x_4		-3	1	1	0		-6	2
	x_6		-6	1		-3	1	0	2
	$-z$		0	1		0		0	2

2.6 Revised Simplex Method

In performing the simplex method, all the information contained in the tableau is not necessarily used. Only the following items are needed:

Information Needed for Updating the Simplex Tableau

(i) Using the relative cost coefficients $\bar{c}_j$, find the index s such that

$$\min_{\bar{c}_j < 0} \bar{c}_j = \bar{c}_s.$$

(ii) Assuming $\bar{c}_s < 0$, we require the elements of the sth column (pivot column)

$$\bar{p}_s = (\bar{a}_{1s}, \bar{a}_{2s}, \ldots, \bar{a}_{ms})^T$$

and the values of the basic variables

$$\bar{b} = (\bar{b}_1, \bar{b}_2, \ldots, \bar{b}_m)^T.$$

By using these values, the quotients

$$\frac{\bar{b}_r}{\bar{a}_{rs}} = \min_{\bar{a}_{is} > 0} \frac{\bar{b}_i}{\bar{a}_{is}}$$

are calculated for finding the index r. Then, a pivot operation is performed on $\bar{a}_{rs}$ for updating the tableau.

From the above discussion, note that only one nonbasic column $\bar{p}_s$ in the current tableau is required. Since there are many more columns than rows in a linear programming problem, dealing with all the columns $\bar{p}_j$ wastes much computation time and computer storage. A more efficient procedure is to calculate first the relative cost coefficients $\bar{c}_j$ and then the pivot column $\bar{p}_s$ from the data of the original problem. The revised simplex method does precisely this, and the inverse of the current basis matrix is what is needed to calculate them.

We assume again that the $m \times n$ rectangular matrix $A = [p_1 \; p_2 \; \cdots \; p_n]$ for the constraints has the rank of m and $n > m$. Moreover, we assume that a linear programming problem in the standard form is feasible. A basis matrix B is defined as an $m \times m$ nonsingular submatrix formed by selecting some m linearly independent columns from the n columns of matrix A. Note that matrix A contains at least one basis matrix B due to rank$(A) = m$ and $n > m$.

For notational simplicity, without loss of generality, assume that the basis matrix B is formed by selecting the first m columns of matrix A, i.e.,

$$B = [p_1 \; p_2 \; \cdots \; p_m]. \tag{2.69}$$

Let

$$x_B = (x_1, x_2, \ldots, x_m)^T \text{ and } c_B = (c_1, c_2, \ldots, c_m) \tag{2.70}$$

be the corresponding vectors of basic variables and coefficients of the objective function, respectively. Note that c_B is a row vector. The vector x_B satisfies

$$Bx_B = b, \tag{2.71}$$

and one finds that

$$x_B = B^{-1}b = \bar{b}. \tag{2.72}$$

Assume that the basis matrix B is feasible, i.e.,

$$x_B \geq 0. \tag{2.73}$$

As shown earlier, it is convenient to deal with the objective function z as the $(m + 1)$th equation and keep the variable $-z$ in the basis. This augmented system can be written in column form as follows:

$$\sum_{j=1}^{n} \begin{pmatrix} p_j \\ c_j \end{pmatrix} x_j + \begin{pmatrix} 0 \\ 1 \end{pmatrix}(-z) = \begin{pmatrix} b \\ 0 \end{pmatrix}. \tag{2.74}$$

By using the corresponding basis $x_B = (x_1, x_2, \ldots, x_m)^T$, $(-z)$ and the nonbasis $x_N = (x_{m+1}, \ldots, x_n)^T$, (2.74) is also rewritten as

2.6 Revised Simplex Method

$$\begin{bmatrix} p_1 & p_2 & \cdots & p_m & 0 \\ c_1 & c_2 & \cdots & c_m & 1 \end{bmatrix} \begin{pmatrix} x_B \\ -z \end{pmatrix} + \begin{bmatrix} p_{m+1} & \cdots & p_n \\ c_{m+1} & \cdots & c_n \end{bmatrix} x_N = \begin{pmatrix} b \\ 0 \end{pmatrix}. \quad (2.75)$$

Since the basis matrix B is feasible, the $(m+1) \times (m+1)$ matrix

$$\hat{B} = \begin{bmatrix} p_1 & p_2 & \cdots & p_m & 0 \\ c_1 & c_2 & \cdots & c_m & 1 \end{bmatrix} = \begin{bmatrix} B & 0 \\ c_B & 1 \end{bmatrix} \quad (2.76)$$

is also a feasible basis matrix for the enlarged system (2.74). It is easily verified by direct matrix multiplication that the inverse of $\hat{B}$ is

$$\hat{B}^{-1} = \begin{bmatrix} B^{-1} & 0 \\ -c_B B^{-1} & 1 \end{bmatrix}. \quad (2.77)$$

Such an $(m+1) \times (m+1)$ matrix $\hat{B}$ is called an enlarged basis matrix and its inverse $\hat{B}^{-1}$ is called an enlarged basis inverse matrix.

Introducing a simplex multiplier vector

$$\pi = (\pi_1, \pi_2, \ldots, \pi_m) = c_B B^{-1} \quad (2.78)$$

associated with the basis matrix B, the enlarged basis inverse matrix $\hat{B}^{-1}$ is written more compactly as

$$\hat{B}^{-1} = \begin{bmatrix} B^{-1} & 0 \\ -\pi & 1 \end{bmatrix}. \quad (2.79)$$

Premultiplying the enlarged system (2.75) by $\hat{B}^{-1}$, (2.75) becomes as

$$\begin{bmatrix} I & 0 \\ 0^T & 1 \end{bmatrix} \begin{pmatrix} x_B \\ -z \end{pmatrix} + \hat{B}^{-1} \begin{bmatrix} p_{m+1} & \cdots & p_n \\ c_{m+1} & \cdots & c_n \end{bmatrix} x_N = \hat{B}^{-1} \begin{pmatrix} b \\ 0 \end{pmatrix} \quad (2.80)$$

which results in the following canonical form:

$$\begin{pmatrix} x_B \\ -z \end{pmatrix} + \sum_{j=m+1}^{n} \begin{pmatrix} \bar{p}_j \\ \bar{c}_j \end{pmatrix} x_j = \begin{pmatrix} \bar{b} \\ -\bar{z} \end{pmatrix}, \quad (2.81)$$

or equivalently

$$\left.\begin{array}{c}\begin{bmatrix}x_1\\x_2\\&\ddots\\&&x_m\end{bmatrix}+\sum_{j=m+1}^{n}\bar{p}_j x_j=\begin{pmatrix}\bar{b}_1\\\bar{b}_2\\\vdots\\\bar{b}_m\end{pmatrix}\\-z+\sum_{j=m+1}^{n}\bar{c}_j x_j=-\bar{z},\end{array}\right\} \quad (2.82)$$

where

$$\begin{pmatrix}\bar{p}_j\\\bar{c}_j\end{pmatrix}=\hat{B}^{-1}\begin{pmatrix}p_j\\c_j\end{pmatrix}=\begin{bmatrix}B^{-1} & 0\\ -\pi & 1\end{bmatrix}\begin{pmatrix}p_j\\c_j\end{pmatrix}=\begin{pmatrix}B^{-1}p_j\\c_j-\pi p_j\end{pmatrix}, \quad (2.83)$$

$$\begin{pmatrix}\bar{b}\\-\bar{z}\end{pmatrix}=\hat{B}^{-1}\begin{pmatrix}b\\0\end{pmatrix}=\begin{bmatrix}B^{-1} & 0\\-\pi & 1\end{bmatrix}\begin{pmatrix}b\\0\end{pmatrix}=\begin{pmatrix}B^{-1}b\\-\pi b\end{pmatrix}. \quad (2.84)$$

In particular, from (2.83), the updated column vector of the coefficients is represented as

$$\bar{p}_j = B^{-1} p_j, \quad (2.85)$$

and the relative cost coefficient is

$$\bar{c}_j = c_j - \pi p_j. \quad (2.86)$$

These two formulas are fundamental to calculation in the revised simplex method, and as seen in (2.85) and (2.86), $\bar{c}_j$ and $\bar{p}_j$, which are used in each iteration of the simplex method, can be calculated by using the original coefficients c_j and p_j given in the initial simplex tableau, provided that the enlarged basis inverse matrix $\hat{B}^{-1}$, or equivalently the basis inverse matrix B^{-1} and the simplex multipliers π are given.

Now, assume that the smallest $\bar{c}_s$ is found among the relative cost coefficients $\bar{c}_j$ calculated by (2.86) for all nonbasic variables and the corresponding column vector $\bar{p}_s = (\bar{a}_{1s}, \ldots, \bar{a}_{ms})^T$ is obtained by (2.85). If the vector $\bar{b}$ of the values of the basic variables is known at the beginning of each cycle, the pivot element $\bar{a}_{rs}$ is immediately determined. Furthermore, however, we need the inverse matrix $\hat{B}^{*-1}$ of the revised enlarged basis matrix $\hat{B}^*$ corresponding to a new basis made by replacing x_r in the current basis with the new basic variable x_s. From the current $(m+1) \times (m+1)$ enlarged basis matrix

$$\hat{B} = \begin{bmatrix}p_1 & \cdots & p_{r-1} & p_r & p_{r+1} & \cdots & p_m & 0\\c_1 & \cdots & c_{r-1} & c_r & c_{r+1} & \cdots & c_m & 1\end{bmatrix}, \quad (2.87)$$

by removing p_r and c_r and entering p_s and c_s instead, the new enlarged basis matrix

2.6 Revised Simplex Method

$$\hat{B}^* = \left[\begin{array}{cccccc|c} p_1 & \cdots & p_{r-1} & p_s & p_{r+1} & \cdots & p_m & 0 \\ \hline c_1 & \cdots & c_{r-1} & c_s & c_{r+1} & \cdots & c_m & 1 \end{array}\right] \quad (2.88)$$

is obtained.

It can be shown that, without the direct calculations of inverse matrices,[5] the new enlarged basis inverse matrix $\hat{B}^{*-1}$ can be obtained from $\hat{B}^{-1}$ by performing the pivot operation on $\bar{a}_{rs}$ in $\hat{B}^{-1}$. Since $\hat{B}^{*-1}$ is the same as $\hat{B}^{-1}$ except the rth column, the product of $\hat{B}^{-1}$ and $\hat{B}^*$ can be represented as

$$\hat{B}^{-1}\hat{B}^* = \begin{bmatrix} 1 & & & \bar{a}_{1s} & & & \\ & \ddots & & \vdots & & & \\ & & & \bar{a}_{rs} & & & \\ & & & \vdots & \ddots & & \\ & & & \bar{a}_{ms} & & 1 & \\ \hline & & & \bar{c}_s & & & 1 \end{bmatrix}, \quad (2.89)$$

where the rth column $(\bar{a}_{1s}, \ldots, \bar{a}_{rs}, \ldots, \bar{a}_{ms}, \bar{c}_s)^T = \left(\dfrac{\bar{p}_s}{\bar{c}_s}\right)$ is $\hat{B}^{-1}\left(\dfrac{p_s}{c_s}\right)$, and the i column ($i \neq r$) is the $(m+1)$ dimensional unit vectors such that the ith element is one.

After introducing the $(m+1) \times (m+1)$ nonsingular square matrix

$$\hat{E} = \begin{bmatrix} 1 & & & -\bar{a}_{1s}/\bar{a}_{rs} & & & \\ & \ddots & & \vdots & & & \\ & & & 1/\bar{a}_{rs} & & & \\ & & & \vdots & \ddots & & \\ & & & -\bar{a}_{ms}/\bar{a}_{rs} & & 1 & \\ \hline & & & -\bar{c}_s/\bar{a}_{rs} & & & 1 \end{bmatrix} \quad (2.90)$$

that differs from the $(m+1) \times (m+1)$ unit matrix $\hat{I}$ in only the rth column, premultiplying (2.89) by $\hat{E}$ yields

$$\hat{E}\hat{B}^{-1}\hat{B}^* = \hat{I}. \quad (2.91)$$

Hence, by postmultiplying both sides of (2.91) by $\hat{B}^{*-1}$, we have the new enlarged basis inverse matrix

---

[5] Annoying calculations of the inverse matrix make no sense of the revised method.

$$\hat{B}^{*-1} = \hat{E}\hat{B}^{-1}. \tag{2.92}$$

Setting

$$\hat{B}^{*-1} = \left[\begin{array}{c|c} \beta_{ij}^* & 0 \\ \hline -\pi_j^* & 1 \end{array}\right], \quad \hat{B}^{-1} = \left[\begin{array}{c|c} \beta_{ij} & 0 \\ \hline -\pi_j & 1 \end{array}\right] \tag{2.93}$$

and calculating the right-hand side of (2.92), we have

$$\left.\begin{array}{l} \beta_{rj}^* = \dfrac{1}{\bar{a}_{rs}}\beta_{rj}, \quad j = 1, 2, \ldots, m, \\[2mm] \beta_{ij}^* = \beta_{ij} - \dfrac{\bar{a}_{is}}{\bar{a}_{rs}}\beta_{rj}, \quad i = 1, 2, \ldots, m, \ i \neq r, \ j = 1, 2, \ldots, m, \\[2mm] -\pi_j^* = -\pi_j - \dfrac{\bar{c}_s}{\bar{a}_{rs}}\beta_{rj}, \quad j = 1, 2, \ldots, m. \end{array}\right\} \tag{2.94}$$

This means that performing the pivot operation on $\bar{a}_{rs}$ to the current enlarged basis inverse matrix $\hat{B}^{-1}$ gives the new enlarged basis inverse matrix $\hat{B}^{*-1}$.

By adding the superscript $*$ to the values of $\bar{b}$ and $\bar{z}$ for the new enlarged basis matrix $\hat{B}^*$ and premultiplying the enlarged system (2.75) corresponding to $\hat{B}^*$ by $\hat{B}^{*-1}$, the right-hand side becomes as

$$\left(\begin{array}{c} \bar{b}^* \\ \hline -\bar{z}^* \end{array}\right) = \hat{B}^{*-1}\left(\begin{array}{c} b \\ \hline 0 \end{array}\right) = \hat{E}\hat{B}^{-1}\left(\begin{array}{c} b \\ \hline 0 \end{array}\right) = \hat{E}\left(\begin{array}{c} \bar{b} \\ \hline -\bar{z} \end{array}\right). \tag{2.95}$$

Each element of (2.95) is represented by

$$\left.\begin{array}{l} \bar{b}_r^* = \dfrac{\bar{b}_r}{\bar{a}_{rs}}, \\[2mm] \bar{b}_i^* = \bar{b}_i - \dfrac{\bar{a}_{is}}{\bar{a}_{rs}}\bar{b}_r, \quad (i \neq r) \\[2mm] -\bar{z}^* = -\bar{z} - \dfrac{\bar{c}_s}{\bar{a}_{rs}}\bar{b}_r. \end{array}\right\} \tag{2.96}$$

This also means that performing the pivot operation on $\bar{a}_{rs}$ to the current $\bar{b}$ and $-\bar{z}$ gives the new constants $\bar{b}^*$ and $-\bar{z}^*$. As just described, since premultiplying the current enlarged basis inverse matrix $\hat{B}^{-1}$ and the right-hand side of (2.81) by $\hat{E}$ corresponds to a pivot operation, $\hat{E}$ is called a pivot matrix or an elementary matrix.

Now, the procedure of the revised simplex method, starting with an initial basic feasible solution, can be summarized as follows.

2.6 Revised Simplex Method

Procedure of the Revised Simplex Method

Assume that the coefficients A, b, and c of the initial feasible canonical form and the inverse matrix B^{-1} of the initial feasible basis are available.

Step 0 By using B^{-1}, calculate

$$\pi = c_B B^{-1}, \quad x_B = \bar{b} = B^{-1}b, \quad \bar{z} = \pi b,$$

and put them in the revised simplex tableau shown in Table 2.12.

Step 1 Calculate the relative cost coefficients $\bar{c}_j$ for all indices j of the nonbasic variables by

$$\bar{c}_j = c_j - \pi p_j.$$

If all of the relative cost coefficients are nonnegative, i.e., $\bar{c}_j \geq 0$, then the current solution is optimal, and stop. Otherwise, find the index s such that

$$\min_{\bar{c}_j < 0} \bar{c}_j = \bar{c}_s.$$

Step 2 Calculate

$$\bar{p}_s = B^{-1} p_s.$$

If all of the elements of $\bar{p}_s = (\bar{a}_{1s}, \bar{a}_{2s}, \ldots, \bar{a}_{ms})^T$ are nonpositive, i.e., $\bar{a}_{is} \leq 0$ for all indices i of the basic variables, then the optimal value is unbounded, and stop.

Step 3 If some of $\bar{a}_{is}$ are positive, put the values of $\hat{\bar{p}}_s = (\bar{p}_s, \bar{c}_s)^T$ in the revised simplex tableau of Table 2.12, and find the index r such that

$$\frac{\bar{b}_r}{\bar{a}_{rs}} = \min_{\bar{a}_{is} > 0} \frac{\bar{b}_i}{\bar{a}_{is}}.$$

Step 4 Perform the pivot operation on $\bar{a}_{rs}$ to B^{-1}, $-\pi$, $\bar{b}$, and $-\bar{z}$ of Table 2.12, and replace x_r with x_s as a new basic variable. The pivot operation for B^{-1} and $-\pi$ is given by (2.94), and that for $\bar{b}$ and $-\bar{z}$ is also given by (2.96). After updating the revised simplex tableau in Table 2.12, return to step 1.

It should be noted here that in the procedure of the revised simplex method, since the $(m+1)$th column of the enlarged basis inverse matrix $\hat{B}^{-1}$ is always $\begin{pmatrix} 0 \\ \cdots \\ 1 \end{pmatrix}$, it is recommended to neglect it and to use the revised simplex tableau without the $(m+1)$th column of $\hat{B}^{-1}$ as given in Table 2.12.

Table 2.12 Revised simplex tableau

Basis	Basis inverse matrix			Constants	$\hat{\bar{p}}_s$
x_1					
⋮					
x_r		B^{-1}		$\bar{b}$	$\bar{p}_s$
⋮					
x_m					
$-z$		$-\pi$		$-\bar{z}$	$\bar{c}_s$

When starting the revised simplex method with phase I of the two-phase method, the enlarged basis inverse matrix $\hat{B}^{-1}$ can be considered as

$$\hat{B}^{-1} = \begin{bmatrix} B^{-1} & 0 & 0 \\ -\pi & 1 & 0 \\ -\sigma & 0 & 1 \end{bmatrix}, \tag{2.97}$$

where $\sigma = (\sigma_1, \sigma_2, \ldots, \sigma_m)$ is a vector of the simplex multipliers for the objective function w of phase I, and the initial enlarged basis inverse matrix $\hat{B}^{-1}$ is an $(m + 2) \times (m + 2)$ unit matrix.

In phase I, the relative cost coefficients $\bar{d}_j$ are calculated by

$$\bar{d}_j = d_j - \sigma p_j, \quad (j : \text{nonbasic}) \tag{2.98}$$

and the pivot column is determined by $\min_{\bar{d}_j < 0} \bar{d}_j$. Including the row of $-w$, the pivot operations are performed according to step 4 in the procedure of the revised simplex method. After eliminating the row of $-w$, i.e., the $(m + 2)$th row at the beginning of phase II, the above-mentioned revised simplex method is continued.

Example 2.12 (Revised simplex method for production planning problem of Example 1.1). Using the revised simplex method, solve the standard form of the production planning problem of Example 1.1.

$$\begin{aligned}
\text{minimize } z &= -3x_1 - 8x_2 \\
\text{subject to } \quad & 2x_1 + 6x_2 + x_3 = 27 \\
& 3x_1 + 2x_2 \quad\quad + x_4 = 16 \\
& 4x_1 + x_2 \quad\quad\quad + x_5 = 18 \\
& x_j \geq 0, \quad j = 1, 2, 3, 4, 5.
\end{aligned}$$

Employing the slack variables x_3, x_4, and x_5 as basic variables, one finds that the basis matrix B is a 3×3 unit matrix and its inverse B^{-1} is also the same unit matrix. Thus, from (2.78) and (2.84), it follows that

$$\pi = c_B B^{-1} = (0, 0, 0), \quad \bar{b} = B^{-1}b = b = (27, 16, 18)^T, \quad \bar{z} = \pi b = 0.$$

2.6 Revised Simplex Method

Table 2.13 Revised simplex tableau of Example 1.1

Cycle	Basis	Basis inverse matrix			Constants	$\hat{\bar{p}}_s$
0	x_3	1			27	[6]
	x_4		1		16	2
	x_5			1	18	1
	$-z$				0	-8
1	x_2	1/6			4.5	1/3
	x_4	$-1/3$	1		7	[7/3]
	x_5	$-1/6$		1	13.5	11/3
	$-z$	4/3			36	$-1/3$
2	x_2	3/14	$-1/7$		3.5	
	x_1	$-1/7$	3/7		3	
	x_5	5/14	$-11/7$	1	2.5	
	$-z$	9/7	1/7		37	

Putting these values in the revised simplex tableau at cycle 0 of Table 2.13.

After calculating $\bar{c}_j$ for nonbasic variables, the minimum of them is calculated in order to select a new basic variable as follows:

$$\bar{c}_1 = c_1 - \pi p_1 = -3 - (0,0,0)\begin{pmatrix}2\\3\\4\end{pmatrix} = -3,$$

$$\bar{c}_2 = c_2 - \pi p_2 = -8 - (0,0,0)\begin{pmatrix}6\\2\\1\end{pmatrix} = -8,$$

$$\min_{\bar{c}_j < 0} \bar{c}_j = (-3, -8) = \bar{c}_2 = -8 < 0.$$

Since $\bar{c}_2$ is the minimum, x_2 becomes a new basic variable. The corresponding coefficient column vector $\bar{p}_2$ is calculated as

$$\bar{p}_2 = B^{-1}p_2 = \begin{bmatrix}1 & 0 & 0\\0 & 1 & 0\\0 & 0 & 1\end{bmatrix}\begin{pmatrix}6\\2\\1\end{pmatrix} = \begin{pmatrix}6\\2\\1\end{pmatrix},$$

and then it is filled in on the rightmost column of the revised simplex tableau. Since

$$\min\left(\frac{\bar{b}_3}{\bar{a}_{32}}, \frac{\bar{b}_4}{\bar{a}_{42}}, \frac{\bar{b}_5}{\bar{a}_{52}}\right) = \min\left(\frac{27}{6}, \frac{16}{2}, \frac{18}{1}\right) = \frac{27}{6} = 4.5,$$

x_3 becomes a nonbasic variable, and it follows that the pivot element is $\bar{a}_{32} = 6$, which is bracketed by [ ] in Table 2.13. After replacing x_3 with x_2 as a new basic variable, the pivot operation on $\bar{a}_{32} = 6$ is performed to the basis inverse matrix and the constants at cycle 0 of Table 2.13, and the result of cycle 1 is obtained. These values become B^{-1}, $-\pi$, b, and $-\bar{z}$ for the new basis matrix $B = [p_2 \ p_4 \ p_5]$ and the procedure returns to step 1.

From

$$\bar{c}_1 = c_1 - \pi p_1 = -3 - \left(-\frac{4}{3}, 0, 0\right) \begin{pmatrix} 2 \\ 3 \\ 4 \end{pmatrix} = -\frac{1}{3}$$

$$\bar{c}_3 = c_3 - \pi p_3 = 0 - \left(-\frac{4}{3}, 0, 0\right) \begin{pmatrix} 1 \\ 0 \\ 0 \end{pmatrix} = \frac{4}{3}$$

$$\min_{\bar{c}_j < 0} \bar{c}_j = \bar{c}_1 = -\frac{1}{3} < 0,$$

x_1 becomes a new basic variable. By using B^{-1} at cycle 1, $\bar{p}_1$ is calculated as

$$\bar{p}_1 = B^{-1} p_1 = \begin{bmatrix} 1/6 & 0 & 0 \\ -1/3 & 1 & 0 \\ -1/6 & 0 & 1 \end{bmatrix} \begin{pmatrix} 2 \\ 3 \\ 4 \end{pmatrix} = \begin{pmatrix} 1/3 \\ 7/3 \\ 11/3 \end{pmatrix},$$

and it is filled in on the rightmost column of the revised simplex tableau. Since

$$\min\left(\frac{\bar{b}_2}{\bar{a}_{21}}, \frac{\bar{b}_4}{\bar{a}_{41}}, \frac{\bar{b}_5}{\bar{a}_{51}}\right) = \min\left(\frac{4.5}{1/3}, \frac{7}{7/3}, \frac{13.5}{11/3}\right) = \frac{7}{7/3} = 3,$$

x_4 becomes a nonbasic variable. After replacing x_4 with x_1, the pivot operation on $\bar{a}_{41} = 7/3$ bracketed by [ ] is performed to the basis inverse matrix and the constants at cycle 1. This yields the result at cycle 2 in Table 2.13. The next basis matrix becomes as $B = [p_2 \ p_1 \ p_5]$, and since

$$\bar{c}_3 = c_3 - \pi p_3 = 0 - \left(-\frac{9}{7}, -\frac{1}{7}, 0\right) \begin{pmatrix} 1 \\ 0 \\ 0 \end{pmatrix} = \frac{9}{7} > 0$$

$$\bar{c}_4 = c_4 - \pi p_4 = 0 - \left(-\frac{9}{7}, -\frac{1}{7}, 0\right) \begin{pmatrix} 0 \\ 1 \\ 0 \end{pmatrix} = \frac{1}{7} > 0,$$

an optimal solution

$$x_1 = 3, \quad x_2 = 3.5 \quad (x_3 = x_4 = 0, x_5 = 2.5), \quad z = -37$$

is obtained.

2.7 Duality

The notion of duality is one of the most important concepts in linear programming. Basically, associated with a linear programming problem (we may call it the primal problem), defined by the constraint matrix A, the right-hand side constant vector b, and the cost coefficient vector c, there is the corresponding linear programming problem (called the dual problem) which is specified by the same set of coefficients A, b, and c. These two problems bear interesting and useful relationships to one another.

Consider the standard form of linear programming

$$\left. \begin{array}{l} \text{minimize } z = cx \\ \text{subject to} \quad Ax = b \\ \phantom{\text{subject to}} \quad x \geq 0, \end{array} \right\} \quad (2.99)$$

where c is an n dimensional row vector, x is an n dimensional column vector, A is an $m \times n$ matrix A, and b is an n dimensional column vector. Introducing an m dimensional row vector $\pi = (\pi_1, \pi_2, \ldots, \pi_m)$, we define an associated linear programming problem:

$$\left. \begin{array}{l} \text{maximize } v = \pi b \\ \text{subject to} \quad \pi A \leq c. \end{array} \right\} \quad (2.100)$$

It should be noted here that problem (2.100) is a maximization problem with m unrestricted variables and n inequality constraints. The roles of the variables and constraints are somewhat reversed in problems (2.99) and (2.100). Usually, the original problem (2.99) is called the primal problem and the related problem (2.100) is called the dual problem. The two problems make a primal–dual pair. Similarly, an element of the vector x is called a primal variable, and that of the vector π is called a dual variable.

The constraint of the dual problem (2.100), which is a product of the m dimensional row vector π and the $m \times n$ constraint matrix A, is alternatively expressed as

$$\left. \begin{array}{l} a_{11}\pi_1 + a_{21}\pi_2 + \cdots + a_{m1}\pi_m \leq c_1 \\ a_{12}\pi_1 + a_{22}\pi_2 + \cdots + a_{m2}\pi_m \leq c_2 \\ \cdots\cdots\cdots \\ a_{1n}\pi_1 + a_{2n}\pi_2 + \cdots + a_{mn}\pi_m \leq c_n, \end{array} \right\} \quad (2.101)$$

which implies that the coefficients of the system of inequalities (2.101) are given by the transposed matrix A^T of A.

Any primal problem can be changed into a linear programming problem in a different format by using the following devices: (i) replace an unconstrained variable with the difference of two nonnegative variables; (ii) replace an equality constraint

Table 2.14 Primal–dual relationships

Minimization problem		Maximization problem
Constraints		Variables
$\geq$	$\Leftrightarrow$	≥ 0
$\leq$	$\Leftrightarrow$	≤ 0
$=$	$\Leftrightarrow$	Unrestricted
Variables		Constraints
≥ 0	$\Leftrightarrow$	$\leq$
≤ 0	$\Leftrightarrow$	$\geq$
Unrestricted	$\Leftrightarrow$	$=$

with two opposing inequalities; and (iii) replace an inequality constraint with an equality by adding a slack or surplus variable.

For example, consider the following linear programming problem involving not only equality constraints but also inequality constraints and free variables:

$$\left.\begin{aligned}
\text{minimize } z = {}& c^1 x^1 + c^2 x^2 + c^3 x^3 \\
\text{subject to } & A_{11} x^1 + A_{12} x^2 + A_{13} x^3 \geq b^1 \\
& A_{21} x^1 + A_{22} x^2 + A_{23} x^3 \leq b^2 \\
& A_{31} x^1 + A_{32} x^2 + A_{33} x^3 = b^3 \\
& x^1 \geq \mathbf{0},\ x^2 \leq \mathbf{0}.
\end{aligned}\right\} \quad (2.102)$$

By converting this problem to its standard form by introducing slack and surplus variables, and substituting $x^2 = -x^{2+}$ ($x^{2+} \geq \mathbf{0}$) and $x^3 = x^{3+} - x^{3-}$ ($x^{3+} \geq \mathbf{0}$, $x^{3-} \geq \mathbf{0}$), it can be easily understood that its dual becomes

$$\left.\begin{aligned}
\text{maximize } v = {}& \pi^1 b^1 + \pi^2 b^2 + \pi^3 b^3 \\
\text{subject to } & \pi^1 A_{11} + \pi^2 A_{21} + \pi^3 A_{31} \leq c^1 \\
& \pi^1 A_{12} + \pi^2 A_{22} + \pi^3 A_{32} \geq c^2 \\
& \pi^1 A_{13} + \pi^2 A_{23} + \pi^3 A_{33} = c^3 \\
& \pi^1 \geq \mathbf{0},\ \pi^2 \leq \mathbf{0}.
\end{aligned}\right\} \quad (2.103)$$

Carefully comparing this dual problem (2.103) with the primal problem (2.102) gives the relationships between the primal and dual pair summarized in Table 2.14. For example, an unrestricted variable corresponds to an equality constraint.

By utilizing the relationships in Table 2.14, it is possible to write the dual problem for a given linear programming problem without going through the intermediate step of converting the problem to the standard form. From Table 2.14, the symmetric primal-dual pair given in Table 2.15 is immediately obtained. In a symmetric form, it is especially easy to see that the dual of the dual is the primal.

The relationship between the primal and dual problems is called duality. The following theorem, sometimes called the weak duality theorem, is easily proven and gives us an important relationship between the two problems. In the following, it is convenient to deal with a primal problem in the standard form.

2.7 Duality

Table 2.15 Symmetric primal–dual pair

Primal		Dual	
Minimize	$z = cx$	Maximize	$v = \pi b$
Subject to	$Ax \geq b$	Subject to	$\pi A \leq c$
	$x \geq 0$		$\pi \geq 0$

Theorem 2.5 (Weak duality theorem). *If $\bar{x}$ and $\bar{\pi}$ are feasible primal and dual solutions, then*

$$\bar{z} = c\bar{x} \geq \bar{\pi} b = \bar{v}. \tag{2.104}$$

Proof. From the dual feasibility of $\bar{\pi}$ and the primal feasibility of $\bar{x}$, we have

$$c \geq \bar{\pi} A, \text{ and } A\bar{x} = b, \bar{x} \geq 0,$$

which implies

$$c\bar{x} \geq \bar{\pi} A\bar{x} = \bar{\pi} b.$$

□

This theorem shows that the primal (minimization) problem is always bounded below by the dual (maximization) problem and the dual (maximization) problem is always bounded above by the primal (minimization) problem if they are feasible.

From the weak duality theorem, several corollaries can be immediately obtained.

Corollary 2.1. *If $\bar{x}^o$ and $\bar{\pi}^o$ are feasible primal and dual solutions and $cx^o = \pi^o b$ holds, then $\bar{x}^o$ and $\bar{\pi}^o$ are optimal solutions to their respective problems.*

This corollary implies that if a pair of feasible solutions can be found to the primal and dual problems with the same objective value, then they are both optimal.

Corollary 2.2. *If the primal problem is unbounded below, then the dual problem is infeasible.*

Corollary 2.3. *If the dual problem is unbounded above, then the primal problem is infeasible.*

With these results, the following duality theorem, sometimes called the strong duality theorem, can be established as a stronger result.

Theorem 2.6 (Strong duality theorem).

(i) *If either the primal or the dual problem has a finite optimal solution, then so does the other, and the corresponding values of the objective functions are the same.*

(ii) *If one problem has an unbounded objective value, then the other problem has no feasible solution.*

Proof. (i) It is sufficient, in proving the first statement, to assume that the primal has a finite optimal solution, and then we show that the dual has a solution with the same value of the objective function.

To show that the optimal values are the same, let x^o solve the primal. Since the primal must have a basic optimal solution, we may as well assume x^o as the basic, with the optimal basis matrix B^o, and the vector of basic variables $x^o_{B^o}$. Thus

$$B^o x^o_{B^o} = b, \quad x^o_{B^o} \geq 0.$$

The simplex multiplier vector associated with B^o is

$$\pi^o = c_{B^o}(B^o)^{-1},$$

where c_{B^o} is the vector of cost coefficients of basic variables. Since x^o is optimal, the relative cost coefficients $\bar{c}_j$ given by (2.86) are nonnegative:

$$\bar{c}_j = c_j - \pi^o p_j \geq 0, \quad j = 1, \ldots, n,$$

or, in matrix form,

$$\pi^o A \leq c.$$

Thus, π^o satisfies the dual constraints, and the corresponding objective value is

$$v^o = \pi^o b = c_{B^o}(B^o)^{-1} b = c_{B^o} x^o_{B^o} = z^o.$$

Hence, from Corollary 2.1, it directly follows that π^o is an optimal solution to the dual problem.

(ii) The second statement is an immediate consequence of Corollaries 2.2 and 2.3.
□

The preceding proof illustrates some important points.

(i) The constraints of the dual problem exactly represent the optimality conditions of the primal problem, and the relative cost coefficients $\bar{c}_j$ can be interpreted as slack variables in them.
(ii) The simplex multiplier vector π^o associated with a primal optimal basis solves the corresponding dual problem. Since, as shown in the previous section, the vector $-\pi$ is contained in the bottom row of the revised simplex tableau for the primal problem, the optimal revised simplex tableau inherently provides a dual optimal solution.

For interpreting the relationships between the primal and dual problems, recall the two-variable diet problem of Example 2.3. Associated with this problem, we examine the following problem, though it is somewhat intentional.

2.7 Duality

Example 2.13 (Dual problem for the diet problem of Example 2.3). A drug company wants to maximize the total profit by producing three pure tablets V_1, V_2, and V_3 which contain exactly one mg (milligram) of the nutrients N_1, N_2, and N_3, respectively. To do so, the company attempts to determine the prices of three tablets which compare favorably with those of the two foods F_1 and F_2. Let π_1, π_2, and π_3 denote the prices in yens of one tablet of V_1, V_2, and V_3, respectively.

One gram of the food F_1 provides 1, 1, and 2 mg of N_1, N_2, and N_3 and costs 4 yen. If the housewife replaces one gram of this food F_1 with tablets of V_1, V_2, and V_3, one tablet of V_1, one tablet of V_2, and two tablets of V_3 are needed. This would cost $\pi_1+\pi_2+2\pi_3$, which should be less than or equal to the price of one gram of the food F_1, i.e., $\pi_1 + \pi_2 + 2\pi_3 \leq 4$. Similarly, one gram of the food F_2 provides 1, 1, and 2 mg of N_1, N_2, and N_3 and costs 3 yen. Thus, the inequality $3\pi_1 + 2\pi_2 + \pi_3 \leq 3$ is imposed. Since the housewife understands that the daily requirements of the nutrients N_1, N_2, and N_3 are 12, 10, and 15 mg, respectively, the cost of meeting these requirements by using the tablets would be $v = 12\pi_1 + 10\pi_2 + 15\pi_3$. Thus, the company should determine the prices of the tablets V_1, V_2, and V_3 so as to maximize this function subject to the above two inequalities. That is, the company determines the prices of the three tablets which maximize the profit function

$$v = 12\pi_1 + 10\pi_2 + 15\pi_3$$

subject to the constraints

$$\pi_1 + \pi_2 + 2\pi_3 \leq 4$$
$$3\pi_1 + 2\pi_2 + \pi_3 \leq 3$$
$$\pi_1 \geq 0, \ \pi_2 \geq 0, \ \pi_3 \geq 0.$$

It should be noted here that this linear programming problem is precisely the dual of the original diet problem of Example 2.3. ◊

As thus far discussed, the dual variables, corresponding to the constraints of the primal problem, coincide with the simplex multipliers for the optimal basic solution of the primal problem. Consider the economic interpretation of the simplex multiplier. Let

$$x^o = (x_1^o, x_2^o, \ldots, x_n^o)^T \text{ and } \pi^o = (\pi_1^o, \pi_2^o, \ldots, \pi_m^o)$$

be the optimal solutions of the primal and dual problems, respectively. From the strong duality theorem, it follows that

$$z^o = c_1 x_1^o + c_2 x_2^o + \cdots + c_n x_n^o = \pi_1^o b_1 + \pi_2^o b_2 + \cdots + \pi_m^o b_m = v^o.$$

In this relation, one finds that

$$z^o = \pi_1^o b_1 + \pi_2^o b_2 + \cdots + \pi_m^o b_m, \tag{2.105}$$

and then it can be intuitively understood that when one unit of the right-hand side constant b_i of the ith constraint $a_{i1}x_1 + a_{i2}x_2 + \cdots + a_{in}x_n = b_i$ of the primal problem is changed from b_i to $b_i + 1$, the value of the objective function will increase by π_i^o as long as the basis does not change.

To be more precise, from (2.105), the amount of change in the objective function z for a small change in b_i is obtained by partially differentiating z with respect to the right-hand side b_i, i.e.,

$$\pi_i^o = \frac{\partial z^o}{\partial b_i}, \quad i = 1, \ldots, m. \tag{2.106}$$

Thus, the simplex multiplier π_i indicates how much the value of the objective function varies for a small change in the right-hand side of the constraint, and therefore it is referred to as the shadow price or the marginal price.

Using the duality theorem, the following result, known as Farkas's theorem concerning systems of linear equalities and inequalities, can be easily proven.

Theorem 2.7 (Farkas's theorem). *One and only one of the following two alternatives holds.*

(i) *There exists a solution $x \geq 0$ such that $Ax = b$.*
(ii) *There exists a solution π such that $\pi A \leq 0^T$ and $\pi b > 0$.*

Proof. Consider the (primal) linear problem

$$\left. \begin{array}{ll} \text{minimize } z = 0^T x \\ \text{subject to} \quad Ax = b \\ \quad\quad\quad\quad\quad x \geq 0, \end{array} \right\} \tag{2.107}$$

and its dual

$$\left. \begin{array}{ll} \text{maximize } v = \pi b \\ \text{subject to} \quad \pi A \leq 0^T. \end{array} \right\} \tag{2.108}$$

If the statement (i) holds, the primal problem is feasible. Since the value of the objective function z is always zero, any feasible solution is optimal. From the strong duality theorem, the value of the objective function v of the dual is zero. Thus, the statement (ii) does not hold.

Conversely, if the statement (ii) holds, the dual problem has a feasible solution such that the objective function v is positive. From the weak duality theorem, this implies that the objective function z of the primal is positive, and therefore the primal problem has no feasible solution. □

Associated with Farkas's theorem, Gordon's theorem also plays an important role for deriving the optimality conditions of nonlinear programming.

2.8 Dual Simplex Method

Theorem 2.8 (Gordon's theorem). *One and only one of the following two alternatives holds.*

(i) *There exists a solution $x \geq 0$, $x \neq 0$ such that $Ax = 0$.*
(ii) *There exists a solution π such that $\pi A < 0^T$.*

The following theorem, relating the primal and dual problems, is often useful.

Theorem 2.9 (Complementary slackness theorem). *Let x be a feasible solution to the primal problem (2.99) and π be a feasible solution to the dual problem (2.100). Then they are respectively optimal if and only if the complementary slackness condition*

$$(c - \pi^o A)x^o = 0 \tag{2.109}$$

is satisfied.

2.8 Dual Simplex Method

There are a number of algorithms for linear programming which start with an infeasible solution to the primal and iteratively force a sequence of solutions to become feasible as well as optimal. The most prominent among such methods is the dual simplex method (Lemke 1954). Operationally, its procedure still involves a sequence of pivot operations, but with different rules for choosing the pivot element.

Consider a primal problem in the standard form

$$\left. \begin{array}{ll} \text{minimize } z = & cx \\ \text{subject to} & Ax = b \\ & x \geq 0, \end{array} \right\}$$

and its dual

$$\left. \begin{array}{ll} \text{maximize } v = & \pi b \\ \text{subject to} & \pi A \leq c. \end{array} \right\}$$

Consider the canonical form of the primal problem starting with the basis $(x_1, x_2, \ldots, x_m)$ expressed as

$$\left. \begin{array}{c} \begin{bmatrix} x_1 \\ & x_2 \\ & & \ddots \\ & & & x_m \end{bmatrix} + \sum_{j=m+1}^{n} \bar{p}_j x_j = \begin{pmatrix} \bar{b}_1 \\ \bar{b}_2 \\ \vdots \\ \bar{b}_m \end{pmatrix} \\ -z + \sum_{j=m+1}^{n} \bar{c}_j x_j = -\bar{z}, \end{array} \right\} \tag{2.110}$$

where not all right-hand side constants $\bar{b}_i$ may be nonnegative, i.e., for some i, $\bar{b}_i \geq 0$ may not hold.

In this canonical form, if $\bar{c}_j = c_j - \pi p_j \geq 0$ for all $j = m+1, \ldots, n$, which can be alternatively expressed as $\pi A \leq c$ in a vector-matrix form, π is a feasible solution to the dual problem. Thus, the canonical form of the primal problem (2.110) satisfying $\bar{c}_j \geq 0$, $j = m+1, \ldots, n$ is called the dual feasible canonical one. Obviously, if the dual feasible canonical form is also feasible to the primal problem, i.e., for all i, $\bar{b}_i \geq 0$ hold, then it is an optimal canonical form.

Now, in a quite similar way to the selection rule of $\bar{c}_s$ in the simplex method, find the pivot row by

$$\min_{\bar{b}_i < 0} \bar{b}_i = \bar{b}_r.$$

It should be noted that if $\bar{b}_r \geq 0$ for all r, it follows that an optimal solution is obtained.

If $\bar{a}_{rj} \geq 0$ for all j, from $\bar{b}_r < 0$, in the rth equation

$$x_r = \bar{b}_r - \sum_{j=m+1}^{n} \bar{a}_{rj} x_j,$$

the right-hand side is negative for $x_j \geq 0$, $j = m+1, \ldots, n$, which implies that the value of the basic variable x_r is negative, i.e., $x_r < 0$ for all the nonnegative nonbasic variables x_j. This means that the primal problem is infeasible, and then the following theorem is obtained.

Theorem 2.10 (Infeasibility of primal problem). *In the rth row of the canonical form (2.110), if*

$$\bar{b}_r < 0, \quad \bar{a}_{rj} \geq 0, \quad j = m+1, m+2, \ldots, n, \qquad (2.111)$$

then the primal problem is infeasible.

Now, in the dual feasible canonical form (2.110), let the rth row be the pivot one. Moreover, assume that $\bar{b}_r$ is negative and for some j at least one $\bar{a}_{rj}$ is negative. Then, if the pivot column is found by

$$\min_{\bar{a}_{rj} < 0} \frac{\bar{c}_j}{-\bar{a}_{rj}} = \frac{\bar{c}_s}{-\bar{a}_{rs}} = \Delta, \qquad (2.112)$$

$\bar{a}_{rs}$ is chosen as the pivot element.

By performing the pivot operation on $\bar{a}_{rs}$ which means that x_r is replaced with x_s as a new basic variable, as shown in Table 2.3, the resulting new relative cost coefficients $\bar{c}_j^*$ for nonbasic variables, which are discriminated by adding the superscript $*$, are represented as

2.8 Dual Simplex Method

$$\bar{c}_j^* = \bar{c}_j - \bar{c}_s \bar{a}_{rj}^* = \bar{c}_j - \bar{c}_s \frac{\bar{a}_{rj}}{\bar{a}_{rs}}.$$

Obviously, for any column index j of a nonbasic variable such that $\bar{a}_{rj} \geq 0$ holds, from $\bar{c}_s > 0$ and $\bar{a}_{rs} < 0$, it directly follows that its relative cost coefficient is nonnegative, i.e.,

$$\bar{c}_j^* = \bar{c}_j - \bar{c}_s \frac{\bar{a}_{rj}}{\bar{a}_{rs}} \geq \bar{c}_j \geq 0.$$

For any column index j of a nonbasic variable such that $\bar{a}_{rj} < 0$ holds, from (2.112), it follows that its relative cost coefficient is also nonnegative, i.e.,

$$\bar{c}_j^* = \bar{a}_{rj} \left(\frac{\bar{c}_s}{-\bar{a}_{rs}} - \frac{\bar{c}_j}{-\bar{a}_{rj}} \right) \geq 0.$$

Hence, it holds that $\bar{c}_j^* \geq 0$ for all j, the resulting new canonical form (tableau) is a dual feasible canonical form.

Moreover, by the pivot operation on $\bar{a}_{rs}$, we also have the updated value of the objective function

$$\bar{z}^* = \bar{z} + \bar{c}_s \frac{\bar{b}_r}{\bar{a}_{rs}} = \bar{z} - \bar{b}_r \Delta,$$

and from $\bar{b}_r < 0$ and $\Delta \geq 0$, the value is increased by $|\bar{b}_r \Delta|$ compared to the previous value of $\bar{z}$.[6]

After starting with the dual feasible canonical form, the dual simplex method improves feasible solutions of the dual problems through a series of pivot operations in order to seek for an optimal solution. Although the dual simplex method uses the pivot operations in a similar way to the simplex method, it employs a different rule for choosing the pivot element and the value of the objective function increases with the number of iterations. The procedure of the dual simplex method, starting with the dual feasible canonical form, can be summarized as follows.

Procedure of the Dual Simplex Method

Start with the dual feasible canonical form. That is, assume that $\bar{c}_j \geq 0$ for all j.

Step 1 If $\bar{b}_i \geq 0$ for all indices i of the basic variables, then the current solution is optimal, and stop. Otherwise, choose the index r for the pivot row such that

---

[6]If $\bar{c}_s = 0$ and dual degeneracy occurs, it is possible to avoid cycling by utilizing the similar anticycling rule in the simplex method.

$$\min_{\bar{b}_i < 0} \bar{b}_i = \bar{b}_r.$$

Step 2 If $\bar{a}_{rj} \geq 0$ for all indices j of the nonbasic variables, then the primal problem is infeasible, and stop.

Step 3 If some of $\bar{a}_{rj}$ are negative, find the index s for the pivot column such that

$$\min_{\bar{a}_{rj} < 0} \frac{\bar{c}_j}{-\bar{a}_{rj}} = \frac{\bar{c}_s}{-\bar{a}_{rs}} = \Delta.$$

Step 4 Perform the pivot operation on $\bar{a}_{rs}$ for obtaining a new dual feasible canonical form with x_s replacing x_r as a basic variable. Return to step 1.

Example 2.14 (Dual simplex method for the diet problem of Example 2.3). Using the dual simplex method, solve the diet problem in the standard form given in Example 2.3:

$$\begin{aligned}
\text{minimize } z &= 4x_1 + 3x_2 \\
\text{subject to } \quad x_1 + 3x_2 - x_3 &= 12 \\
x_1 + 2x_2 \quad\quad - x_4 &= 10 \\
2x_1 + x_2 \quad\quad\quad - x_5 &= 15 \\
x_j &\geq 0, \quad j = 1, 2, 3, 4, 5.
\end{aligned}$$

Multiplying both sides of the three equations of the constraints by -1 yields the dual feasible canonical form

$$\begin{aligned}
-x_1 - 3x_2 + x_3 \quad\quad\quad\quad &= -12 \\
-x_1 - 2x_2 \quad\quad + x_4 \quad\quad &= -10 \\
-2x_1 - x_2 \quad\quad\quad\quad + x_5 &= -15. \\
4x_1 + 3x_2 \quad\quad\quad\quad\quad - z &= 0 \\
x_j \geq 0, \quad j = 1, 2, 3, 4, 5 &
\end{aligned}$$

Since $\bar{c}_1 = 4 > 0$ and $\bar{c}_2 = 3 > 0$, this canonical form with basic variables x_3, x_4, and x_5 is dual feasible. However, it is not primal feasible because $\bar{b}_1 = -12 < 0$, $\bar{b}_2 = -10 < 0$ and $\bar{b}_3 = -9 < 0$.

At cycle 0 in Table 2.16, from

$$\min(\bar{b}_3, \bar{b}_4, \bar{b}_5) = \min(-12, -10, -15) = -15 < 0,$$

x_5 becomes a nonbasic variable in the next cycle. From

$$\min\left(\frac{\bar{c}_1}{-\bar{a}_{51}}, \frac{\bar{c}_2}{-\bar{a}_{52}}\right) = \min\left(\frac{4}{2}, \frac{3}{1}\right) = \frac{4}{2},$$

2.8 Dual Simplex Method

Table 2.16 Simplex tableau of Example 2.3 (dual simplex method)

Cycle	Basis	x_1	x_2	x_3	x_4	x_5	Constants
0	x_3	-1	-3	1			-12
	x_4	-1	-2		1		-10
	x_5	$[-2]$	-1			1	-15
	$-z$	4	3				0
1	x_3		$[-2.5]$	1		-0.5	-4.5
	x_4		-1.5		1	-0.5	-2.5
	x_1	1	0.5			-0.5	7.5
	$-z$		1			2	-30
2	x_2		1	-0.4		0.2	1.8
	x_4			-0.6	1	-0.2	0.2
	x_1	1		0.2		-0.6	6.6
	$-z$			0.4		1.8	-31.8

x_1 becomes a basic variable in the next cycle, and the pivot element is determined at $\bar{a}_{51} = -2$ bracketed by [ ] in Table 2.16. After performing the pivot operation on $\bar{a}_{51} = -2$, the tableau at cycle 1 is obtained. At cycle 1, from $\bar{b}_1 > 0$ and

$$\min(\bar{b}_3, \bar{b}_4) = \min(-4.5, -2.5) = -4.5 < 0$$

x_3 becomes a nonbasic variable in the next cycle. From

$$\min\left(\frac{\bar{c}_2}{-\bar{a}_{32}}, \frac{\bar{c}_5}{-\bar{a}_{35}}\right) = \min\left(\frac{1}{2.5}, \frac{2}{0.5}\right) = \frac{1}{2.5}$$

x_2 becomes a basic variable in the next cycle, and the pivot element is determined at $\bar{a}_{32} = -2.5$ bracketed by [ ]. After performing the pivot operation on $\bar{a}_{32} = -2.5$, the tableau at cycle 2 is obtained. At cycle 2, all of the constants $\bar{b}_i$ become positive, and an optimal solution

$$x_1 = 6.6,\ x_2 = 1.8\ (x_3 = 0, x_4 = 0.2, x_5 = 0),\quad z = 31.8$$

is obtained. Observe that the tableau of cycle 2 in Table 2.16 coincides with that of cycle 3 in Table 2.7 when the row of $-w$ is dropped. ◊

It should be noted here that the idea of the revised simplex method can be employed in the discussion of the dual simplex method. In the dual simplex method, in addition to the data of the initial feasible canonical form A, b, and c, the coefficients $\bar{a}_{rj}$ for all indices j of the nonbasic variables with respect to x_r left from the basis and the relative cost coefficients $\bar{c}_j$ for all indices j of the nonbasic variables are required, where $\bar{c}_j$ can be computed by the formula $\bar{c}_j = c_j - \pi p_j$ of the revised simplex method. Hence, if the formula for calculating $\bar{a}_{rj}$ for all indices j of the nonbasic variables through the basis inverse matrix B^{-1} is given, the dual simplex method can be expressed in a style followed by the revised simplex method. Since the coefficient $\bar{a}_{rj}$ is the rth element of $\bar{p}_j$, by using the rth row vector of B^{-1}, denoted by $[B^{-1}]_{r\cdot}$, it can be calculated just as

$$\bar{a}_{rj} = [B^{-1}]_{r\cdot}\, p_j, \quad j : \text{nonbasic}. \tag{2.113}$$

With the above discussion, the procedure of the revised dual simplex method can be summarized as follows.

Procedure of the Revised Dual Simplex Method

Assume that the coefficients A, b, and c of the initial dual feasible canonical form and the inverse matrix B^{-1} of the initial dual feasible basis are available.

Step 0   Using B^{-1}, calculate

$$\pi = c_B B^{-1}, \ x_B = \bar{b} = B^{-1}b, \ \bar{z} = \pi b$$

and put them in the revised simplex tableau shown in Table 2.12.

Step 1   If $\bar{b}_i \geq 0$ for all indices i of the basic variables, then the current solution is optimal, and stop. Otherwise, choose the index r for the pivot row such that

$$\min_{\bar{b}_i < 0} \bar{b}_i = \bar{b}_r.$$

Step 2   For all indices j of the nonbasic variables, calculate

$$\bar{a}_{rj} = [B^{-1}]_{r\cdot}\, p_j.$$

If $\bar{a}_{rj} \geq 0$ for all indices j of the nonbasic variables, then the primal problem is infeasible, and stop.

Step 3   If some of $\bar{a}_{rj}$ are negative, calculate

$$\bar{c}_j = c_j - \pi p_j$$

and find the index s for the pivot column such that

$$\min_{\bar{a}_{rj}<0} \frac{\bar{c}_j}{-\bar{a}_{rj}} = \frac{\bar{c}_s}{-\bar{a}_{rs}} = \Delta.$$

In Table 2.12, replace x_r with x_s as a basic variable.

Step 4   Calculate

$$\bar{p}_s = B^{-1} p_s$$

2.8 Dual Simplex Method

Table 2.17 Revised dual simplex tableau of Example 2.3

Cycle	Basis	Basis inverse matrix			Constants	$\hat{p}_s$
0	x_3	1			−12	−1
	x_4		1		−10	−1
	x_5			1	−15	[−2]
	$-z$					4
1	x_3	1		−1/2	−9/2	[−5/2]
	x_4		1	−1/2	−5/2	−3/2
	x_1			−1/2	15/2	1/2
	$-z$			2	−30	1
2	x_2	−2/5		1/5	9/5	
	x_4	−3/5	1	−1/5	1/5	
	x_1	1/5		−3/5	33/5	
	$-z$	2/5		9/5	−159/5	

and put the values of $\hat{p}_s = (\bar{p}_s, \bar{c}_s)^T$ in the column $\hat{p}_s$ of Table 2.12. Perform the pivot operation on $\bar{a}_{rs}$ to $B^{-1}, -\pi, \bar{b}, -\bar{z}$ of Table 2.12, and return to step 1.

Example 2.15 (Revised dual simplex method for the diet problem of Example 2.3). The canonical form

$$\begin{aligned}
-x_1 - 3x_2 + x_3 &= -12 \\
-x_1 - 2x_2 + x_4 &= -10 \\
-2x_1 - x_2 + x_5 &= -15 \\
4x_1 + 3x_2 - z &= 0 \\
x_j \geq 0, \quad j = 1, 2, 3, 4, 5,
\end{aligned}$$

for the diet problem discussed in Examples 2.3 and 2.14, where x_3, x_4, and x_5 are basic variables, is dual feasible because $\bar{c}_1 = 4 > 0$ and $\bar{c}_2 = 3 > 0$. However, since $\bar{b}_1 = -12 < 0$, $\bar{b}_2 = -10 < 0$, and $\bar{b}_3 = -9 < 0$, the primal problem is not feasible. The initial basis matrix B is the 3×3 unit matrix and its inverse B^{-1} is also the same unit matrix. Hence, from (2.78) and (2.84), it follows that

$$\pi = c_B B^{-1} = (0,0,0), \quad \bar{b} = B^{-1}b = b = (-12, -10, -15)^T, \quad \bar{z} = \pi b = 0.$$

Putting these values in the revised dual simplex tableau at cycle 0 of Table 2.17.
At cycle 0 in Table 2.17, since

$$\min(\bar{b}_3, \bar{b}_4, \bar{b}_5) = \min(-12, -10, -15) = -15 < 0,$$

x_5 becomes a nonbasic variable in the next cycle and the index r of the variable leaving the basis is determined as $r = 3$.

According to (2.113), we calculate the coefficients $\bar{a}_{rj}$, $r = 3$, $j = 1, 2$ for the nonbasic variables. That is, using the third row $[B^{-1}]_3$. of B^{-1}, p_1, and p_2, we have

$$\bar{a}_{31} = [B^{-1}]_3 \cdot p_1 = (0, 0, 1)(-1, -1, -2)^T = -2$$
$$\bar{a}_{32} = [B^{-1}]_3 \cdot p_2 = (0, 0, 1)(-3, -2, -1)^T = -1.$$

Also the relative cost coefficients $\bar{c}_j$, $j = 1, 2$ are calculated as

$$\bar{c}_1 = c_1 - \pi p_1 = 4 + (0, 0, 0)(-1, -1, -2)^T = 4$$
$$\bar{c}_2 = c_2 - \pi p_2 = 3 + (0, 0, 0)(-3, -2, -1)^T = 3.$$

From

$$\min_{\bar{a}_{rj}<0} \frac{\bar{c}_j}{-\bar{a}_{rj}} = \min\left(\frac{\bar{c}_1}{-\bar{a}_{31}}, \frac{\bar{c}_2}{-\bar{a}_{32}}\right) = \min\left(\frac{4}{2}, \frac{3}{1}\right) = \frac{4}{2},$$

x_1 becomes a basic variable in the next cycle. We calculate $\bar{p}_1$ as

$$\bar{p}_1 = B^{-1} p_1 = \begin{bmatrix} 1 & 0 & 0 \\ 0 & 1 & 0 \\ 0 & 0 & 1 \end{bmatrix} \begin{pmatrix} -1 \\ -1 \\ -2 \end{pmatrix} = \begin{pmatrix} -1 \\ -1 \\ -2 \end{pmatrix}$$

and put $\bar{p}_1$ and $\bar{c}_1 = 4$ in the column of $\hat{\bar{p}}_s$ at cycle 0 in Table 2.17. Since $r = 3$, the pivot element is -2 bracketed by [ ]. By performing the pivot operation on -2 at cycle 0, the tableau at cycle 1 is obtained.

At cycle 1, the variables x_3, x_4, and x_1 are basic variables, and in Table 2.17, since $\bar{b}_1 > 0$ and

$$\min(\bar{b}_3, \bar{b}_4) = \min(-9/2, -5/2) = -9/2 < 0,$$

x_3 becomes a nonbasic variable in the next cycle, and the index r of the variable leaving the basis is determined as $r = 1$.

From (2.113), we calculate the coefficients $\bar{a}_{rj}$, $r = 1$, $j = 2, 5$ for nonbasic variables. Using the first row $[B^{-1}]_1$. of B^{-1}, p_2, and p_5, we have

$$\bar{a}_{12} = [B^{-1}]_1 \cdot p_2 = (1, 0, -1/2)(-3, -2, -1)^T = -5/2$$
$$\bar{a}_{15} = [B^{-1}]_1 \cdot p_5 = (1, 0, -1/2)(0, 0, 1)^T = -1/2.$$

Also the relative cost coefficients $\bar{c}_j$, $j = 2, 5$ are calculated as

$$\bar{c}_2 = c_2 - \pi p_2 = 3 + (0, 0, 2)(-3, -2, -1)^T = 1$$
$$\bar{c}_5 = c_5 - \pi p_5 = 0 + (0, 0, 2)(0, 0, 1)^T = 2.$$

2.8 Dual Simplex Method

From

$$\min_{\bar{a}_{rj}<0} \frac{\bar{c}_j}{-\bar{a}_{rj}} = \min\left(\frac{\bar{c}_2}{-\bar{a}_{12}}, \frac{\bar{c}_5}{-\bar{a}_{15}}\right) = \min\left(\frac{1}{5/2}, \frac{2}{1/2}\right) = \frac{1}{5/2},$$

x_2 becomes a basic variable in the next cycle. We calculate $\bar{p}_2$ as

$$\bar{p}_2 = B^{-1}p_2 = \begin{bmatrix} 1 & 0 & 0 \\ 0 & 1 & 0 \\ 0 & 0 & 1 \end{bmatrix} \begin{pmatrix} -3 \\ -2 \\ -1 \end{pmatrix} = \begin{pmatrix} -5/2 \\ -3/2 \\ 1/2 \end{pmatrix}$$

and put $\bar{p}_2$ and $\bar{c}_2 = 1$ in the column of $\hat{\bar{p}}_s$ at cycle 1 in Table 2.17. Since $r = 1$, the pivot element is $-5/2$ bracketed by [ ]. By performing the pivot operation on $-5/2$ at cycle 1, the tableau at cycle 2 is obtained.

At cycle 2, the variables x_2, x_4 and x_1 are basic variables. Since all of the constants $\bar{b}_i$ are positive, an optimal solution

$$x_1 = \frac{33}{5}, \ x_2 = \frac{9}{5}\left(x_3 = 0, x_4 = \frac{1}{5}, x_5 = 0\right), \quad z = \frac{159}{5}$$

is obtained. ◊

Finally, consider the sensitivity analysis, which examines the effects of small changes in the parameters of a linear programming problem on its optimal solution. In particular, we deal with a case where the right-hand side vector is changed, which is closely related to the dual simplex method.

Assume that in the standard form of linear programming

$$\left.\begin{array}{l} \text{minimize } z = cx \\ \text{subject to} \quad Ax = b \\ \qquad\qquad\quad x \geq 0, \end{array}\right\} \quad (2.114)$$

an optimal basis B is known, and then the corresponding optimal basic solution x_B is

$$x_B = \bar{b} = B^{-1}b. \quad (2.115)$$

Moreover, the corresponding simplex multiplier vector π is

$$\pi = c_B B^{-1}, \quad (2.116)$$

and the value of the objective function $\bar{z}$ is also calculated as

$$\bar{z} = c_B x_B = c_B \bar{b} = \pi b. \quad (2.117)$$

Obviously, the optimality criterion

$$\bar{c}_j = c_j - \pi p_j \geq 0 \quad \text{for all } j \text{ of the nonbasic variables} \tag{2.118}$$

is satisfied.

In discussing changes in the right-hand side vector, assume that b is changed to $b + \Delta b$. Consider the following linear programming problem:

$$\left. \begin{array}{l} \text{minimize } z = cx \\ \text{subject to} \quad Ax = b + \Delta b \\ \phantom{\text{subject to}} \quad x \geq 0. \end{array} \right\} \tag{2.119}$$

Since the simplex multiplier vector π and the relative cost coefficients $\bar{c}_j$ for all indices j of the nonbasic variables do not depend on b as shown in (2.116) and (2.118), they remain the same even if b is changed to $b + \Delta b$. However, the basic solution x_B itself may no longer be feasible.

The new basic solution and the value of the objective function are calculated as

$$x_B^* = B^{-1}(b + \Delta b) = x_B + B^{-1}\Delta b \tag{2.120}$$

and

$$\bar{z}^* = \pi(b + \Delta b) = \bar{z} + \pi \Delta b, \tag{2.121}$$

respectively.

Therefore, the following statements hold:

(i) If $x_B^* \geq 0$ holds, then x_B^* is an optimal solution, and the variation in the objective function is $\pi \Delta b$.
(ii) If $x_B^* \geq 0$ does not hold, since the optimality condition $\bar{c}_j \geq 0$ for all indices j of the nonbasic variables is satisfied, the dual simplex method can be used to find a new optimal solution.

Example 2.16 (Sensitivity analysis for the production planning problem of Example 1.1). In the production planning problem of Example 1.1, we calculate optimal solutions when the total amounts of available materials are changed as follows:

(i) The available amounts of material M_1 is changed from 27 tons to 32 tons.
(ii) The available amounts of material M_2 is changed from 16 tons to 23 tons.

Although the optimal solution to the original problem is given at cycle 2 in the revised simplex method of Table 2.13, for the sake of convenience, we rewrite the initial tableau (cycle 0) and the optimal tableau (cycle 2) in Table 2.18.

From the optimal tableau, one finds that the basic variables are $x_B = (x_2, x_1, x_5)^T$, the basis inverse matrix is

2.8 Dual Simplex Method

Table 2.18 Initial and optimal tableaux of Example 1.1

Cycle	Basis	Basis inverse matrix			Constants	$\hat{p}_s$
Cycle 0 (initial)	x_3	1			27	[6]
	x_4		1		16	2
	x_5			1	18	1
	$-z$				0	-8
Cycle 2 (optimal)	x_2	3/14	$-1/7$		3.5	
	x_1	$-1/7$	3/7		3	
	x_5	5/14	$-11/7$	1	2.5	
	$-z$	9/7	1/7		37	

$$B^{-1} = \begin{bmatrix} 3/14 & -1/7 & 0 \\ -1/7 & 3/7 & 0 \\ 5/14 & -11/7 & 1 \end{bmatrix},$$

and the simplex multiplier vector is

$$\pi = (-9/7, -1/7, 0).$$

(i) Let the amounts of changes be $\Delta b = \begin{pmatrix} 5 \\ 0 \\ 0 \end{pmatrix}$, and from $b = \begin{pmatrix} 27 \\ 16 \\ 18 \end{pmatrix}$, it follows that

$$\begin{aligned} x_B^* &= B^{-1}(b + \Delta b) \\ &= \begin{bmatrix} 3/14 & -1/7 & 0 \\ -1/7 & 3/7 & 0 \\ 5/14 & -11/7 & 1 \end{bmatrix} \begin{pmatrix} 32 \\ 16 \\ 18 \end{pmatrix} = \begin{pmatrix} 32/7 \\ 16/7 \\ 30/7 \end{pmatrix} \end{aligned}$$

$$\bar{z}^* = \pi(b + \Delta b) = (-9/7, -1/7, 0) \begin{pmatrix} 32 \\ 16 \\ 18 \end{pmatrix} = -304/7.$$

Since $x_B^* \geq 0$ holds, x_B^* is an optimal basic solution, and then an optimal solution

$$x_2 = 32/7, \ x_1 = 16/7, \ x_5 = 30/7, \ (x_3 = x_4 = 0) \quad z = -304/7$$

is obtained.

Table 2.19 Revised simplex tableau after change of $b_2 = 23$

Cycle	Basis	Basis inverse matrix			Constants	$\hat{p}_s$
1	x_2	3/14	−1/7		5/2	−1/7
	x_1	−1/7	3/7		6	3/7
	x_5	5/14	−11/7	1	−17/2	[−11/7]
	−z	9/7	1/7		38	1/7
2	x_2	2/11		1/11	36/11	
	x_1	−1/22		3/11	81/22	
	x_4	−5/22	1	−7/11	119/22	
	−z	29/22		1/11	819/22	

(ii) Let the amounts of changes be $\Delta b = \begin{pmatrix} 0 \\ 7 \\ 0 \end{pmatrix}$, and from $b = \begin{pmatrix} 27 \\ 16 \\ 18 \end{pmatrix}$, it follows that

$$x_B^* = B^{-1}(b + \Delta b) = B^{-1} \begin{pmatrix} 27 \\ 23 \\ 18 \end{pmatrix} = \begin{pmatrix} 5/2 \\ 6 \\ -17/2 \end{pmatrix}$$

$$\bar{z}^* = \pi(b + \Delta b) = (-9/7, -1/7, 0) \begin{pmatrix} 27 \\ 23 \\ 18 \end{pmatrix} = -38.$$

Since the negative component $-17/2$ appears in x_B^*, using the revised dual simplex method, we can obtain an optimal tableau shown in Table 2.19.

That is, using the third row $[B^{-1}]_3$. of B^{-1}, p_3, and p_4, we have

$$\bar{a}_{33} = [B^{-1}]_3 \cdot p_3 = (5/14, -11/7, 1)(1, 0, 0)^T = 5/14,$$

$$\bar{a}_{34} = [B^{-1}]_3 \cdot p_4 = (5/14, -11/7, 1)(0, 1, 0)^T = -11/7.$$

Thus, x_4 becomes a basic variable in the next cycle. The relative cost coefficients $\bar{c}_4$ is calculated as

$$\bar{c}_4 = c_4 - \pi p_4 = 0 - (-9/7, -1/7, 0)(0, 1, 0)^T = 1/7.$$

We calculate $\bar{p}_4$ as

$$\begin{bmatrix} 3/14 & -1/7 & 0 \\ -1/7 & 3/7 & 0 \\ 5/14 & -11/7 & 1 \end{bmatrix} \begin{pmatrix} 0 \\ 1 \\ 0 \end{pmatrix} = \begin{pmatrix} -1/7 \\ 3/7 \\ -11/7 \end{pmatrix}.$$

2.8 Dual Simplex Method

These values are put in the column of $\hat{p}_s$ in the tableau. By performing the pivot operation on $[-11/7]$, a new tableau is obtained.

In this example, after the only one pivot operation, an optimal solution

$$x_2 = 36/11, \quad x_1 = 81/22 \quad (x_4 = 119/22, \ x_3 = x_5 = 0), \quad z = -819/22$$

is obtained. ◊

When the coefficients of the objective function are changed, since only the changes in the cost coefficients c affect the optimality criterion and the value of the objective function, the (revised) simplex method is used for finding the new optimal solution only when some relative cost coefficients become negative, i.e., $\bar{c}_j < 0$ for some j.

Problems

2.1 Convert the following problems to the standard form of linear programming:

(i) (Absolute value problem)

$$\text{minimize} \quad z = \sum_{j=1}^{n} c_j |x_j|$$
$$\text{subject to} \quad \sum_{j=1}^{n} a_{ij} x_j = b_i, \quad i = 1, 2, \ldots, m,$$

where $c_j > 0$, $j = 1, 2, \ldots, n$, and x_j, $j = 1, 2, \ldots, n$ are free variables.

(ii) (Fractional programming problem)

$$\text{minimize} \quad z = \frac{\sum_{j=1}^{n} c_j x_j + c_0}{\sum_{j=1}^{n} d_j x_j + d_0}$$
$$\text{subject to} \quad \sum_{j=1}^{n} a_{ij} x_j = b_i, \quad i = 1, 2, \ldots, m$$
$$x_j \geq 0, \quad j = 1, 2, \ldots, n,$$

where $\sum_{j=1}^{n} d_j x_j + d_0 > 0$ holds for all feasible solutions.

(iii) (Minimax problem)

$$\begin{aligned}
\text{minimize} \quad & z = \max\left(\sum_{j=1}^{n} c_j^1 x_j, \sum_{j=1}^{n} c_j^2 x_j, \ldots, \sum_{j=1}^{n} c_j^L x_j\right) \\
\text{subject to} \quad & \sum_{j=1}^{n} a_{ij} x_j = b_i, \quad i = 1, 2, \ldots, m \\
& x_j \geq 0, \quad j = 1, 2, \ldots, n.
\end{aligned}$$

2.2 Formulate the following problems as linear programming ones.

(i) A manufacturing company produces two products A and B. There are 40 h of labor available each day, and 1 kg (kilogram) of product A requires 2 h of labor, whereas 1 kg of product B requires 5 h. There are up to 30 machine-hours available per day, and machine processing time for 1 kg of product A is 3 h and for 1 kg of product B is 1 h. There are 39 kg of raw material available each day, and 1 kg of product A requires 3 kg of the material, whereas 1 kg of product B requires 4 kg. The daily profit for product A is 30 thousand yen per 1 kg, while B is 20 thousand yen per 1 kg, and the manager wishes to maximize the daily profit.

(ii) A firm manufactures cattle feed by mixing two ingredients A and B. Each ingredient contains three nutrients C, D, and E. Each 1 g (gram) of the ingredient A contains 9 mg (milligram) of C, 1 mg of D, and 1 mg of E. Each 1 g of the ingredient B contains 2 mg of C, 5 mg of D, and 1 mg of E. Each 1 g of the feed must contain at least 54 g, 25 g, and 13 g of C, D, and E, respectively. The costs per gram of the ingredients A and B are 9 yen and 15 yen, respectively, and the manager wishes to find the best feed mix that has the minimum cost per gram.

2.3 Assume that all $x^l = (x_1^l, x_2^l, \ldots, x_n^l)^T$, $l = 1, 2, \ldots, L$ are optimal solutions to a certain linear programming problem. Show that $x^* = \sum_{l=1}^{L} \lambda_l x^l$ is also an optimal solution to the problem, where λ_l, $l = 1, \ldots, L$ are nonnegative constants satisfying $\sum_{l=1}^{L} \lambda_l = 1$.

2.4 For a linear programming problem involving a free variable x_k, assume that we substitute the difference of two nonnegative variables $x_k^+ - x_k^-$, $x_k^+ \geq 0$, $x_k^- \geq 0$ for x_k. Explain why both x_k^+ and x_k^- cannot be in the same basis simultaneously.

2.5 Consider the two linear programming problems

$$\begin{array}{ll}
\text{minimize } z = cx & \text{minimize } z = (\mu c)x \\
\text{subject to } \quad Ax = b \quad \text{and} & \text{subject to } \quad Ax = (\lambda b) \\
\qquad x \geq 0, & \qquad x \geq 0,
\end{array}$$

where λ and μ are positive real numbers. Explain the relationships between these two problems. What happens if either λ or μ is negative?

2.8 Dual Simplex Method

2.6 Solve the following problems using the simplex method:

(i) Minimize $-2x_1 - 5x_2$
subject to $2x_1 + 6x_2 \leq 27$
$8x_1 + 6x_2 \leq 45$
$3x_1 + x_2 \leq 15$
$x_j \geq 0, \ j = 1, 2$

(ii) Minimize $-3x_1 - 2x_2$
subject to $2x_1 + 5x_2 \leq 130$
$6x_1 + 3x_2 \leq 110$
$x_j \geq 0, \ j = 1, 2$

(iii) Minimize $-3x_1 - 4x_2$
subject to $3x_1 + 12x_2 \leq 400$
$6x_1 + 3x_2 \leq 600$
$8x_1 + 7x_2 \leq 800$
$x_j \geq 0, \ j = 1, 2$

(iv) Minimize $-2.5x_1 - 5x_2 - 3.4x_3$
subject to $-5x_1 + 10x_2 + 6x_3 \leq 425$
$2x_1 - 5x_2 + 4x_3 \leq 400$
$3x_1 - 10x_2 + 8x_3 \leq 600$
$x_j \geq 0, \ j = 1, 2, 3$

(v) minimize $-12x_1 - 18x_2 - 8x_3 - 40x_4$
subject to $2x_1 + 5.5x_2 + 6x_3 + 10x_4 \leq 80$
$4x_1 + x_2 + 4x_3 + 20x_4 \leq 50$
$x_j \geq 0, \ j = 2, 3, 4; \ x_1$: a free variable

(vi) Minimize $2x_1 - 3x_2 - x_3 + 2x_4$
subject to $-3x_1 + 2x_2 - x_3 + 3x_4 = 2$
$-x_1 + 2x_2 + x_3 + 2x_4 = 3$
$x_j \geq 0, \ j = 1, 2, 3, 4$

2.7 Solve the following problems using the simplex method:

(i) Minimize $|x_1| + 4|x_2| + 2|x_3|$
subject to $2x_1 + x_2 \leq 3$
$x_1 + 2x_2 + x_3 = 5$

(ii) Minimize $\dfrac{-x_1 + 4x_2 + x_3 + 1}{x_1 + 2x_2 + x_3 + 1}$
subject to $2x_1 - 2x_2 + x_3 \leq 1$
$x_1 + 2x_2 - x_3 \geq 1.5$
$x_j \geq 0, \ j = 1, 2, 3$

(iii) Minimize max $(-x_1 + 2x_2 - x_3, -2x_1 + 3x_2 - 2x_3, x_1 - x_2 - 2x_3)$
subject to $2x_1 + x_2 + x_3 \leq 5$
$2x_1 + 2x_2 + 5x_3 \leq 10$
$x_j \geq 0, \ j = 1, 2, 3$

2.8 Prove by contradiction that the use of Bland's rule prevents cycling in the following way.

(i) Let T be the index set of all variables that enter the basis during cycling, and let q be the largest index in T, i.e., $q = \max\{j \mid j \in T\}$. The variable x_q enters the basis during cycling, and then x_q must also leave the basis. Let I be the index set of basic variables before x_q enters the basis, and let $J = \{1, 2, \cdots, n\} - I$ be the index set of nonbasic variables. The corresponding canonical form is represented by

$$x_i + \sum_{j \in J} \bar{a}_{ij} x_j = \bar{b}_i, \ i \in I, \qquad -z + \sum_{j \in J} \bar{c}_j x_j = -z.$$

Furthermore, let I' be the index set of basic variables when x_q leaves the basis, and let $J' = \{1, 2, \cdots, n\} - I'$ be the index set of nonbasic variables. The corresponding canonical form is represented by

$$x_i + \sum_{j \in J'} \bar{a}'_{ij} x_j = \bar{b}_i, \ i \in I', \qquad -z + \sum_{j \in J'} \bar{c}'_j x_j = -z.$$

Let $t \in J'$ be the index of the basic variable that enters I' instead of x_q. By the definitions of q and t, it follows that $\bar{c}_q < 0, \bar{c}'_t < 0, \bar{a}'_{qt} > 0, t \in T$, $t < q$. In the canonical form for I' and J', assume that $x_t = -1$ for $t \in J'$ and $x_j = 0$ for all $j \in J' - \{t\}$. Explain that the relation $-\bar{c}'_t = \sum_{j \in J} \bar{c}_j x_j$ holds.

(ii) From $\bar{c}'_t < 0$, there must be a positive term in $\sum_{j \in J} \bar{c}_j x_j$ of the above relation. Let the term be $\bar{c}_r x_r > 0, r \in J$. Show that $r < q$.

(iii) Show $x_r = \bar{a}'_{rt} > 0$ and derive the contradiction.

2.9 Apply the standard simplex method to the following linear programming problem due to E.M.L. Beale, starting with x_5, x_6, and x_7 as the initial basic variables, and verify that the procedure of the simplex method cycles:

$$\begin{aligned}
\text{minimize} \quad & (-3/4)x_1 + 150x_2 - (1/50)x_3 + 6x_4 \\
\text{subject to} \quad & (1/4)x_1 - 60x_2 - (1/25)x_3 + 9x_4 + x_5 = 0 \\
& (1/2)x_1 - 90x_2 - (1/50)x_3 + 3x_4 + x_6 = 0 \\
& x_3 + x_7 = 1 \\
& x_j \geq 0, \ j = 1, 2, \ldots, 7.
\end{aligned}$$

Solve the problem using the simplex method incorporating Bland's rule.

2.8 Dual Simplex Method

2.10 A vector π of the simplex multipliers can also be defined as follows: Multiply π by a vector b of the right-hand side constants of the original equation system, and subtract it from the objective function z. Then, π_i is determined such that the coefficient of a basic variable x_i is zero. Explain that the above definition and the original definition of the simplex multiplier are equivalent.

2.11 Solve problem 2.6 by the revised simplex method.

2.12 Show that the dual to the linear programming problem

$$\begin{aligned}
\text{minimize} \quad & x_1 + x_2 + x_3 \\
\text{subject to} \quad & -x_2 + x_3 \geq -1 \\
& x_1 \phantom{{}+x_2} - x_3 \geq -1 \\
& -x_1 + x_2 \phantom{{}-x_3} \geq -1 \\
& x_j \geq 0, \quad j = 1, 2, 3
\end{aligned}$$

is equivalent to the primal problem. Such a pair of linear programming is known as self-dual. Assuming A is a square matrix, derive the conditions for c, A, and b for which the linear programming problem

$$\begin{aligned}
\text{minimize} \quad & cx \\
\text{subject to} \quad & Ax \geq b \\
& x \geq 0
\end{aligned}$$

is self-dual.

2.13 Prove the complementary slackness theorem.

2.14 Prove Gordon's theorem.

2.15 Solve the following problems using the dual simplex method:

(i) Minimize $4x_1 + 3x_2$
subject to $x_1 + 3x_2 \geq 12$
$\phantom{\text{subject to }} x_1 + 2x_2 \geq 10$
$\phantom{\text{subject to }} 2x_1 + x_2 \geq 9$
$\phantom{\text{subject to }} x_j \geq 0, \; j = 1, 2$

(ii) Minimize $3x_1 + 5x_2$
subject to $2x_1 + 3x_2 \geq 20$
$\phantom{\text{subject to }} 2x_1 + 5x_2 \geq 22$
$\phantom{\text{subject to }} 5x_1 + 3x_2 \geq 25$
$\phantom{\text{subject to }} x_j \geq 0, \; j = 1, 2$

(iii) Minimize $4x_1 + 2x_2 + 3x_3$
subject to $5x_1 + 3x_2 - 2x_3 \geq 10$
$\phantom{\text{subject to }} 3x_1 + 2x_2 + 4x_3 \geq 8$
$\phantom{\text{subject to }} x_j \geq 0, \; j = 1, 2, 3$

(iv) Minimize $4x_1 + 8x_2 + 3x_3$
subject to $2x_1 + 5x_2 + 3x_3 \geq 185$
$3x_1 + 2.5x_2 + 8x_3 \geq 155$
$8x_1 + 10x_2 + 4x_3 \geq 600$
$x_j \geq 0, \ j = 1, 2, 3$

2.16 In the production planning problem of Example 1.1, assume that the total amounts of available materials are changed as follows:

(i) The total amount of M_1 is changed from 27 tons to 33 tons.
(ii) The total amount of M_2 is changed from 16 tons to 21 tons.

In each case, find a new optimal solution starting from the last optimal tableau.

2.17 In the linear programming problem solved in problem 2.6 (i), assume that the right-hand side constants are changed as follows:

(i) The right-hand side constant 27 is changed to 30.
(ii) The right-hand side constant 45 is changed to 51.

In each case, find a new optimal solution starting from the last optimal tableau.

Chapter 3
Multiobjective Linear Programming

The problem to optimize multiple conflicting linear objective functions simultaneously under the given linear constraints is called the multiobjective linear programming problem. This chapter begins with a discussion of fundamental notions and methods of multiobjective linear programming. After introducing the notion of Pareto optimality, several methods for characterizing Pareto optimal solutions including the weighting method, the constraint method, and the weighted minimax method are explained, and goal programming and compromise programming are also introduced. Extensive discussions of interactive multiobjective linear programming conclude this chapter.

3.1 Problem Formulation and Solution Concepts

Recall the production planning problem discussed in Example 1.1.

Example 3.1 (Production planning problem). A manufacturing company desires to maximize the total profit from producing two products P_1 and P_2 utilizing three different materials M_1, M_2, and M_3. The company knows that to produce 1 ton of product P_1 requires 2 tons of material M_1, 3 tons of material M_2, and 4 tons of material M_3, while to produce 1 ton of product P_2 requires 6 tons of material M_1, 2 tons of material M_2, and 1 ton of material M_3. The total amounts of available materials are limited to 27, 16, and 18 tons for M_1, M_2, and M_3, respectively. It also knows that product P_1 yields a profit of 3 million yen per ton, while P_2 yields 8 million yen. Given these limited materials, the company is trying to figure out how many units of products P_1 and P_2 should be produced to maximize the total profit. This production planning problem can be formulated as the following linear programming problem:

$$\begin{aligned}
\text{minimize } \quad & z_1 = -3x_1 - 8x_2 \\
\text{subject to} \quad & 2x_1 + 6x_2 \leq 27 \\
& 3x_1 + 2x_2 \leq 16 \\
& 4x_1 + x_2 \leq 18 \\
& x_1 \geq 0, \quad x_2 \geq 0.
\end{aligned} \tag{3.1}$$

◊

Example 3.2 (Production planning with environmental consideration). Unfortunately, however, in production process, it is pointed out that product P_1 yields 5 units of pollution per ton and product P_2 yields 4 units of pollution per ton. Thus, the manager should not only maximize the total profit but also minimize the amount of pollution.

For simplicity, assume that the amount of pollution is a linear function of two variables x_1 and x_2 such as

$$5x_1 + 4x_2,$$

where x_1 and x_2 denote the numbers of tons produced of products P_1 and P_2, respectively.

Considering environmental quality, the production planning problem can be reformulated as the following two-objective linear programming problem:

$$\begin{aligned}
\text{minimize } \quad & z_1 = -3x_1 - 8x_2 \\
\text{minimize } \quad & z_2 = 5x_1 + 4x_2 \\
\text{subject to} \quad & 2x_1 + 6x_2 \leq 27 \\
& 3x_1 + 2x_2 \leq 16 \\
& 4x_1 + x_2 \leq 18 \\
& x_1 \geq 0, \quad x_2 \geq 0.
\end{aligned} \tag{3.2}$$

◊

The problem to optimize such multiple conflicting linear objective functions simultaneously under the given linear constraints is called the multiobjective linear programming problem and can be generalized as follows:

$$\begin{aligned}
\text{minimize } \quad & z_1(x) = c_1 x \\
\text{minimize } \quad & z_2(x) = c_2 x \\
& \dots\dots\dots\dots\dots \\
\text{minimize } \quad & z_k(x) = c_k x \\
\text{subject to} \quad & Ax \leq b \\
& x \geq 0,
\end{aligned} \tag{3.3}$$

3.1 Problem Formulation and Solution Concepts

where

$$c_i = (c_{i1}, \ldots, c_{in}), \quad i = 1, \ldots, k$$

$$x = \begin{pmatrix} x_1 \\ x_2 \\ \vdots \\ x_n \end{pmatrix}, \quad A = \begin{bmatrix} a_{11} & a_{12} & \cdots & a_{1n} \\ a_{21} & a_{22} & \cdots & a_{2n} \\ \vdots & \vdots & \ddots & \vdots \\ a_{m1} & a_{m2} & \cdots & a_{mn} \end{bmatrix}, \quad b = \begin{pmatrix} b_1 \\ b_2 \\ \vdots \\ b_m \end{pmatrix}.$$

Such a multiobjective linear programming problem is sometimes expressed as the following vector minimization problem:

$$\left. \begin{array}{ll} \text{minimize} & z(x) = (z_1(x), z_2(x), \ldots, z_k(x))^T \\ \text{subject to} & Ax \leq b \\ & x \geq 0, \end{array} \right\} \quad (3.4)$$

where $z(x) = (z_1(x), \ldots, z_k(x))^T = (c_1 x, \ldots, c_k x)^T$ is a k-dimensional vector. Let the feasible region of the problem be denoted by

$$X = \left\{ x \in \mathbb{R}^n \mid Ax \leq b, \, x \geq 0 \right\}. \quad (3.5)$$

Introducing a $k \times n$ matrix $C = (c_1, c_2, \ldots, c_k)^T$ of the coefficients of the objective functions, we can express the multiobjective linear programming problem (3.4) in a more compact form:

$$\left. \begin{array}{ll} \text{minimize} & z(x) = Cx \\ \text{subject to} & x \in X. \end{array} \right\} \quad (3.6)$$

If we directly apply the notion of optimality for single-objective linear programming to this multiobjective linear programming, we arrive at the following notion of a complete optimal solution.

Definition 3.1 (Complete optimal solution). A point x^* is said to be a complete optimal solution if and only if there exists $x^* \in X$ such that $z_i(x^*) \leq z_i(x), i = 1, \ldots, k$ for all $x \in X$.

However, in general, such a complete optimal solution that simultaneously minimizes all of the multiple objective functions does not always exist when the objective functions conflict with each other. Thus, instead of a complete optimal solution, a new solution concept, called Pareto optimality, is introduced in multiobjective linear programming.

Fig. 3.1 Feasible region and Pareto optimal solutions for Example 3.2 in x_1-x_2 plane

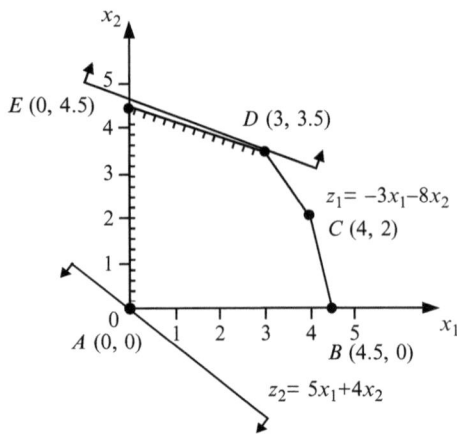

Definition 3.2 (Pareto optimal solution). A point x^* is said to be a Pareto optimal solution if and only if there does not exist another $x \in X$ such that $z_i(x) \leq z_i(x^*)$ for all i and $z_j(x) \neq z_j(x^*)$ for at least one j.

As can be seen from the definition, a Pareto optimal solution consists of an infinite number of points. A Pareto optimal solution is sometimes called a noninferior solution since it is not inferior to other feasible solutions.

In addition to Pareto optimality, the following weak Pareto optimality is defined as a slightly weaker solution concept than Pareto optimality.

Definition 3.3 (Weak Pareto optimal solution). A point x^* is said to be a weak Pareto optimal solution if and only if there does not exist another $x \in X$ such that $z_i(x) < z_i(x^*), i = 1, \ldots, k$.

For notational convenience, let X^{CO}, X^P, and X^{WP} denote the complete optimal, Pareto optimal, and weak Pareto optimal solution sets, respectively. From their definitions, it can be easily understood that the following relation holds:

$$X^{CO} \subseteq X^P \subseteq X^{WP}. \tag{3.7}$$

Example 3.3 (Pareto optimal solutions to production planning of Example 3.2). To understand the notion of Pareto optimal solutions in multiobjective linear programming, consider the two-objective linear programming problem given in Example 3.2. The feasible region X for this problem in the x_1-x_2 plane becomes the boundary lines and interior points of the convex pentagon $ABCDE$ in Fig. 3.1. Among the five extreme points A, B, C, D, and E, observe that z_1 is minimized at the extreme point $D(3, 3.5)$ while z_2 is minimized at the extreme point $A(0, 0)$. These two extreme points A and D are obviously Pareto optimal solutions since they cannot improve respective objective functions z_1 and z_2 anymore. In addition to the extreme points A and D, the extreme point E and all of the points of the

3.2 Scalarization Methods

Fig. 3.2 Feasible region and Pareto optimal solutions for Example 3.2 in z_1-z_2 plane

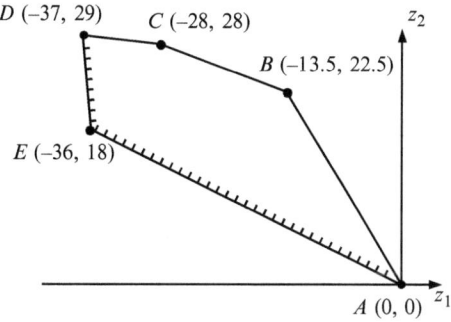

segments AE and ED are Pareto optimal solutions since they can be improved only at the expense of either z_1 or z_2. However, all of the remaining feasible points are not Pareto optimal since there always exist other feasible points which improve both objective functions or at least one of them without sacrificing the other.

This situation can be more easily understood by observing the feasible region

$$Z = \{(z_1(x), z_2(x)) \mid x \in X\} \tag{3.8}$$

in the z_1-z_2 plane as depicted in Fig. 3.2.

It is not hard to check that if the objective functions of this problem are changed to

$$z_1 = -x_1 - 4x_2, \quad z_2 = 2x_1 - x_2,$$

there exists a complete optimal solution $(z_1, z_2) = (-18, -4.5)$.

Moreover, if they are changed to

$$z_1 = -4x_1 - 5x_2, \quad z_2 = 2x_1 - x_2,$$

there exist weak Pareto optimal solutions.

3.2 Scalarization Methods

Several computational methods have been proposed for characterizing Pareto optimal solutions depending on the different scalarizations of the multiobjective linear programming problems. Among the many possible ways of scalarizing the multiobjective linear programming problems, the weighting method, the constraint method, and the weighted minimax method have been studied as a means of characterizing Pareto optimal solutions of the multiobjective linear programming problems.

3.2.1 Weighting Method

The weighting method for obtaining a Pareto optimal solution is to solve the weighting problem formulated by taking the weighted sum of all of the objective functions in the original multiobjective linear programming problem (Kuhn and Tucker 1951; Zadeh 1963). Thus, the weighting problem is defined by

$$\min_{x \in X} \ wz(x) = \sum_{i=1}^{k} w_i z_i(x), \quad (3.9)$$

where $w = (w_1, \ldots, w_k)$ is a vector of weighting coefficients assigned to the objective functions and assumed to be

$$w = (w_1, \ldots, w_k) \geq 0, \quad w \neq 0.$$

The relationships between the optimal solution x^* of the weighting problem and the Pareto optimality concept of the multiobjective linear programming problems can be characterized by the following theorems.

Theorem 3.1. *If $x^* \in X$ is an optimal solution of the weighting problem for some $w > 0$, then x^* is a Pareto optimal solution of the multiobjective linear programming problem.*

Proof. If an optimal solution x^* of the weighting problem is not a Pareto optimal solution of the multiobjective linear programming problem, then there exists $x \in X$ such that $z_j(x) < z_j(x^*)$ for some j and $z_i(x) \leq z_i(x^*)$, $i = 1, \ldots, k; i \neq j$. From $w = (w_1, \ldots, w_k) > 0$, this implies $\sum_{i=1}^{k} w_i z_i(x) < \sum_{i=1}^{k} w_i z_i(x^*)$. Thus, it does not follow that x^* is an optimal solution of the weighting problem for some $w > 0$. □

It should be noted here that the condition of Theorem 3.1 can be replaced with a unique optimal solution of the weighting problem for $w \geq 0, w \neq 0$.

Theorem 3.2. *If $x^* \in X$ is a Pareto optimal solution to a multiobjective linear programming problem, then x^* is an optimal solution to the weighting problem for some $w = (w_1, \ldots, w_k) \geq 0, w \neq 0$.*

Proof. First we prove that x^* is an optimal solution of the linear programming problem

$$\left.\begin{array}{ll} \text{minimize} & \mathbf{1}^T C x \\ \text{subject to} & C x \leq C x^* \\ & A x \leq b, \quad x \geq 0, \end{array}\right\} \quad (3.10)$$

where $\mathbf{1}$ is a k-dimensional column vector whose elements are all ones.

3.2 Scalarization Methods

If x^* is not an optimal solution of this problem, then there exists $x \in X$ such that $Cx \leq Cx^*$ and $\mathbf{1}^T Cx < \mathbf{1}^T Cx^*$. This means that there exists $x \in X$ such that

$$\mathbf{1}^T Cx = \sum_{i=1}^{k} c_i x < \sum_{i=1}^{k} c_i x^* = \mathbf{1}^T Cx^* \quad \text{and} \quad c_i x \leq c_i x^*, \; i = 1, \ldots, k$$

or equivalently

$$c_j x < c_j x^* \text{ for some } j \quad \text{and} \quad c_i x \leq c_i x^*, \; i = 1, \ldots, k; i \neq j$$

which contradicts the fact that x^* is a Pareto optimal solution. Hence, x^* is an optimal solution of this problem.

Now consider the dual problem

$$\left. \begin{array}{ll} \text{maximize} & (-b^T, -x^{*T}C^T)y \\ \text{subject to} & (-A^T, -C^T)y \leq C^T \mathbf{1} \\ & y \geq 0 \end{array} \right\} \quad (3.11)$$

of the linear programming problem (3.10).

From the (strong) duality theorem of linear programming, it holds that

$$(-b^T, -x^{*T}C^T)y^* = \mathbf{1}^T Cx^*$$

and thus

$$(\mathbf{1}^T + y_2^{*T})Cx^* = -b^T y_1^*, \quad y^{*T} = (y_1^{*T}, y_2^{*T}).$$

Letting

$$w = \mathbf{1}^T + y_2^{*T},$$

we have

$$\sum_{i=1}^{k} w_i c_i x = w^T Cx = (\mathbf{1} + y_2^*)^T Cx \quad \forall x \in X.$$

Next observe that the dual problem of the linear programming problem

$$\left. \begin{array}{ll} \text{minimize} & (\mathbf{1} + y_2^*)^T Cx \\ \text{subject to} & Ax \leq b, \quad x \geq 0 \end{array} \right\} \quad (3.12)$$

becomes

$$\left. \begin{array}{ll} \text{maximize} & -b^T u \\ \text{subject to} & -A^T u \leq C^T (\mathbf{1} + y_2^*) \\ & u \geq 0. \end{array} \right\} \quad (3.13)$$

Fig. 3.3 Weighting method for Example 3.2

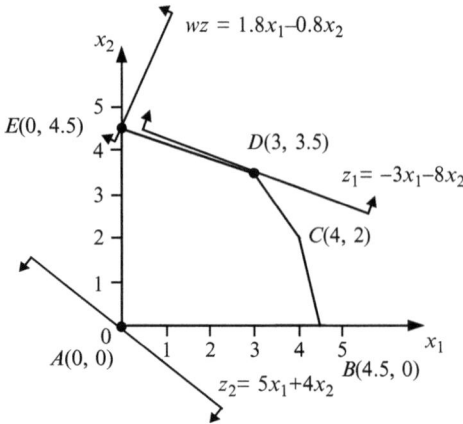

Then, $x = x^*$ and $u = y_1^*$ are feasible solutions for the corresponding linear programming problems, and each value of the objective functions is equal. Hence, for any $x \in X$, it follows that

$$\sum_{i=1}^{k} w_i c_i x^* = (1+y_2^*)^T C x^* = \min_{x \in X}(1+y_2^*)^T C x \leq (1+y_2^*)^T C x = \sum_{i=1}^{k} w_i c_i x.$$

This implies that x^* is an optimal solution of the weighting problem for $w = (w_1, \ldots, w_k) \geq \mathbf{0}$. This completes the proof of the theorem. □

The weighting coefficients of the weighting problem give the trade-off information between the objective functions. They show how many units of value of one objective function have to be given up in order to obtain one additional unit of value of the other objective function. This fact may be intuitively understood as follows:

Geometrically, in the k-dimensional $z = (z_1, \ldots, z_k)$ space,

$$W = w_1 z_1(x) + w_2 z_2(x) + \cdots + w_k z_k(x) = c \quad \text{(constant)} \tag{3.14}$$

represents the hyperplane (note that in the case of two objectives it is a line, and in the case of three objectives, a plane) with the normal vector $w = (w_1, \ldots, w_k)$. Solving the weighting problem for the given weighting coefficients $w^* > \mathbf{0}$ yields the minimum c such that this hyperplane has at least one common point with the feasible region Z in the $z = (z_1, \ldots, z_k)$ space, and the corresponding Pareto optimal solution x^* is obtained as in Fig. 3.3.

The hyperplane for this minimum c is the supporting hyperplane of the feasible region Z at the point $z(x^*)$ on the Pareto optimal surface. The condition for the small displacement from the point $z(x^*)$ belonging to this supporting hyperplane is $\Delta W = 0$, i.e.,

3.2 Scalarization Methods

$$w_1^* \Delta z_1 + w_2^* \Delta z_2 + \cdots + w_k^* \Delta z_k = 0. \quad (3.15)$$

For fixed values of $\Delta z_j = 0$, $j = 2, 3, \ldots, k$, $j \neq i$ except z_1 and z_i, we have

$$w_1^* \Delta z_1 + w_i^* \Delta z_i = 0. \quad (3.16)$$

Hence, it holds that

$$-\frac{\partial z_1}{\partial z_i} = \frac{w_i^*}{w_1^*}. \quad (3.17)$$

Therefore, the ratio of the weighting coefficients w_i^*/w_1^* gives a trade-off rate between the two-objective functions at $z(x^*)$.

Example 3.4 (Weighting method for production planning of Example 3.2). To illustrate the weighting method, consider the problem of Example 3.2. The corresponding weighting problem becomes as follows:

$$\begin{aligned}
\text{minimize} \quad & wz(x) = w_1(-3x_1 - 8x_2) + w_2(5x_1 + 4x_2) \\
\text{subject to} \quad & 2x_1 + 6x_2 \leq 27 \\
& 3x_1 + 2x_2 \leq 16 \\
& 4x_1 + x_2 \leq 18 \\
& x_1 \geq 0, \ x_2 \geq 0.
\end{aligned}$$

For this problem, for example, if we choose $w_1 = 0.4$ and $w_2 = 0.6$, we obtain

$$wz(x) = 1.8x_1 - 0.8x_2.$$

As depicted in Fig. 3.3, it can be easily understood that solving the corresponding weighting problem yields the extreme point $E(0, 4.5)$ as a Pareto optimal solution. Also, as two extreme cases, if we set $w_1 = 1$, $w_2 = 0$ and $w_1 = 0$, $w_2 = 1$, from Fig. 3.3, the optimal solutions of the corresponding weighting problems become the extreme points $D(3, 3.5)$ and $A(0, 0)$, respectively. In these cases, although the condition $w > 0$ of Theorem 3.1 is not satisfied, from Fig. 3.3, it can be seen that these two extreme points are Pareto optimal solutions. ◊

3.2.2 Constraint Method

The constraint method for characterizing Pareto optimal solutions is to solve the constraint problem formulated by taking one objective function of a multiobjective linear programming problem as the objective function of the constraint problem and letting all the other objective functions be inequality constraints (Haimes and Hall 1974; Haimes et al. 1971). The constraint problem is defined by

$$\begin{rcases} \text{minimize} & z_j(x) \\ \text{subject to} & z_i(x) \leq \varepsilon_i, \ i = 1, 2, \ldots, k; \ i \neq j \\ & x \in X. \end{rcases} \qquad (3.18)$$

The relationships between the optimal solution x^* to the constraint problem and Pareto optimality of a multiobjective linear programming problem can be characterized by the following theorems.

Theorem 3.3. *If $x^* \in X$ is a unique optimal solution to the constraint problem for some ε_i, $i = 1, \ldots, k; \ i \neq j$, then x^* is a Pareto optimal solution to a multiobjective linear programming problem.*

Proof. If a unique optimal solution x^* to the constraint problem is not a Pareto optimal solution to the multiobjective linear programming problem, then there exists $x \in X$ such that $z_l(x) < z_l(x^*)$ for some l and $z_i(x) \leq z_i(x^*)$, $i = 1, \ldots, k; i \neq l$. This means either

$$z_i(x) \leq z_i(x^*) \leq \varepsilon_i, \ i = 1, \ldots, k; \ i \neq j, \ z_j(x) < z_j(x^*)$$

or

$$z_i(x) \leq z_i(x^*) \leq \varepsilon_i, \ i = 1, \ldots, k; \ i \neq j, \ z_j(x) = z_j(x^*)$$

which contradicts the assumption that x^* is a unique optimal solution of the constraint problem for some ε_i, $i = 1, \ldots, k; \ i \neq j$. □

As can be easily understood from the proof of this theorem, in the absence of the uniqueness of a solution in the theorem, only weak Pareto optimality is guaranteed.

Theorem 3.4. *If $x^* \in X$ is a Pareto optimal solution to a multiobjective linear programming problem, then x^* is an optimal solution of the constraint problem for some ε_i, $i = 1, \ldots, k; \ i \neq j$.*

Proof. If a Pareto optimal solution $x^* \in X$ of the multiobjective linear programming problem is not an optimal solution to the constraint problem for $\varepsilon_i, i = 1, \ldots, k; i \neq j$, then there exists $x \in X$ such that

$$z_j(x) < z_j(x^*) \quad z_i(x) \leq \varepsilon_i = z_i(x^*), \ i = 1, \ldots, k; \ i \neq j,$$

which contradicts the fact that x^* is a Pareto optimal solution to the multiobjective linear programming problem. □

Example 3.5 (Constraint method for production planning of Example 3.2). To illustrate the constraint method, consider Example 3.2. The constraint problem for $j = 1$ becomes

3.2 Scalarization Methods

Fig. 3.4 Constraint method for Example 3.2

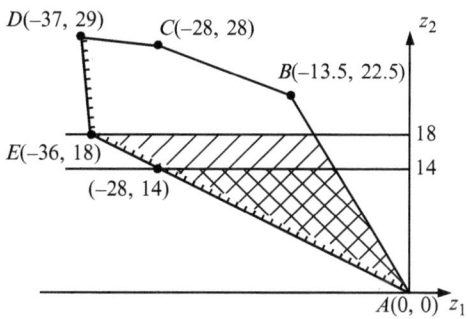

$$\begin{array}{ll}
\text{minimize} & z_1 = -3x_1 - 8x_2 \\
\text{subject to} & 2x_1 + 6x_2 \leq 27 \\
& 3x_1 + 2x_2 \leq 16 \\
& 4x_1 + x_2 \leq 18 \\
& z_2 = 5x_1 + 4x_2 \leq \varepsilon_2 \\
& x_1 \geq 0, \; x_2 \geq 0.
\end{array}$$

Here, for example, if we choose $\varepsilon_2 = 18$, as illustrated in Fig. 3.4, it can be understood that the optimal solution to this constraint problem occurs at the extreme point $E(-36, 18)$ and, hence, yields a Pareto optimal solution. Also, if we choose $\varepsilon_2 = 14$, we can obtain a Pareto optimal solution such that $(z_1, z_2) = (-28, 14)$ as in Fig. 3.4. ◊

3.2.3 Weighted Minimax Method

The weighted minimax method for characterizing Pareto optimal solutions is to solve the following weighted minimax problem (Bowman 1976):

$$\underset{x \in X}{\text{minimize}} \; \max_{i=1,\ldots,k} \{w_i z_i(x)\}, \qquad (3.19)$$

or equivalently

$$\left. \begin{array}{ll}
\text{minimize} & v \\
\text{subject to} & w_i z_i(x) \leq v, \quad i = 1, 2, \ldots, k \\
& x \in X,
\end{array} \right\} \qquad (3.20)$$

where v is an auxiliary variable.

Here, without loss of generality, it can be assumed that $z_i(x) > 0$, $i = 1, \ldots, k$ for all $x \in X$. Because, for the objective functions not satisfying $z_i(x) > 0$ for all $x \in X$, using their individual minima $z_i^{\min} = \min_{x \in X} z_i(x)$ and setting

Fig. 3.5 Graphical interpretation of weighted minimax method

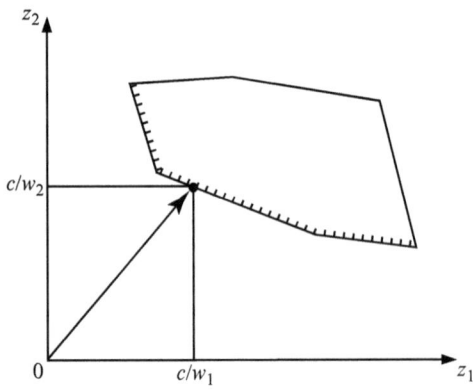

$$\hat{z}_i(x) = z_i(x) - z_i^{\min},$$

it holds that $\hat{z}_i(x) > 0$ for all $x \in X$.

Geometrically, in the weighted minimax problem, the contours of $\max\{w_i z_i\} = c$ (constant) in the objective function space become two edges of rectangles corresponding to the given weighting coefficients. Hence, solving the weighted minimax problem yields Pareto optimal solutions such that these rectangles support the feasible region

$$Z = \{z(x) \mid x \in X\}$$

as depicted in Fig. 3.5.

The relationships between the optimal solution x^* of the weighted minimax problem and Pareto optimality of a multiobjective linear programming problem can be characterized by the following theorems.

Theorem 3.5. *If $x^* \in X$ is a unique optimal solution of the weighted minimax problem for some $w = (w_1, \ldots, w_k) \geq 0$, then x^* is a Pareto optimal solution of the multiobjective linear programming.*

Proof. If a unique optimal solution x^* of the weighted minimax problem for some $w = (w_1, \ldots, w_k) \geq 0$ is not a Pareto optimal solution, then there exists $x \in X$ such that $z_j(x) < z_j(x^*)$ for some j and $z_i(x) \leq z_i(x^*)$, $i = 1, \ldots, k$; $i \neq j$. In view of $w = (w_1, \ldots, w_k) \geq 0$, it follows

$$w_i z_i(x) \leq w_i z_i(x^*), \quad i = 1, \ldots, k.$$

Hence,

$$\max_{i=1,\ldots,k} w_i z_i(x) \leq \max_{i=1,\ldots,k} w_i z_i(x^*).$$

3.2 Scalarization Methods

This contradicts the fact that x^* is a unique optimal solution to the weighted minimax problem for $w = (w_1, \ldots, w_k) \geq 0$. □

From the proof of this theorem, in the absence of the uniqueness of a solution in Theorem 3.5, only weak Pareto optimality is guaranteed.

Theorem 3.6. *If $x^* \in X$ is a Pareto optimal solution to a multiobjective linear programming problem, then x^* is an optimal solution of the weighted minimax problem for some $w = (w_1, \ldots, w_k) > 0$.*

Proof. For a Pareto optimal solution $x^* \in X$ of the multiobjective linear programming problem, choose $w^* = (w_1^*, \ldots, w_k^*) > 0$ such that $w_i^* z_i(x^*) = v$, $i = 1, \ldots, k$. Now assume that x^* is not an optimal solution of the weighted minimax problem, then there exists $x \in X$ such that

$$w_i^* z_i(x) < w_i^* z_i(x^*) = v^*, \ i = 1, \ldots, k.$$

Noting $w^* = (w_1^*, \ldots, w_k^*) > 0$, this implies the existence of $x \in X$ such that

$$z_i(x) < z_i(x^*), \ i = 1, \ldots, k,$$

which contradicts the assumption that x^* is a Pareto optimal solution. □

Example 3.6 (Weighted minimax method for production planning of Example 3.2). To illustrate the weighted minimax method, consider the problem of Example 3.2. Observe that the individual minima of $z_1(x)$ and $z_2(x)$ are $\min_{x \in X} z_1(x) = -37$ and $\min_{x \in X} z_2(x) = 0$, respectively. If we substitute $\hat{z}_1(x) = z_1(x) - (-37)$ and choose $w_1 = 0.8$, $w_2 = 0.4$, then the weighted minimax problem becomes

$$\left.\begin{aligned}
&\text{minimize} \quad \max(-2.4x_1 - 6.4x_2 + 29.6, 2x_1 + 1.6x_2) \\
&\text{subject to} \quad 2x_1 + 6x_2 \leq 27 \\
&\phantom{\text{subject to}} \quad 3x_1 + 2x_2 \leq 16 \\
&\phantom{\text{subject to}} \quad 4x_1 + x_2 \leq 18 \\
&\phantom{\text{subject to}} \quad x_1 \geq 0, \ x_2 \geq 0,
\end{aligned}\right\}$$

or equivalently

$$\left.\begin{aligned}
&\text{minimize} \quad v \\
&\text{subject to} \quad 2x_1 + 6x_2 \leq 27 \\
&\phantom{\text{subject to}} \quad 3x_1 + 2x_2 \leq 16 \\
&\phantom{\text{subject to}} \quad 4x_1 + x_2 \leq 18 \\
&\phantom{\text{subject to}} \quad -2.4x_1 - 6.4x_2 + 29.6 \leq v \\
&\phantom{\text{subject to}} \quad 2x_1 + 1.6x_2 \leq v \\
&\phantom{\text{subject to}} \quad x_1 \geq 0, \ x_2 \geq 0.
\end{aligned}\right\}$$

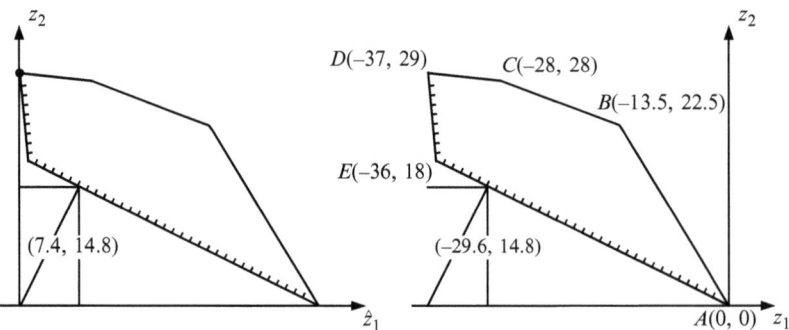

Fig. 3.6 Weighted minimax method for Example 3.2

Noting that $\hat{z}_1 = z_1 + 37$, as illustrated in Fig. 3.6, it can be understood that a vector (7.4, 14.8) of the objective function values of this problem is a point moved from the point (−29.6, 14.8) of the original problem by 37 along the z_1 axis. Hence, the point (−29.6, 14.8) is a vector of the original objective function values which corresponds to a Pareto optimal solution $(x_1, x_2) = (0, 3.7)$. ◊

From Theorems 3.3 and 3.5, if the uniqueness of the optimal solution x^* for the scalarizing problem is not guaranteed, it is necessary to perform the Pareto optimality test of x^*. The Pareto optimality test for x^* can be performed by solving the following linear programming problem with the decision variables $x = (x_1, \ldots, x_n)^T$ and $\varepsilon = (\varepsilon_1, \ldots, \varepsilon_k)^T$:

$$\left. \begin{aligned} \text{maximize} \quad & \sum_{i=1}^{k} \varepsilon_i \\ \text{subject to} \quad & z_i(x) + \varepsilon_i = z_i(x^*), \quad i = 1, \ldots, k \\ & x \in X, \quad \varepsilon = (\varepsilon_1, \ldots, \varepsilon_k) \geq \mathbf{0}. \end{aligned} \right\} \quad (3.21)$$

For an optimal solution $(\bar{x}, \bar{\varepsilon})$ of this linear programming problem, the following theorem holds.

Theorem 3.7. *For an optimal solution $(\bar{x}, \bar{\varepsilon})$ of the Pareto optimality test problem, the following statements hold.*

(i) *If $\bar{\varepsilon}_i = 0$ for all $i = , \ldots, k$, then x^* is a Pareto optimal solution of the multiobjective linear programming problem.*

(ii) *If $\bar{\varepsilon}_i > 0$ for at least one i, then x^* is not a Pareto optimal solution of the multiobjective linear programming problem. Instead of x^*, $\bar{x}$ is the Pareto optimal solution corresponding to the scalarization problem.*

Proof. (i) If x^* is not a Pareto optimal solution to the multiobjective linear programming problem, then there exists $x \in X$ such that $z_j(x) < z_j(x^*)$ for some j and $z_i(x) \leq z_i(x^*)$, $i = 1, \ldots, k$; $i \neq j$, which contradicts the assumption that $\bar{\varepsilon}_i = 0$ for all $i = , \ldots, k$.

(ii) If at least one $\bar{\varepsilon}_i > 0$ and $\bar{x}$ is not a Pareto optimal solution of the multiobjective linear programming problem, then there exists $x \in X$ such that $z_j(x) < z_j(\bar{x})$ for some j and $z_i(x) \le z_i(\bar{x})$, $i = 1, \ldots, k$; $i \ne j$. Hence, there exists $x \in X$ such that $z(x) + \varepsilon' = z(\bar{x})$ for some $\varepsilon' \ge 0$, and then $z(x) + \varepsilon' + \bar{\varepsilon} = z(x^*)$. This contradicts the optimality of $\bar{\varepsilon}$.

□

As discussed above, when an optimal solution x^* is not unique, x^* is not always Pareto optimal, and then for the Pareto optimality test, (3.21) is solved. However, it should be noted here that although the formulation is somewhat complex, by solving the following augmented minimax problem, a Pareto optimal solution is obtainable:

$$\left. \begin{array}{ll} \text{minimize} & v \\ \text{subject to} & w_i z_i(x) + \rho \sum_{i=1}^{k} w_i z_i(x) \le v, \quad i = 1, 2, \ldots, k \\ & x \in X, \end{array} \right\} \quad (3.22)$$

where ρ is a sufficiently small positive number.

3.3 Linear Goal Programming

The term "goal programming" first appeared in the 1961 text by Charnes and Cooper to deal with multiobjective linear programming problems that assumed the decision maker (DM) could specify goals or aspiration levels for the objective functions. Subsequent works on goal programming have been numerous, including texts on goal programming by Ijiri (1965), Lee (1972), and Ignizio (1976, 1982) and survey papers by Charnes and Cooper (1977) and Ingnizio (1983).

The key idea behind goal programming is to minimize the deviations from goals or aspiration levels set by the DM. Goal programming therefore, in most cases, seems to yield a satisficing solution in the same spirit as March and Simon (1958) rather than an optimal solution.

As discussed in the previous subsection, in general, the multiobjective linear programming problem can be formulated as follows:

$$\underset{x \in X}{\text{minimize}} \; z(x) = (z_1(x), \ldots, z_k(x))^T, \quad (3.23)$$

where $z_1(x) = c_1 x, \ldots, z_k(x) = c_k x$ are k distinct objective functions of the decision variable vector x and

$$X = \{x \in \mathbb{R}^n \mid Ax \le b, \; x \ge 0\} \quad (3.24)$$

is the linearly constrained feasible region.

For linear goal programming, however, a set of k goals is specified by the DM for the k objective functions $z_i(x)$, $i = 1,\ldots,k$ and the multiobjective linear programming problem is converted into the problem of coming "as close as possible" to the set of specified goals which may not be simultaneously attainable.

The general formulation of goal programming thus becomes

$$\underset{x \in X}{\text{minimize}} \ d(z(x), \hat{z}), \tag{3.25}$$

where $\hat{z} = (\hat{z}_1, \ldots, \hat{z}_k)$ is the goal vector specified by the DM and $d(z(x), \hat{z})$ represents the distance between $z(x)$ and $\hat{z}$ in some selected norm.

The simplest version of (3.25), where the absolute value or the ℓ_1 norm is used, is

$$\underset{x \in X}{\text{minimize}} \ d_1(z(x), \hat{z}) = \sum_{i=1}^{k} |c_i x - \hat{z}_i|. \tag{3.26}$$

More generally, using the ℓ_1 norm with weights (the weighted ℓ_1 norm), it becomes

$$\underset{x \in X}{\text{minimize}} \ d_1^w(z(x), \hat{z}) = \sum_{i=1}^{k} w_i |c_i x - \hat{z}_i|, \tag{3.27}$$

where w_i is a nonnegative weight to the ith objective function.

This linear goal programming problem can easily be converted to an equivalent linear programming problem by introducing the auxiliary variables

$$d_i^+ = \frac{1}{2}\{|z_i(x) - \hat{z}_i| + (z_i(x) - \hat{z}_i)\} \tag{3.28}$$

and

$$d_i^- = \frac{1}{2}\{|z_i(x) - \hat{z}_i| - (z_i(x) - \hat{z}_i)\} \tag{3.29}$$

for each $i = 1,\ldots,k$. Thus, the equivalent linear goal programming formulation to the problem (3.27) becomes

$$\left. \begin{aligned} &\text{minimize} \sum_{i=1}^{k} w_i(d_i^+ + d_i^-) \\ &\text{subject to } z_i(x) - d_i^+ + d_i^- = \hat{z}_i \quad i = 1,\ldots,k \\ &\quad Ax \leq b, \ x \geq 0 \\ &\quad d_i^+ \cdot d_i^- = 0, \ i = 1,\ldots,k \\ &\quad d_i^+ \geq 0, \ d_i^- \geq 0, \ i = 1,\ldots,k. \end{aligned} \right\} \tag{3.30}$$

3.3 Linear Goal Programming

It is appropriate to consider here the practical significance of d_i^+ and d_i^-. From the definition of d_i^+ and d_i^-, it can be easily understood that

$$d_i^+ = \begin{cases} z_i(x) - \hat{z}_i & \text{if } z_i(x) \geq \hat{z}_i \\ 0 & \text{if } z_i(x) < \hat{z}_i \end{cases} \quad (3.31)$$

and

$$d_i^- = \begin{cases} \hat{z}_i - z_i(x) & \text{if } \hat{z}_i \geq z_i(x) \\ 0 & \text{if } \hat{z}_i < z_i(x). \end{cases} \quad (3.32)$$

Thus, d_i^+ and d_i^- represent, respectively, the overachievement and underachievement of the ith goal and, hence, are called deviational variables. Obviously, overachievement and underachievement can never occur simultaneously. When $d_i^+ > 0$, then d_i^- must be zero, and vice versa. This fact is reflected by the third constraint, $d_i^+ \cdot d_i^- = 0$, $i = 1, \ldots, k$, of (3.30) which is automatically satisfied at every iteration of the simplex method of linear programming because d_i^+ and d_i^- never become basic variables simultaneously. The third constraint of (3.30) is always satisfied in the simplex method, and consequently, it is clear that the simplex method can be applied to solve this type of linear goal programming problem.

Depending on the decision situations, the DM may be sometimes concerned only with either the overachievement or underachievement of a specified goal. Such a situation can be incorporated into the goal programming formulation by assigning the over- and underachievement weights w_i^+ and w_i^- to d_i^+ and d_i^-, respectively. For example, if each $z_i(x)$ is a cost-type objective function with its goal $\hat{z}_i$, the overachievement is not desirable. For this case, we set $w_i^+ = 1$ and $w_i^- = 0$, and the problem (3.30) is modified as follows:

$$\left.\begin{array}{ll} \text{minimize} & \sum_{i=1}^{k} w_i^+ d_i^+ \\ \text{subject to} & z_i(x) - d_i^+ + d_i^- = \hat{z}_i \quad i = 1, \ldots, k \\ & Ax \leq b, \; x \geq 0 \\ & d_i^+ \cdot d_i^- = 0, \; i = 1, \ldots, k \\ & d_i^+ \geq 0, \; d_i^- \geq 0, \; i = 1, \ldots, k. \end{array}\right\} \quad (3.33)$$

Conversely, for a benefit-type objective function, the underachievement is not desirable. For such a case, we set $w_i^- = 1$ and $w_i^+ = 0$ to replace $\sum_{i=1}^{k} w_i^+ d_i^+$ with $\sum_{i=1}^{k} w_i^- d_i^-$ as the objective function of (3.33). This particular goal programming is called one-sided goal programming.

The linear goal programming formulation can also be modified into a more general form by introducing the preemptive priorities P_i in place of, or together with, the numerical weights $w_i^+, w_i^- \geq 0$. When the objective functions $z_1(x), \ldots, z_k(x)$ are divided into L ordinal ranking classes having the preemptive priorities $P_1, \ldots, P_L$ in decreasing order, it may be convenient to write

$$P_l \gg P_{l+1}, \quad l = 1, \ldots, L - 1 \tag{3.34}$$

to mean that no real number t, however large, can produce

$$tP_{l+1} \geq P_l, \quad l = 1, \ldots, L - 1, \tag{3.35}$$

where $1 \leq L \leq k$.

By incorporating such preemptive priorities P_l together with the over- and underachievement weights w_i^+ and w_i^-, the general linear goal programming formulation takes on the following form:

$$\left.\begin{aligned} \text{minimize} \quad & \sum_{l=1}^{L} P_l \left(\sum_{i \in I_l} (w_i^+ d_i^+ + w_i^- d_i^-) \right) \\ \text{subject to} \quad & z_i(x) - d_i^+ + d_i^- = \hat{z}_i \quad i = 1, \ldots, k \\ & Ax \leq b, \, x \geq 0 \\ & d_i^+ \cdot d_i^- = 0, \, i = 1, \ldots, k \\ & d_i^+ \geq 0, \, d_i^- \geq 0, \, i = 1, \ldots, k, \end{aligned}\right\} \tag{3.36}$$

where $I_l (\neq \emptyset)$ is the index set of objective functions in the lth priority class. Observe that when there are k distinct ordinal ranking classes with the ith objective function $c_i x$ belonging to the ith priority class, i.e., $L = k$, the objective function of (3.36) then becomes simply

$$\sum_{i=1}^{k} P_i (w_i^+ d_i^+ + w_i^- d_i^-). \tag{3.37}$$

To solve this type of linear goal programming problems, we begin by trying to achieve the goals of all objective functions in the first priority class. Having done that, we try to satisfy the goals in the second priority class, keeping the goals in the first class satisfied. The process is repeated until either a unique solution is obtained at some stage or all priority classes are considered. This is equivalent to solving, at most, L linear programming problems sequentially, for which the simplex method can easily be applied with some modifications.

Further details concerning the algorithm, extensions, and applications can be found in the text of Lee (1972) and Ignizio (1976, 1982).

Example 3.7 (Production planning problem with goals). To illustrate the linear goal programming method, assume that a manager of the manufacturing company establishes the following goals P_1, P_2, and P_3 for the production planning problem

3.3 Linear Goal Programming

Fig. 3.7 Graphical solution for Example 3.7

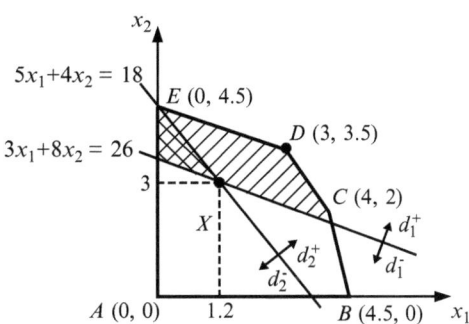

incorporating environmental quality in Example 3.2. Here, to avoid confusion in the notation, let Q and R denote the product names instead of P_1 and P_2 used in the previous examples.

P_1: To achieve at least 26 million yen of total profit.
P_2: To keep the pollution level below 18 units.
P_3: To produce at least 2 tons of Q and 3 tons of R. However, assign the weight ratio of 3–8 for Q and R by considering the profit contribution ratio of these two products.

The corresponding linear goal programming problem can be formulated as follows:

$$\begin{aligned}
\text{minimize} \quad & P_1 d_1^- + P_2 d_2^+ + P_3(3 d_3^- + 8 d_4^-) \\
\text{subject to} \quad & 3x_1 + 8x_2 - d_1^+ + d_1^- = 26 \\
& 5x_1 + 4x_2 - d_2^+ + d_2^- = 18 \\
& x_1 \qquad\quad - d_3^+ + d_3^- = 2 \\
& \qquad x_2 - d_4^+ + d_4^- = 3 \\
& 2x_1 + 6x_2 \qquad\qquad \le 27 \\
& 3x_1 + 2x_2 \qquad\qquad \le 16 \\
& 4x_1 + x_2 \qquad\qquad \le 18 \\
& x_1 \ge 0, \; x_2 \ge 0 \\
& d_i^+ \ge 0, \; d_i^- \ge 0, \; i = 1, 2, 3, 4.
\end{aligned}$$

To graphically obtain an optimal solution for this simple example in the x_1-x_2 plane, the two priority goals are depicted as straight lines together with the original feasible region in Fig. 3.7. Although only the decision variables x_1 and x_2 are used in this graph, the effect of increasing either d_i^+ or d_i^- is reflected by the arrow signs. The region which satisfies both the original constraints and the first priority goal, i.e., $d_1^+ \ge 0$ and $d_1^- = 0$, is shown as the cross-hatched region. To achieve the second priority goal without degrading the achievement of the first priority goal, the area of feasible solution should be limited to the crisscross-hatched area in Fig. 3.7. However, as can been seen, concerning the third priority goals, d_3^+ cannot increase to be positive. As just described, the final solution of this problem occurs at the point $(x_1, x_2) = (1.2, 3)$ in which only the first and second priority goals are satisfied. ◇

3.4 Compromise Programming

A well-known extension of the goal programming approach is obtained if the goal vector $\hat{z} = (\hat{z}_1, \ldots, \hat{z}_k)$ is replaced with the so-called ideal or utopia vector $z^{\min} = (z_1^{\min}, \ldots, z_k^{\min})$, where $z_i^{\min} = \min_{x \in X} z_i(x)$, $i = 1, \ldots, k$. The resulting problem can be interpreted as an attempt to minimize the deviation from the ideal or utopia vector (point). Realizing that the ideal point is generally infeasible in most of multiobjective linear programming problems with conflicting objective functions, Yu (1973) and Zeleny (1973) introduced the concept of compromise solutions. Geometrically, the compromise solution defined by Yu and Zeleny is a solution which is the closest to the ideal point.

To be more specific mathematically, given a weighting vector w, x_w^p is called a compromise solution of the multiobjective linear programming problem with respect to the parameter p of the norm if and only if it solves

$$\underset{x \in X}{\text{minimize}} \quad d_p^w(z(x), z^{\min}) = \left(\sum_{i=1}^{k} w_i |z_i(x) - z_i^{\min}|^p \right)^{1/p} \tag{3.38}$$

or equivalently, for $1 \leq p < \infty$,

$$\underset{x \in X}{\text{minimize}} \quad \tilde{d}_p^w(z(x), z^{\min}) = \sum_{i=1}^{k} w_i (z_i(x) - z_i^{\min})^p, \tag{3.39}$$

and for $p = \infty$,

$$\underset{x \in X}{\text{minimize}} \quad \tilde{d}_\infty^w(z(x), z^{\min}) = \max_{i=1,\ldots,k} w_i (z_i(x) - z_i^{\min}). \tag{3.40}$$

Observe that for $p = 1$, all deviations from $z_i^{\min}$ are taken into account in direct proportion to their magnitudes, while for $2 \leq p < \infty$, the largest deviation has the greatest influence. Ultimately for $p = \infty$, only the largest deviation is taken into account.

It should be noted here that any solution of (3.39) for any $1 \leq p < \infty$ or a unique solution of (3.40) with $w_i > 0$ for all $i = 1, \ldots, k$ is a Pareto optimal solution of the multiobjective linear programming problem.

The compromise set C_w, given the weighting vector w, is defined as the set of all compromise solutions x_w^p, $1 \leq p \leq \infty$. To be more explicit,

3.4 Compromise Programming

$$C_w = \{x \in X \mid x \text{ solves (3.39) or (3.40) given } w \text{ for all } 1 \le p \le \infty\}. \quad (3.41)$$

In the context of linear programming problems, Zeleny (1973) suggested that the compromise set C_w can be approximated by the Pareto optimal solutions of the following two-objective problem:

$$\underset{x \in X}{\text{minimize}} \; (\tilde{d}_1^w(z(x), z^{\min}), \tilde{d}_\infty^w(z(x), z^{\min})). \quad (3.42)$$

Although it can be seen that the compromise solution set C_w is a subset of the set of Pareto optimal solutions, C_w may still be too large to select the final solution and, hence, should be reduced further.

Zeleny (1973, 1976) suggests several methods to reduce the compromise solution set C_w. One possible reduction method without the DM's aid is to generate another compromise solution set $\bar{C}_w$ similar to C_w by maximizing the distance from the so-called anti-ideal point $z^{\max} = (z_1^{\max}, \ldots, z_k^{\max})$, where $z_i^{\max} = \underset{x \in X}{\max}\, z_i(x)$. The problem to be solved thus becomes

$$\underset{x \in X}{\text{maximize}} \; \left(\sum_{i=1}^{k} w_i |z_i^{\max} - z_i(x)|^p \right)^{1/p} \quad (3.43)$$

or equivalently, for $1 \le p < \infty$,

$$\underset{x \in X}{\text{maximize}} \; \sum_{i=1}^{k} w_i (z_i^{\max} - z_i(x))^p, \quad (3.44)$$

and for $p = \infty$,

$$\underset{x \in X}{\text{maximize}} \; \underset{i=1,\ldots,k}{\min} \; w_i (z_i^{\max} - z_i(x)). \quad (3.45)$$

The compromise solution set $\bar{C}_w$ is defined by

$$\bar{C}_w = \{x \in X \mid x \text{ solves (3.44) or (3.45) given } w \text{ for all } 1 \le p \le \infty\}. \quad (3.46)$$

The compromise solution set C_w based on the ideal point is not identical with the compromise solution set $\bar{C}_w$ based on the anti-ideal point. Zeleny (1976) suggests using this fact to further reduce the compromise solution set by considering the intersection $C_w \cap \bar{C}_w$.

An interactive strategy for reducing the compromise solution set proposed by Zeleny (1976) is based on the concept of the so-called displaced ideal and thus is called the method of the displaced ideal. In this approach, the ideal point with respect to the new C_w displaces the previous ideal point, and the (reduced) compromise solution set eventually encloses the new ideal point, terminating the process.

The procedure of the method of displaced ideal is summarized as follows.

Procedure of the Method of Displaced Ideal

Step 1 Let $C_w^{(0)} = X$, and initialize the iteration index, $r = 1$.

Step 2 Find the ideal point $z^{\min(r)}$ by solving

$$\underset{x \in C_w^{(r-1)}}{\text{minimize}} \; z_i(x), \; i = 1, \ldots, k.$$

Step 3 Construct the compromise solution set $C_w^{(r)}$ by finding the Pareto optimal solution set of

$$\underset{x \in C_w^{(r-1)}}{\text{minimize}} \; (\tilde{d}_1^w(z(x), z^{\min(r)}), \tilde{d}_\infty^w(z(x), z^{\min(r)})).$$

Step 4 If the DM can select the final solution from $C_w^{(r)}$, or if $C_w^{(r)}$ contains $z^{\min(r)}$, stop. Otherwise, set $r = r + 1$ and return to step 2.

It should be noted here that the method of displaced ideal can be viewed as the best ideal-seeking process, not the ideal itself. Further refinements and details can be found in Zeleny (1976, 1982).

Example 3.8 (Displaced ideal method for production planning of Example 3.2). To illustrate the method of displaced ideal, consider the problem of Example 3.2:

$$\begin{aligned}
\text{minimize} \quad & z_1 = -3x_1 - 8x_2 \\
\text{minimize} \quad & z_2 = 5x_1 + 4x_2 \\
\text{subject to} \quad & 2x_1 + 6x_2 \le 27 \\
& 3x_1 + 2x_2 \le 16 \\
& 4x_1 + x_2 \le 18 \\
& x_1 \ge 0, \quad x_2 \ge 0.
\end{aligned}$$

Let $w_1 = w_2 = 1$ and $C_w^{(0)} = X = \{(x_1, x_2) \in \mathbb{R}^2 \mid 2x_1 + 6x_2 \le 27, 3x_1 + 2x_2 \le 16, 4x_1 + x_2 \le 18, x_1 \ge 0, x_2 \ge 0\}$. From the definition, we have

$$z_1^{\min(1)} = \min_{x \in C_w^{(0)}} z_1(x) = -37, \quad z_2^{\min(1)} = \min_{x \in C_w^{(0)}} z_2(x) = 0.$$

In step 3, the following two-objective programming problem is formulated:

$$\underset{x \in C_w^{(0)}}{\text{minimize}} \; (\tilde{d}_1^w(z(x), z^{\min(1)}), \tilde{d}_\infty^w(z(x), z^{\min(1)})),$$

where

$$\tilde{d}_1^w(z(x), z^{\min(1)}) = (-3x_1 - 8x_2 - (-37)) + (5x_1 + 4x_2 - 0),$$

3.4 Compromise Programming

Fig. 3.8 Displaced ideal method for Example 3.2

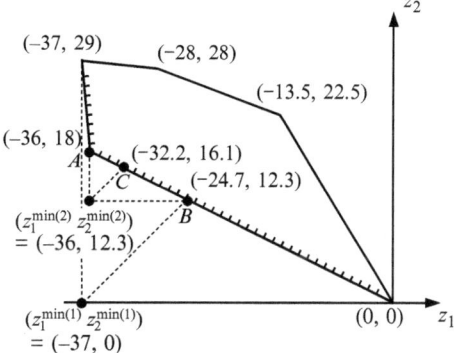

$$\tilde{d}_\infty^w(z(x), z^{\min(1)}) = \max\{(-3x_1 - 8x_2 - (-37)), (5x_1 + 4x_2 - 0)\}.$$

From this problem, we have the compromise solution set $C_w^{(1)}$ which is a straight-line segment between points $A(-36, 18)$ and $B(-24.667, 12.333)$ shown in Fig. 3.8, where point A corresponding to a solution $x = (0, 4.5)$ minimizes $\tilde{d}_1^w(z(x), z^{\min(1)})$ and point B corresponding to a solution $x = (0, 3.0833)$ minimizes $\tilde{d}_\infty^w(z(x), z^{\min(1)})$.

Suppose that the DM cannot select the final solution. In step 2, we have

$$z_1^{\min(2)} = \min_{x \in C_w^{(1)}} z_1(x) = -36, \quad z_2^{\min(2)} = \min_{x \in C_w^{(1)}} z_2(x) = 12.333.$$

In step 3, the following two-objective programming problem is reformulated:

$$\underset{x \in C_w^{(1)}}{\text{minimize}} \ (\tilde{d}_1^w(z(x), z^{\min(2)}), \tilde{d}_\infty^w(z(x), z^{\min(2)})),$$

where

$$\tilde{d}_1^w(z(x), z^{\min(2)}) = (-3x_1 - 8x_2 - (-36)) + (5x_1 + 4x_2 - 12.333),$$
$$\tilde{d}_\infty^w(z(x), z^{\min(2)}) = \max\{(-3x_1 - 8x_2 - (-36)), (5x_1 + 4x_2 - 12.333)\}.$$

From this problem, we have the revised compromise solution set $C_w^{(2)}$ which is a straight-line segment between points $A(-36, 18)$ and $C(-32.222, 16.111)$ shown in Fig. 3.8, where point A minimizes $\tilde{d}_1^w(z(x), z^{\min(2)})$ and point C corresponding to a solution $x^* = (0, 4.028)$ minimizes $\tilde{d}_\infty^w(z(x), z^{\min(2)})$.

One finds that the compromise solution set diminishes from $C_w^{(1)}$ to $C_w^{(2)}$. If the DM still cannot select the final solution in $C_w^{(2)}$, it follows that the procedure continues. ◇

3.5 Interactive Multiobjective Linear Programming

The STEP method (STEM) proposed by Benayoun et al. (1971) seems to be known as one of the first interactive multiobjective linear programming techniques, and there have been some modifications and extensions [see, for example, Choo and Atkins (1980); Fichefet (1976)]. Essentially, the STEM algorithm consists of two major steps. Step 1 seeks a Pareto optimal solution that is near to the ideal point in the minimax sense. Step 2 requires the DM to compare a vector of the objective function values with the ideal point and to indicate which objectives can be sacrificed, and by how much, in order to improve the current levels of unsatisfactory objectives. The STEM algorithm is quite simple to understand and implement, in the sense that the DM is required to give only the amounts to be sacrificed of some satisfactory objectives until all objectives become satisfactory. However, the DM will never arrive at the final solution if the DM is not willing to sacrifice any of the objectives. Moreover, in many practical situations, the DM will probably want to indicate directly the aspiration level for each objective rather than just specify the amount by which satisfactory objectives can be sacrificed.

Wierzbicki (1980) developed a relatively practical interactive method called the reference point method by introducing the concept of a reference point suggested by the DM which reflects in some sense the desired values of the objective functions. The basic idea behind the reference point method is that the DM can specify reference values for the objective functions and change the reference objective levels interactively due to learning or improved understanding during the solution process. In this procedure, when the DM specifies a reference point, the corresponding scalarization problem is solved for generating a Pareto optimal solution which is, in a sense, close to the reference point or better than that if the reference point is attainable. Then the DM either chooses the current Pareto optimal solution or modifies the reference point to find a satisficing solution.

Since then, some similar interactive multiobjective programming methods have been developed along this line [see, for example, Steuer and Choo (1983)]. However, it is important to point out here that for dealing with the fuzzy goals of the DM for the objective functions of the multiobjective linear programming problem, Sakawa et al. (1987) developed the extended fuzzy version of the reference point method that supplies the DM with the trade-off information. Although the details of the method will be discussed in the next chapter, it would certainly be appropriate to discuss here the reference point method with trade-off information rather than the reference point method proposed by Wierzbicki.

Consider the following multiobjective linear programming problem:

$$\left. \begin{array}{ll} \text{minimize} & z_1(x) = c_1 x \\ \text{minimize} & z_2(x) = c_2 x \\ & \dots\dots\dots \\ \text{minimize} & z_k(x) = c_k x \\ \text{subject to} & x \in X = \{x \in \mathbb{R}^n \mid Ax \leq b,\ x \geq 0\}, \end{array} \right\} \quad (3.47)$$

3.5 Interactive Multiobjective Linear Programming

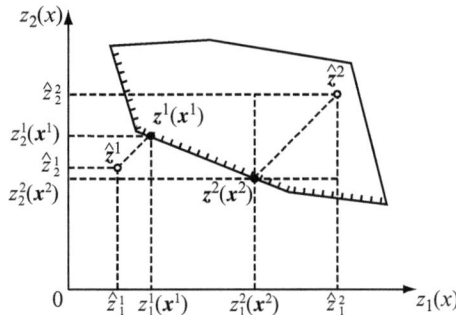

Fig. 3.9 Graphical interpretation of minimax method

where $z_1(x) = c_1 x, \ldots, z_k(x) = c_k x$ are k distinct objective functions of the decision variable vector x and X is the linearly constrained feasible region.

For each of the multiple conflicting objective functions $z(x) = (z_1(x), \ldots, z_k(x))^T$, assume that the DM can specify the so-called reference point $\hat{z}l = (\hat{z}_1, \ldots, \hat{z}_k)^T$ which reflects in some sense the desired values of the objective functions for the DM. Also assume that the DM can change the reference point interactively due to learning or improved understanding during the solution process. When the DM specifies the reference point $\hat{z} = (\hat{z}_1, \ldots, \hat{z}_k)^T$, the corresponding Pareto optimal solution, which is, in the minimax sense, the nearest to the reference point or better than that if the reference point is attainable, is obtained by solving the following minimax problem:

$$\left.\begin{array}{l} \text{minimize} \quad \max_{i=1,\ldots,k} \{z_i(x) - \hat{z}_i\} \\ \text{subject to} \quad x \in X, \end{array}\right\} \quad (3.48)$$

or equivalently

$$\left.\begin{array}{l} \text{minimize} \quad v \\ \text{subject to} \quad z_i(x) - \hat{z}_i \leq v, \quad i = 1, \ldots, k \\ \phantom{\text{subject to}} \quad x \in X, \end{array}\right\} \quad (3.49)$$

where v is an auxiliary variable.

The case of the two-objective functions in the z_1-z_2 plane is shown geometrically in Fig. 3.9. For two reference points $\hat{z}^1 = (\hat{z}_1^1, \hat{z}_2^1)^T$ and $\hat{z}^2 = (\hat{z}_1^2, \hat{z}_2^2)^T$ specified by the DM, the respective Pareto optimal solutions $z^1(x^1)$ and $z^2(x^2)$ are obtained by solving the minimax problems with $\hat{z}^1$ and $\hat{z}^2$.

The relationships between the optimal solutions of the minimax problem and Pareto optimality of a multiobjective linear programming problem can be characterized by the following two theorems.

Theorem 3.8. *If $x^* \in X$ is a unique optimal solution of the minimax problem for any reference point $\hat{z}$, then x^* is a Pareto optimal solution of the multiobjective linear programming problem.*

Proof. If a unique optimal solution x^* of the minimax problem is not a Pareto optimal solution of the multiobjective linear programming problem, then there exists $x \in X$ such that $z_i(x) \le z_i(x^*)$, $i = 1, \ldots, k$; $i \ne j$, and $z_j(x) < z_j(x^*)$ for some j. Hence, it follows that

$$\max_{i=1,\ldots,k} \{z_i(x) - \hat{z}_i\} \le \max_{i=1,\ldots,k} \{z_i(x^*) - \hat{z}_i\}.$$

This contradicts the assumption that x^* is a unique optimal solution of the minimax problem. □

From the proof of this theorem, in the absence of the uniqueness of a solution in the theorem, only weak Pareto optimality is guaranteed.

Theorem 3.9. *If x^* is a Pareto optimal solution of the multiobjective linear programming problem, then x^* is an optimal solution of the minimax problem for some reference point $\hat{z}$.*

Proof. For a Pareto optimal solution $x^* \in X$ of the multiobjective linear programming, choose a reference point $\hat{z} = (\hat{z}_1, \ldots, \hat{z}_k)^T$ such that $z_i(x^*) - \hat{z}_i = v^*$, $i = 1, \ldots, k$. For this reference point, if x^* is not an optimal solution of the minimax problem, then there exists $x \in X$ such that

$$z_i(x) - \hat{z}_i < z_i(x^*) - \hat{z}_i = v^*, \quad i = 1, \ldots, k.$$

This implies the existence of $x \in X$ such that

$$z_i(x) < z_i(x^*), \quad i = 1, \ldots, k,$$

which contradicts the fact that x^* is a Pareto optimal solution. □

If an optimal solution x^* to the minimax problem is not unique, then, as discussed in the previous section, the Pareto optimality test for x^* can be performed by solving the following problem:

$$\left. \begin{array}{l} \text{maximize} \quad \sum_{i=1}^{k} \varepsilon_i \\ \text{subject to} \quad z_i(x) + \varepsilon_i = z_i(x^*), \quad i = 1, \ldots, k \\ \phantom{\text{subject to}} \quad x \in X, \quad \boldsymbol{\varepsilon} = (\varepsilon_1, \ldots, \varepsilon_k) \ge \mathbf{0}. \end{array} \right\} \quad (3.50)$$

For an optimal solution $(\bar{x}, \bar{\boldsymbol{\varepsilon}})$ of this linear programming problem, as was shown in Theorem 3.7, (i) if $\bar{\varepsilon}_i = 0$ for all $i = 1, \ldots, k$, then x^* is a Pareto optimal solution of the multiobjective linear programming problem, and (ii) if $\bar{\varepsilon}_i > 0$ for at least

3.5 Interactive Multiobjective Linear Programming

one i, then not x^* but $\bar{x}$ is a Pareto optimal solution of the multiobjective linear programming problem.

Now, given a Pareto optimal solution for the reference point specified by the DM by solving the corresponding minimax problem, the DM must either be satisfied with the current Pareto optimal solution or modify the reference point. To help the DM express a degree of preference, trade-off information between a standing objective function $z_1(x)$ and each of the other objective functions is very useful. Such a trade-off between $z_1(x)$ and $z_i(x)$ for each $i = 2, \ldots, k$ is easily obtainable since it is closely related to the strict positive simplex multipliers of the minimax problem (3.49). Let the simplex multipliers associated with be denoted by π_i, $i = 1, \ldots, k$. If all $\pi_i > 0$ for each i, it can be proved that the following expression holds:

$$-\frac{\partial z_i(x)}{\partial z_1(x)} = \frac{\pi_1}{\pi_i}. \qquad (3.51)$$

Geometrically, however, we can understand it as follows.

In the $(z_1, \ldots, z_k, w)$ space, the tangent hyperplane at some point on a Pareto surface can be described by

$$H(z_1, \ldots, z_k, w) = a_1 z_1 + \cdots + a_k z_k + bw = c.$$

The necessary and sufficient condition for the small displacement from this point belonging to this tangent hyperplane is $\Delta H = 0$, i.e.,

$$a_1 \Delta z_1 + \cdots + a_k \Delta z_k + b \Delta w = 0.$$

For fixed values of $\Delta z_j = 0$ ($j = 2, \ldots, k$, $j \neq i$) and $\Delta w = 0$ except Δz_1 and Δz_i, we have

$$a_1 \Delta z_1 + a_i \Delta z_i = 0.$$

Similarly, we have

$$a_i \Delta z_i + b \Delta w = 0, \quad a_1 \Delta z_1 + b \Delta w = 0.$$

It follows from the last two relations that

$$-\frac{\Delta z_i}{\Delta z_1} = \frac{a_1}{a_i} = \frac{-a_1/b}{-a_i/b} = \frac{\Delta w/\Delta z_1}{\Delta w/\Delta z_i}.$$

Consequently, it holds that

$$-\frac{\partial z_i}{\partial z_1} = \frac{\partial w/\partial z_1}{\partial w/\partial z_i}.$$

Using the simplex multipliers π_i, $i = 1, \ldots, k$, associated with all the active constraints of the minimax problem, since $\partial w/\partial z_i = \pi_i$, we obtain (3.51).

It should be stressed here that in order to obtain the trade-off rate from (3.51), all constraints of the minimax problem must be active. Therefore, if there are inactive constraints, it is necessary to replace $\hat{z}_i$ for inactive constraints with $z_i(x^*)$ and to solve the corresponding minimax problem to obtain the simplex multipliers.

We can now give the interactive algorithm to derive the satisficing solution for the DM from the Pareto optimal solution set. The steps marked with an asterisk involve interaction with the DM. Observe that this interactive multiobjective linear programming method can be interpreted as the reference point method with trade-off information.

Procedure of Interactive Multiobjective Linear Programming

Step 0   Calculate the individual minimum $z_i^{\min} = \min_{x \in X} z_i(x)$ and maximum $z_i^{\max} = \max_{x \in X} z_i(x)$ of each objective function under the given constraints.

Step 1*   Ask the DM to select the initial reference point by considering the individual minimum and maximum. If the DM finds it difficult or impossible to identify such a point, $z_i^{\min} = \min_{x \in X} z_i(x)$ can be used for that purpose.

Step 2   For the reference point specified by the DM, solve the corresponding minimax problem to obtain a Pareto optimal solution together with the trade-off rate between the objective functions.

Step 3*   If the DM is satisfied with the current objective function values of the Pareto optimal solution, stop. Then the current Pareto optimal solution is the satisficing solution for the DM. Otherwise, ask the DM to update the reference point by considering the current values of the objective functions together with the trade-off rates between the objective functions and return to Step 2.

It should be stressed to the DM that any improvement of one objective function can be achieved only at the expense of at least one of the other objective functions.

Example 3.9 (Production planning with environmental consideration). To demonstrate interactive multiobjective linear programming, consider the following production planning problem with environmental consideration as previously discussed in Example 3.2:

$$\begin{aligned}
\text{minimize} \quad & z_1 = -3x_1 - 8x_2 \\
\text{minimize} \quad & z_2 = 5x_1 + 4x_2 \\
\text{subject to} \quad & 2x_1 + 6x_2 \le 27 \\
& 8x_1 + 6x_2 \le 45 \\
& 3x_1 + x_2 \le 15 \\
& x_1 \ge 0, \quad x_2 \ge 0.
\end{aligned}$$

Recall that the two objectives in this problem are to minimize both the opposite of the total profit (z_1) and the amount of pollution (z_2).

3.5 Interactive Multiobjective Linear Programming

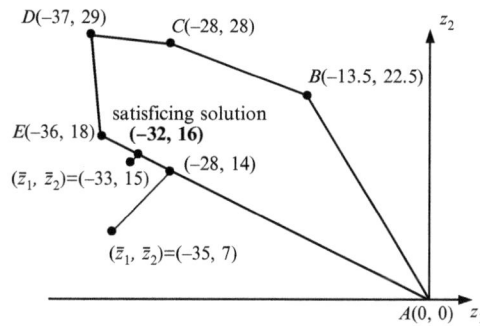

Fig. 3.10 Interactive multiobjective linear programming for Example 3.2

First, observe that the individual minima and maxima for the objective functions are

$$z_1^{\min} = -37, \quad z_1^{\max} = 0, \quad z_2^{\min} = 0, \quad z_2^{\max} = 29.$$

Considering these values, suppose that the DM specifies the reference point as

$$\hat{z}_1 = -35, \quad \hat{z}_2 = 7.$$

For this reference point, as can be easily seen from Fig. 3.10, solving the corresponding minimax problem yields a Pareto optimal solution

$$z_1 = -28, \ z_2 = 14 \quad (x_1 = 0, x_2 = 3.5)$$

and the trade-off rate between the objective functions

$$-\frac{\partial z_2}{\partial z_1} = 0.5.$$

On the basis of such information, suppose that the DM updates the reference point to

$$\hat{z}_1 = -33, \quad \hat{z}_2 = 15$$

in order to improve the satisfaction level of the profit at the expense of that of the pollution amount. For the updated reference point, solving the corresponding minimax problem yields a Pareto optimal solution

$$z_1 = -32, \quad z_2 = 16 \quad (x_1 = 0.75, \ x_2 = 4)$$

and the trade-off rate

$$-\frac{\partial z_2}{\partial z_1} = 0.5.$$

If the DM is satisfied with the current values of the objective functions, the procedure stops. Otherwise, a similar procedure continues in this fashion until the satisficing solution of the DM is derived. ◊

It may be appropriate to point out here that several interactive multiobjective programming methods including the methods presented in this chapter were developed (Changkong and Haimes 1983; Miettinen 1999; Miettinen et al. 2008; Sakawa 1993; Steuer 1986; Vanderpooten and Vincke 1989) from the 1970s to the 1980s, and a basic distinction has been made concerning the underlying approach. Especially, Vanderpooten and Vincke (1989) highlighted an evolution from search-oriented methods to learning-oriented procedures.

The recently published book entitled "Multiple Criteria Decision Making—From Early History to the 21st Century—" (Köksalan et al. 2011), which begins with the early history of Multiple Criteria Decision Making and proceeds to give a decade by decade account of major developments in the field starting from the 1970s until now, would be very useful for interested readers.

Problems

3.1 Graph the following two-objective linear programming problem in the x_1-x_2 plane and z_1-z_2 plane, and find all Pareto optimal solutions.

$$\begin{aligned}
\text{minimize} \quad & z_1 = -x_1 - 3x_2 \\
\text{minimize} \quad & z_2 = -x_1 - x_2 \\
\text{subject to} \quad & x_1 + 8x_2 \leq 112 \\
& x_1 + 2x_2 \leq 34 \\
& 9x_1 + 2x_2 \leq 162 \\
& x_1 \geq 0, \; x_2 \geq 0.
\end{aligned}$$

Verify that if the objective functions are changed to $z_1 = -x_1 - 8x_2$ and $z_2 = -x_1 - x_2$, weak Pareto optimal solutions exist, and a complete optimal solution exists if changed to $z_1 = -2x_1 - 3x_2$ and $z_2 = -x_1 - x_2$.

3.2 Prove that if $x^* \in X$ is an optimal solution of the weighting problem for some $w > 0$, then x^* is a Pareto optimal solution of the multiobjective linear programming problem.

3.3 Graph the following two-objective linear programming problem in the x_1-x_2 plane and z_1-z_2 plane, and find all Pareto optimal solutions.

$$\begin{aligned}
\text{minimize} \quad & z_1 = -2x_1 - 5x_2 \\
\text{minimize} \quad & z_2 = 3x_1 + 2x_2 \\
\text{subject to} \quad & 2x_1 + 6x_2 \leq 27 \\
& 8x_1 + 6x_2 \leq 45 \\
& 3x_1 + x_2 \leq 15 \\
& x_1 \geq 0, \; x_2 \geq 0.
\end{aligned}$$

3.4 For the two-objective linear programming problem discussed in Problem 3.3, solve the following problems.

(i) Obtain a Pareto optimal solution for the weighting method with $w_1 = 0.5$ and $w_2 = 0.5$.
(ii) Obtain a Pareto optimal solution for the constraint method with $\varepsilon_2 = 8$.
(iii) Setting $\hat{z}_1 = z_1 - (-23.5)$, obtain a Pareto optimal solution for the minimax method with $w_1 = 0.5$ and $w_2 = 0.5$.

3.5 Find an optimal solution to the following linear goal programming problem graphically.

$$\begin{aligned}
\text{minimize} \quad & P_1 d_1^- + P_2 d_2^+ + P_3(d_3^- + 2d_4^-) \\
\text{subject to} \quad & x_1 + 2x_2 - d_1^+ + d_1^- = 8 \\
& 3x_1 + 2x_2 - d_2^+ + d_2^- = 10 \\
& x_1 \qquad\quad - d_3^+ + d_3^- = 2 \\
& x_2 - d_4^+ + d_4^- = 3 \\
& 2x_1 + 6x_2 \leq 27 \\
& 8x_1 + 6x_2 \leq 45 \\
& 3x_1 + x_2 \leq 15 \\
& x_1 \geq 0, \, x_2 \geq 0 \\
& d_i^+ \geq 0, \, d_i^- \geq 0, \, i = 1, 2, 3, 4.
\end{aligned}$$

3.6 Setting $w_1 = w_2 = 1$, employ the method of displaced ideal to the two-objective linear programming problem

$$\begin{aligned}
\text{minimize} \quad & z_1 = -5x_1 - 5x_2 \\
\text{minimize} \quad & z_2 = 5x_1 + x_2 \\
\text{subject to} \quad & 5x_1 + 7x_2 \leq 12 \\
& 9x_1 + 1x_2 \leq 10 \\
& -5x_1 + 3x_2 \leq 3 \\
& x_1 \geq 0, \quad x_2 \geq 0.
\end{aligned}$$

3.7 Using the Excel solver, solve Examples 3.2–3.4 and confirm the Pareto optimal solutions.

3.8 Apply the interactive multiobjective linear programming to the two-objective linear programming problem of Problem 3.3.

Chapter 4
Fuzzy Linear Programming

In 1976, Zimmermann first introduced fuzzy set theory into linear programming problems. He considered linear programming problems with a fuzzy goal and fuzzy constraints. Following the fuzzy decision proposed by Bellman and Zadeh (1970) together with linear membership functions, he proved that there exists an equivalent linear programming problem.

In this chapter, after an overview of the fundamentals of basic fuzzy set theory, clear and detailed explanations of fuzzy linear programming and fuzzy multiobjective linear programming are given. Then, interactive fuzzy multiobjective linear programming is discussed in detail. Finally, multiobjective linear programming problems involving fuzzy parameters are formulated, and linear programming-based interactive fuzzy programming for deriving a satisficing solution is presented.

4.1 Fuzzy Sets and Fuzzy Decision

In general, a fuzzy set initiated by Zadeh (1965) is defined as follows.

Definition 4.1 (Fuzzy sets). Let X denote a universal set. Then a fuzzy subset $\tilde{A}$ of X is defined by its membership function

$$\mu_{\tilde{A}} : X \to [0, 1], \qquad (4.1)$$

which assigns a real number $\mu_{\tilde{A}}(x)$ in the interval $[0, 1]$ to each element $x \in X$, where $\mu_{\tilde{A}}(x)$ represents the grade of membership of x in $\tilde{A}$. Thus, the nearer the value of $\mu_{\tilde{A}}(x)$ is unity, the higher the grade of membership of x in $\tilde{A}$.

A fuzzy subset $\tilde{A}$ can be characterized as a set of ordered pairs of element x and its grade $\mu_{\tilde{A}}(x)$ and is often written as

$$\tilde{A} = \{(x, \mu_{\tilde{A}}(x)) \mid x \in X\}. \qquad (4.2)$$

Fig. 4.1 Membership function and characteristic function

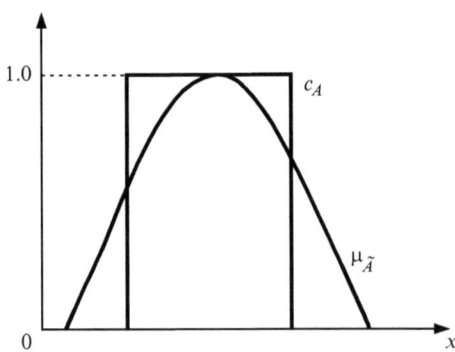

When the membership function $\mu_{\tilde{A}}(x)$ contains only the two values 0 and 1, then $\mu_{\tilde{A}}(x)$ is identical to the characteristic function $c_A : X \to \{0, 1\}$, and hence, $\tilde{A}$ is no longer a fuzzy subset, but an ordinary set A.

As is well known, an ordinary set A is expressed as

$$A = \{x \in X \mid c_A(x) = 1\}, \tag{4.3}$$

through its characteristic function

$$c_A(x) = \begin{cases} 1 & x \subset A \\ 0 & x \notin A. \end{cases} \tag{4.4}$$

Figure 4.1 illustrates the membership function $\mu_{\tilde{A}}(x)$ of a fuzzy subset $\tilde{A}$ together with the characteristic function $c_A(x)$ of an ordinary set A.

Observe that the membership function is an obvious extension of the idea of a characteristic function for an ordinary set because it takes values between 0 and 1, not only 0 and 1.

As can be easily understood from the definition, a fuzzy subset is always defined as a subset of a universal set X. For the concise representation, a fuzzy subset is usually called a fuzzy set by omitting the term "sub." To distinguish an ordinary set from a fuzzy set, an ordinary set is called a nonfuzzy set or a crisp set. A fuzzy set is often denoted by $\tilde{A}, \tilde{B}, \tilde{C}, \ldots$, but it is sometimes written as $A, B, C, \ldots$ for simplicity in the notation.

Example 4.1 (The young). Suppose that ages are denoted by a numerical-valued variable which ranges over the interval $X=[0, \infty)$. Then the set of ages less than or equal to 20 is obviously a crisp set. However, the set of "young ages" has no sharply defined boundaries and can be interpreted as a fuzzy set A on X. It may be subjectively characterized, for example, by the following membership function:

$$\mu_A(x) = (1 + (0.04x)^2)^{-1}. \tag{4.5}$$

Fig. 4.2 Membership function for young ages

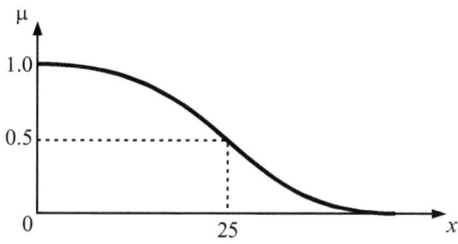

This membership function $\mu_A(x)$ is illustrated in Fig. 4.2. In this case, the degree to which a numerical age, say $x = 25$, is compatible with the concept young is 0.5.

◊

When X is a finite set whose elements are $x_1, x_2, \ldots, x_n$, i.e.,

$$X = \{x_1, x_2, \ldots, x_n\}, \tag{4.6}$$

a fuzzy set A on X is expressed as

$$A = \{(x_1, \mu_A(x_1)), (x_2, \mu_A(x_2)), \ldots, (x_n, \mu_A(x_n))\}. \tag{4.7}$$

In this expression, for simplicity, the pairs with $\mu_A(x) = 0$ are usually omitted.

According to the notation proposed by Zadeh (1965), this fuzzy set A on X is often written as

$$A = \mu_A(x_1)/x_1 + \mu_A(x_2)/x_2 + \cdots + \mu_A(x_n)/x_n \tag{4.8}$$

or more simply

$$A = \sum_{i=1}^{n} \mu_A(x_i)/x_i. \tag{4.9}$$

This expression means that the grade of x_1 is $\mu_A(x_1)$, the grade of x_2 is $\mu_A(x_2), \ldots,$ the grade of x_n is $\mu_A(x_n)$, and the operations "+" and "$\sum$" do not refer to the ordinary addition but the set-theoretic "or."

When X is infinite, in addition to the notation of (4.2), a fuzzy set A is frequently written as

$$A = \int_X \mu_A(x)/x. \tag{4.10}$$

Here, the integral $\int$ can be viewed as a natural extension of $\sum$ in (4.9).

The following basic notions are defined for fuzzy sets:

(i) Support: The support of a fuzzy set A on X, denoted by supp (A), is the set of points in X at which $\mu_A(x)$ is positive, i.e.,

$$\text{supp}(A) = \{x \in X \mid \mu_A(x) > 0\}. \tag{4.11}$$

(ii) Height: The height of a fuzzy set A on X, denoted by hgt(A), is the least upper bound of $\mu_A(x)$, i.e.,

$$\text{hgt}(A) = \sup_{x \in X} \mu_A(x). \tag{4.12}$$

(iii) Normal: A fuzzy set A on X is said to be normal if its height is unity, i.e., if there is $x \in X$ such that $\mu_A(x) = 1$. If it is not normal, a fuzzy set is said to be subnormal.

(iv) Empty: A fuzzy set A on X is empty, denoted by $\emptyset$, if and only if $\mu_A(x) = 0$ for all $x \in X$. Obviously, the universal set X can be viewed as a fuzzy set whose membership function is $\mu_X(x) = 1$ for all $x \in X$.

Observe that a nonempty subnormal fuzzy set A can be normalized by dividing $\mu_A(x)$ by its hgt(A).

Several set-theoretic operations involving fuzzy sets originally proposed by Zadeh (1965) are as follows:

(i) Equality: The fuzzy sets A and B on X are equal, denoted by $A = B$, if and only if their membership functions are equal everywhere on X:

$$A = B \Leftrightarrow \mu_A(x) = \mu_B(x) \text{ for all } x \in X. \tag{4.13}$$

(ii) Containment: The fuzzy set A is contained in B (or a subset of B), denoted by $A \subseteq B$, if and only if the membership function of A is less than or equal to that of B everywhere on X:

$$A \subseteq B \Leftrightarrow \mu_A(x) \leq \mu_B(x) \text{ for all } x \in X. \tag{4.14}$$

(iii) Complementation: The complement of a fuzzy set A on X, denoted by $\bar{A}$, is defined by

$$\mu_{\bar{A}}(x) = 1 - \mu_A(x) \text{ for all } x \in X. \tag{4.15}$$

(iv) Intersection: The intersection of two fuzzy sets A and B on X, denoted by $A \cap B$, is defined by

$$\mu_{A \cap B}(x) = \min\{\mu_A(x), \mu_B(x)\} \text{ for all } x \in X. \tag{4.16}$$

(v) Union: The union of two fuzzy sets A and B on X, denoted by $A \cup B$, is defined by

$$\mu_{A \cup B}(x) = \max\{\mu_A(x), \mu_B(x)\} \text{ for all } x \in X. \tag{4.17}$$

4.1 Fuzzy Sets and Fuzzy Decision

Fig. 4.3 Examples of α-level sets

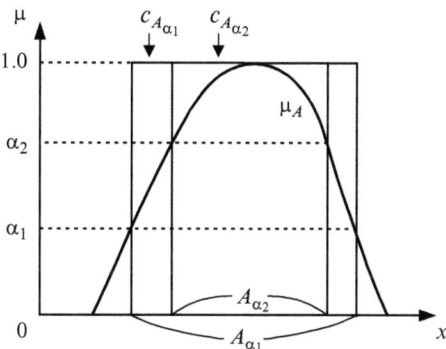

Observe that the intersection $A \cap B$ is the largest fuzzy set which is contained in both A and B and the union $A \cup B$ is the smallest fuzzy set containing both A and B since any fuzzy set C such that $C \subseteq A$ and $C \subseteq B$ satisfies $C \subseteq A \cap B$ and any fuzzy set D such that $D \supseteq A$ and $D \supseteq B$ satisfies $D \supseteq A \cup B$.

Based on the definitions of the set-theoretic operations for fuzzy sets, Zadeh (1965) pointed out that it is possible to extend many of the basic identities which hold for ordinary sets to fuzzy sets.

The concept of α-level sets serves as an important transfer between ordinary sets and fuzzy sets. It also plays an important role in the construction of a fuzzy set by a series of ordinary sets.

Definition 4.2 (α-level set). The α-level set of a fuzzy set A is defined as an ordinary set A_α for which the degree of its membership function exceeds the level α:

$$A_\alpha = \{x \mid \mu_A(x) \geq \alpha\}, \ \alpha \in [0, 1]. \tag{4.18}$$

Observe that the α-level set A_α can be defined by the characteristic function

$$c_{A_\alpha} = \begin{cases} 1 & \text{if } \mu_A(x) \geq \alpha \\ 0 & \text{if } \mu_A(x) < \alpha, \end{cases} \tag{4.19}$$

since it is an ordinary set. Actually, an α-level set is an ordinary set whose elements belong to the corresponding fuzzy set with at least a certain degree α.

It is clear that the following evident property holds for the α-level sets:

$$\alpha_1 \leq \alpha_2 \Leftrightarrow A_{\alpha_1} \supseteq A_{\alpha_2}. \tag{4.20}$$

This relationship is illustrated in Fig. 4.3.

From the definition of the α-level sets, it can be easily understood that the following basic properties hold:

$$(A \cup B)_\alpha = A_\alpha \cup B_\alpha, \tag{4.21}$$

$$(A \cap B)_\alpha = A_\alpha \cap B_\alpha. \qquad (4.22)$$

For example, the first property can be shown as follows:

$$\begin{aligned}(A \cup B)_\alpha &= \{x \mid \max\{\mu_A(x), \mu_B(x)\} \geq \alpha\} \\ &= \{x \mid \mu_A(x) \geq \alpha \text{ or } \mu_B(x) \geq \alpha\} \\ &= \{x \mid \mu_A(x) \geq \alpha\} \cup \{x \mid \mu_B(x) \geq \alpha\} \\ &= A_\alpha \cup B_\alpha.\end{aligned}$$

Using the concept of α-level sets, the relationship between ordinary sets and fuzzy sets can be characterized by the following theorem.

Theorem 4.1 (Decomposition theorem). *A fuzzy set A can be represented by*

$$A = \bigcup_{\alpha \in [0,1]} \alpha A_\alpha \qquad (4.23)$$

where αA_α denotes the algebraic product of a scalar α with the α-level set A_α, i.e., its membership function (characteristic function) is given by

$$\mu_{\alpha A_\alpha}(x) = \alpha \mu_{A_\alpha}(x) = \alpha c_{A_\alpha}(x), \quad \forall x \in X. \qquad (4.24)$$

Actually, since

$$\begin{aligned}\mu_{\bigcup_{\alpha \in [0,1]} \alpha A_\alpha}(x) &= \sup_{\alpha \in [0,1]} \mu_{\alpha A_\alpha}(x) \\ &= \sup_{\alpha \in [0,1]} \alpha \cdot c_{A_\alpha}(x) \\ &= \sup_{\alpha \leq \mu_A(x)} \alpha = \mu_A(x),\end{aligned}$$

the decomposition theorem follows directly.

Theorem 4.1 states that a fuzzy set A can be decomposed into a series of α-level sets A_α by which A can be reconstructed. Thus any fuzzy set can be viewed as a family of ordinary sets.

Before introducing the definition of fuzzy numbers such as "approximately m" and "near zero," consider an extension of ordinary convex sets to fuzzy sets. Naturally, a convex fuzzy set presented by Zadeh (1965) is defined through its membership function in a real n-dimensional Euclidean space $\mathbb{R}^n$.

Definition 4.3 (Convex fuzzy set). A fuzzy set A in $X = \mathbb{R}^n$ is said to be a convex fuzzy set if and only if its α-level sets are convex. An alternative and more direct definition of a convex fuzzy set is as follows:

4.1 Fuzzy Sets and Fuzzy Decision

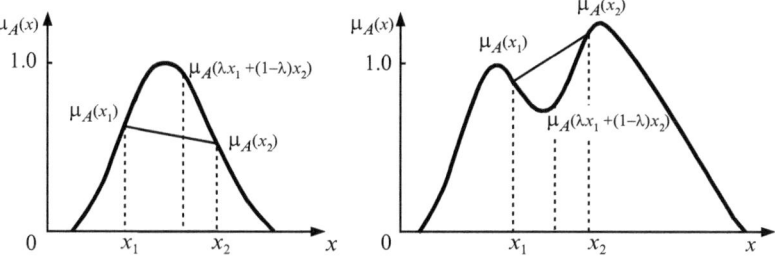

Fig. 4.4 Convex and nonconvex fuzzy sets

A fuzzy set A is convex if and only if

$$\mu_A(\lambda x_1 + (1 - \lambda)x_2) \geq \min\{\mu_A(x_1), \mu_A(x_2)\} \quad (4.25)$$

for all $x_1, x_2 \in X$ and $\lambda \in [0, 1]$.

Observe that this definition does not imply that $\mu_A(x)$ is a convex function of x. The proof of the equivalence between the above definitions follows.

Let A be convex in the sense of the first definition and assume that there exist $x_1', x_2' \in X$ and $\lambda' \in [0, 1]$ such that

$$\mu_A(\lambda' x_1' + (1 - \lambda')x_2') < \min\{\mu_A(x_1'), \mu_A(x_2')\}.$$

Let $\alpha' = \min\{\mu_A(x_1'), \mu_A(x_2')\}$. Then $x_1', x_2' \in A_{\alpha'}$ and $\lambda' x_1' + (1-\lambda')x_2' \notin A_{\alpha'}$, which means $A_{\alpha'}$ is not convex and leads to a contradiction.

Conversely, assume that a fuzzy set A is convex in the sense of the second definition and take $x, y \in A_\alpha$ for any $\alpha \in [0, 1]$. Then $\mu_A(x) \geq \alpha$ and $\mu_A(y) \geq \alpha$. Hence, (4.25) implies $\mu_A(\lambda x + (1 - \lambda)y) \geq \alpha$ for any $\lambda \in [0, 1]$. Thus, we have $\lambda x + (1 - \lambda)y \in A_\alpha$. Thus, the proof of the assertion is established.

From the definition of convexity, it is easy to show that if two fuzzy sets A and B are convex, their intersection $A \cap B$ is also convex. Figure 4.4 illustrates a convex fuzzy set and a nonconvex fuzzy set.

Among fuzzy sets, numbers such as "approximately m" or "about n" can be defined as fuzzy sets on the real line $\mathbb{R}$. Such fuzzy numbers are formally defined as follows (Dubois and Prade 1978, 1980; Zimmermann 1987).

Definition 4.4 (Fuzzy number). A fuzzy number is a convex normalized fuzzy set on the real line $\mathbb{R}$ whose membership function is piecewise continuous.

Frequently, a fuzzy number M is called positive (negative), denoted by $M > 0$ ($M < 0$), if its membership function $\mu_M(x)$ satisfies $\mu_M(x) = 0$, $\forall x < 0$ ($\forall x > 0$).

Example 4.2 (Examples of fuzzy numbers). As membership functions for a fuzzy number M, such as "approximately m," a triangular membership function

$$\mu_M(x) = \max(0, 1 - |x - m|/a), \quad a > 0$$

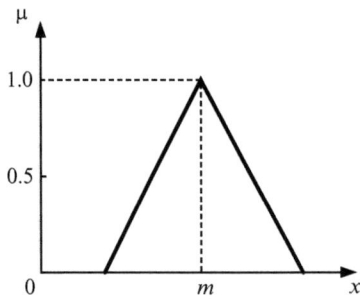
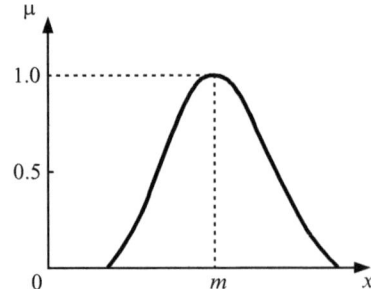

Fig. 4.5 Examples of fuzzy numbers

Fig. 4.6 α-level set of fuzzy number M

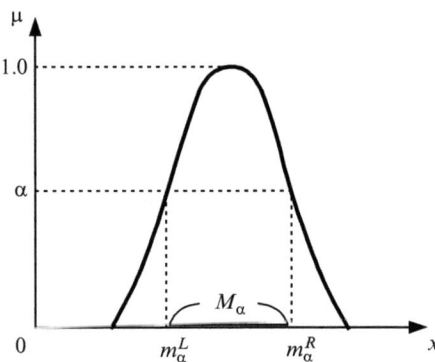

and a bell-shaped membership function

$$\mu_M(x) = e^{-b(x-m)^2}, \quad b \geq 1$$

are widely used. Such membership functions are illustrated in Fig. 4.5.

◊

From the definition of a fuzzy number M, it is significant to note that the α-level set M_α of a fuzzy number M can be represented by the closed interval which depends on the value of α as is shown in Fig. 4.6. Namely,

$$M_\alpha = \{x \in \mathbb{R} \mid \mu_M(x) \geq \alpha\} = [m_\alpha^L, m_\alpha^R] \quad (4.26)$$

where m_α^L and m_α^R represent the left and right extreme points of the α-level set M_α, respectively.

In their 1970 paper "Decision making in a fuzzy environment," Bellman and Zadeh (1970) introduced three basic concepts: fuzzy goal, fuzzy constraint, and fuzzy decision, and explored the application of these concepts to decision-making processes under fuzziness. Let us now introduce the conceptual framework for decision making in a fuzzy environment.

4.1 Fuzzy Sets and Fuzzy Decision

Let X be a given set of possible alternatives which contains the solution of a decision-making problem under consideration. A fuzzy goal G is a fuzzy set on X characterized by its membership function:

$$\mu_G : X \to [0, 1]. \tag{4.27}$$

A fuzzy constraint C is a fuzzy set on X characterized by its membership function:

$$\mu_C : X \to [0, 1]. \tag{4.28}$$

Realizing that both the fuzzy goal and the fuzzy constraint are desired to be satisfied simultaneously, Bellman and Zadeh (1970) defined the fuzzy decision D resulting from the fuzzy goal G and fuzzy constraint C as the intersection of G and C.

To be more explicit, the fuzzy decision of Bellman and Zadeh is the fuzzy set D on X defined as

$$D = G \cap C \tag{4.29}$$

and is characterized by its membership function:

$$\mu_D(x) = \min\{\mu_G(x), \mu_C(x)\}. \tag{4.30}$$

The maximizing decision is then defined as

$$\operatorname*{maximize}_{x \in X} \mu_D(x) = \operatorname*{maximize}_{x \in X} \min\{\mu_G(x), \mu_C(x)\}. \tag{4.31}$$

More generally, the fuzzy decision D resulting from k fuzzy goals $G_1, \ldots, G_k$ and m fuzzy constraints $C_1, \ldots, C_m$ is defined by

$$D = G_1 \cap \ldots \cap G_k \cap C_1 \cap \cdots \cap C_m \tag{4.32}$$

and the corresponding maximizing decision is defined as

$$\operatorname*{maximize}_{x \in X} \mu_D(x) = \operatorname*{maximize}_{x \in X} \min\{\mu_{G_1}(x), \ldots, \mu_{G_k}(x), \mu_{C_1}(x), \ldots, \mu_{C_m}(x)\}. \tag{4.33}$$

It is significant to realize here that in the fuzzy decision defined by Bellman and Zadeh (1970), the fuzzy goals and the fuzzy constraints enter into the expression for D in exactly the same way. In other words, in the definition of the fuzzy decision, there is no longer a difference between the fuzzy goals and the fuzzy constraints.

However, depending on the situations, other aggregation patterns for the fuzzy goal G and the fuzzy constraint C may be worth considering. When fuzzy goals and fuzzy constraints have unequal importance, Bellman and Zadeh (1970) also suggested the convex fuzzy decision defined by

$$\mu_D^{co}(x) = \sum_{i=1}^{k} \alpha_i \mu_{G_i}(x) + \sum_{j=1}^{m} \beta_j \mu_{C_j}(x), \tag{4.34}$$

$$\sum_{i=1}^{k} \alpha_i + \sum_{j=1}^{m} \beta_j = 1, \quad \alpha_i, \beta_j \geq 0, \tag{4.35}$$

where the weighting coefficients reflect the relative importance among the fuzzy goals and constraints.

As an example of an alternative definition of a fuzzy decision, the product fuzzy decision defined by

$$\mu_D^{pr}(x) = \left(\prod_{i=1}^{k} \mu_{G_i}(x) \right) \cdot \left(\prod_{j=1}^{m} \mu_{C_j}(x) \right) \tag{4.36}$$

has been proposed.

For the convex fuzzy decision or the product fuzzy decision, similar to the maximizing decision for the fuzzy decision, the maximizing decision to select x^* such that

$$\mu_D^{co}(x^*) = \max_{x \in X} \left[\sum_{i=1}^{k} \alpha_i \mu_{G_i}(x) + \sum_{j=1}^{m} \beta_j \mu_{C_j}(x) \right] \tag{4.37}$$

or

$$\mu_D^{pr}(x^*) = \max_{x \in X} \left[\left(\prod_{i=1}^{k} \mu_{G_i}(x) \right) \cdot \left(\prod_{j=1}^{m} \mu_{C_j}(x) \right) \right] \tag{4.38}$$

is also defined.

It should be noted here that among these three types of fuzzy decisions $\mu_D^{co}(x)$, $\mu_D^{pr}(x)$, and $\mu_D(x)$, the following relation holds:

$$\mu_D^{pr}(x) \leq \mu_D(x) \leq \mu_D^{co}(x). \tag{4.39}$$

Example 4.3 (Fuzzy decision, convex fuzzy decision, and product fuzzy decision). Let $X = [0, \infty]$ be a set of alternatives. Suppose that we have a fuzzy goal G and a fuzzy constraint C expressed as "x should be much larger than 10" and "x should be substantially smaller than 30" where their membership functions are subjectively defined by

$$\mu_G(x) = \begin{cases} 0 & \text{if } x \leq 10 \\ 1 - (1 + (0.1(x - 10))^2)^{-1} & \text{if } x > 10 \end{cases}$$

$$\mu_C(x) = \begin{cases} 0 & \text{if } x \geq 30 \\ (1 + x(x - 30)^{-2})^{-1} & \text{if } x < 30. \end{cases}$$

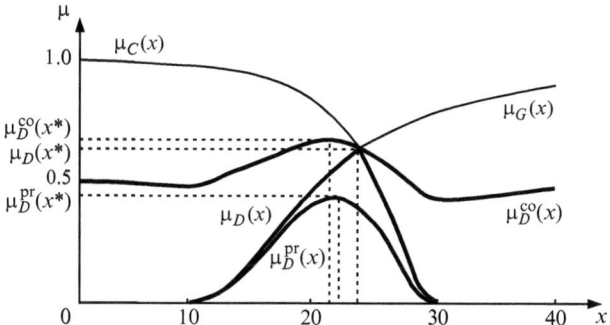

Fig. 4.7 Fuzzy decision, convex fuzzy decision, and product fuzzy decision

The fuzzy decision, the convex fuzzy decision, and the product fuzzy decision for this situation are depicted in Fig. 4.7.

4.2 Fuzzy Linear Programming

Consider the linear programming problem written in the following form[1]:
Minimize the linear objective function

$$z = c_1 x_1 + c_2 x_2 + \cdots + c_n x_n$$

subject to the m linear inequality constraints

$$a_{11} x_1 + a_{12} x_2 + \cdots + a_{1n} x_n \leq b_1$$
$$a_{21} x_1 + a_{22} x_2 + \cdots + a_{2n} x_n \leq b_2$$
$$\cdots\cdots\cdots\cdots\cdots\cdots\cdots$$
$$a_{m1} x_1 + a_{m2} x_2 + \cdots + a_{mn} x_n \leq b_m$$

and nonnegativity conditions for all variables

$$x_j \geq 0, \quad j = 1, 2, \ldots, n$$

where the a_{ij}, b_i, and c_j are given constants.

By introducing an n-dimensional row vector $c = (c_1, \ldots, c_n)$, an n-dimensional column vector $x = (x_1, \ldots, x_n)^T$, an m-dimensional column vector $b = (b_1, \ldots, b_m)^T$, and an $m \times n$ matrix $A = [a_{ij}]$, this problem can be expressed in a more compact vector–matrix form:

---

[1] For convenience, we start from inequality constraints, but similar discussions can be made for equality constraints.

$$\begin{aligned} \text{minimize} \quad & z = cx \\ \text{subject to} \quad & Ax \le b \\ & x \ge 0. \end{aligned} \quad \Biggr\} \tag{4.40}$$

In contrast to conventional linear programming problem, Zimmermann (1976) proposed to soften the rigid requirements of the decision maker (DM) to strictly minimize the objective function and to strictly satisfy the constraints. That is, by considering the imprecision or fuzziness of the DM's judgment, he softened the usual linear programming problem into the following fuzzy version:

$$\left. \begin{aligned} cx &\precsim z_0 \\ Ax &\precsim b \\ x &\ge 0, \end{aligned} \right\} \tag{4.41}$$

where the symbol "$\precsim$" denotes a relaxed or fuzzy version of the ordinary inequality "$\le$." To be more explicit, these fuzzy inequalities representing the DM's fuzzy goal and fuzzy constraints mean that "the objective function cx should be essentially smaller than or equal to an aspiration level z_0 of the DM" and "the constraints Ax should be essentially smaller than or equal to b," respectively.

In the same spirit as the fuzzy decision of Bellman and Zadeh (1970), considering the fuzzy goal $cx \precsim z_0$ and fuzzy constraints $Ax \precsim b$ as equally important, Zimmermann (1976) expressed the problem as follows:

$$\left. \begin{aligned} Bx &\precsim b' \\ x &\ge 0, \end{aligned} \right\} \tag{4.42}$$

where

$$B = \begin{bmatrix} c \\ A \end{bmatrix}, \quad b' = \begin{bmatrix} z_0 \\ b \end{bmatrix}. \tag{4.43}$$

For treating the ith fuzzy inequality $(Bx)_i \precsim b'_i$, $i = 0, \ldots, m$ of the DM's fuzzy inequalities $Bx \precsim b'$, he proposed the following linear membership function:

$$\mu_i((Bx)_i) = \begin{cases} 1 & \text{if } (Bx)_i \le b'_i \\ 1 - \dfrac{(Bx)_i - b'_i}{d_i} & \text{if } b'_i \le (Bx)_i \le b'_i + d_i \\ 0 & \text{if } (Bx)_i \ge b'_i + d_i, \end{cases} \tag{4.44}$$

where each d_i is a subjectively chosen constant expressing the limit of the admissible violation of the ith inequality. It is assumed that the ith membership function should be 1 if the ith constraint is well satisfied, 0 if the ith constraint is violated beyond its limit d_i, and linear from 0 to 1. Such a linear membership function is illustrated in Fig. 4.8.

4.2 Fuzzy Linear Programming

Fig. 4.8 Linear membership function

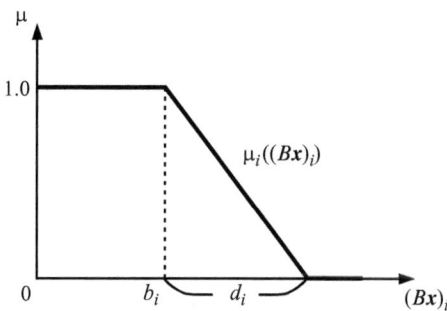

Following the fuzzy decision of Bellman and Zadeh (1970) together with the linear membership functions, the problem of finding the maximizing decision is to choose x^* such that

$$\mu_D(x^*) = \max_{x \geq 0} \min_{i=0,\ldots,m} \{\mu_i((Bx)_i)\}. \tag{4.45}$$

In other words, the problem is to find a solution $x^* \geq 0$ which maximizes the minimum membership function values.

Substituting

$$b_i'' = b_i'/d_i, \quad (B'x)_i = (Bx)_i/d_i$$

the problem is rewritten as

$$\mu_D(x^*) = \max_{x \geq 0} \min_{i=0,\ldots,m} \{1 + b_i'' - (B'x)_i\}. \tag{4.46}$$

By introducing an auxiliary variable λ, this problem can be transformed into the following equivalent conventional linear programming problem:

$$\left. \begin{array}{ll} \text{maximize} & \lambda \\ \text{subject to} & \lambda \leq 1 + b_i'' - (B'x)_i, \quad i = 0, 1, \ldots, m \\ & x \geq 0. \end{array} \right\} \tag{4.47}$$

The fuzzy decision of Bellman and Zadeh is sometimes called the minimum operator, since it is expressed as $\min_{i=0,\ldots,m}\{\mu_i((Bx)_i)\}$.

Example 4.4 (Production planning problem). Recall the production planning problem of Example 1.1:

$$\begin{array}{ll} \text{minimize} & z = -3x_1 - 8x_2 \\ \text{subject to} & 2x_1 + 6x_2 \leq 27 \\ & 3x_1 + 2x_2 \leq 16 \\ & 4x_1 + x_2 \leq 18 \\ & x_1 \geq 0, \quad x_2 \geq 0. \end{array}$$

Table 4.1 Nonfuzzy and fuzzy constraints

	Nonfuzzy	Fuzzy $\mu = 0$	Fuzzy $\mu = 1$
Objective function	−37	−36.5	−38.5
First constraint	27	30	27
Second constraint	16	18	16
Third constraint	18	20	18

The optimal solution to this problem is

$$x_1 = 3, \quad x_2 = 3.5, \quad z = -37$$

as discussed in Example 1.1.

Instead of giving rigid numerical values for the right-hand side constants as in the conventional linear programming problem, assume that the DM has the fuzzy goal and the fuzzy constraints shown in Table 4.1.

Assuming the linear membership functions from $\mu = 0$ to $\mu = 1$ for the fuzzy goal and the fuzzy constraints, the flexible formulation for the original problem in the form of a fuzzy linear programming problem can be transformed into the following equivalent conventional linear programming problem:

$$\begin{aligned}
\text{maximize} \quad & \lambda \\
\text{subject to} \quad & \tfrac{2}{3}x_1 + 2x_2 + \lambda \leq 10 \\
& 1.5x_1 + x_2 + \lambda \leq 9 \\
& 2x_1 + 0.5x_2 + \lambda \leq 10 \\
& 1.5x_1 + 4x_2 - \lambda \geq 18.25 \\
& x_1 \geq 0, \; x_2 \geq 0.
\end{aligned}$$

Solving this problem by the simplex method of linear programming yields the optimal solution

$$x_1 = 3.1047, \quad x_2 = 3.5872, \quad \lambda = 0.7558.$$

In Table 4.2, the optimal solutions corresponding to the fuzzy linear programming problem as well as the nonfuzzy linear programming problem are shown.

In this fuzzy linear programming problem, observe that all of the original rigid constraints are softened by admitting some extent of violations. For example, the first constraint is softened to be "substantially less than or equal to 27" by using a linear membership function with $\mu_1(30) = 0$ and $\mu_1(27) = 1$. Moreover, instead of the minimization of the objective function, the admissible region roughly from −38.5 to −36.5 is considered as the satisfaction level for the objective function. As a result, in this example, it may be natural that the DM has about 2.7% additional total profit at the expense of some additional materials as shown in Table 4.2. Thus,

Table 4.2 Solutions of nonfuzzy and fuzzy production planning problem

		Nonfuzzy	Fuzzy
Solution (x_1, x_2)		(3, 3.5)	(3.1047, 3.5872)
Optimal value z		-37	-38.0116
		Used material	
	Available amount	Nonfuzzy	Fuzzy
Material $M1$	27	27	27.73
Material $M2$	16	16	16.49
Material $M3$	18	15.5	16.01

in fuzzy linear programming, the DM is no longer forced to formulate the problem in precise and rigid form as in linear programming, which seems to be one of the major advantages of fuzzy linear programming. ◊

In 1978, Sommer and Pollastschek (1978) proposed the adoption of the add operator for aggregating the DM's fuzzy goal and fuzzy constraints instead of the minimum operator. Their add operator can be viewed as a special case of the convex fuzzy decision by setting all α_i and β_j equal to 1s.

Using the linear membership functions in (4.44) for representing the DM's fuzzy goal and fuzzy constraints and adopting the add operator instead of the minimum operator, the fuzzy version of the original linear programming problem becomes

$$\text{maximize} \quad \sum_{i=0}^{m} \mu_i((Bx)_i)$$
$$\text{subject to} \quad x \geq 0.$$

Obviously, this is a linear programming problem and can be easily solved by the simplex method of linear programming. Sommer and Pollastschek (1978) also showed that constraints such as $Ax \geq b$ and $Ax = b$ can be softened similar to $Ax \leq b$ and applied their method successfully to the air pollution regulation problem.

4.3 Fuzzy Multiobjective Linear Programming

In 1978, Zimmermann (1978) extended his fuzzy linear programming approach to the following multiobjective linear programming problem with k linear objective functions $z_i(x) = c_i x, i = 1, \ldots, k$:

$$\left. \begin{array}{l} \text{minimize} \quad z(x) = (z_1(x), z_2(x), \ldots, z_k(x))^T \\ \text{subject to} \quad Ax \leq b, \quad x \geq 0, \end{array} \right\} \quad (4.48)$$

Fig. 4.9 Linear membership function

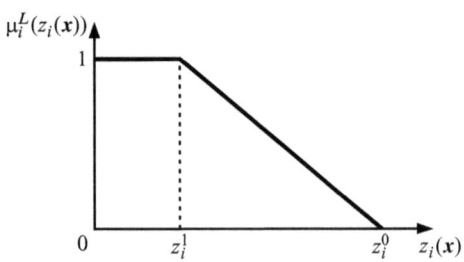

where $c_i = (c_{i1}, \ldots, c_{in})$, $i = 1, \ldots, k$, $x = (x_1, \ldots, x_n)^T$, $b = (b_1, \ldots, b_m)^T$ and $A = [a_{ij}]$ is an $m \times n$ matrix.

For each of the objective functions $z_i(x) = c_i x$, $i = 1, \ldots, k$ of this problem, assume that the DM has a fuzzy goal such as "the objective function $z_i(x)$ should be substantially less than or equal to some value p_i." Then the corresponding linear membership function $\mu_i^L(z_i(x))$ is defined as

$$\mu_i^L(z_i(x)) = \begin{cases} 0 & \text{if } z_i(x) \geq z_i^0 \\ \dfrac{z_i(x) - z_i^0}{z_i^1 - z_i^0} & \text{if } z_i^0 \geq z_i(x) \geq z_i^1 \\ 1 & \text{if } z_i(x) \leq z_i^1, \end{cases} \quad (4.49)$$

where z_i^0 and z_i^1 denote the values of the objective function $z_i(x)$ such that the degrees of the membership function are 0 and 1, respectively. Figure 4.9 illustrates the graph of a possible shape of the linear membership function.

Using such linear membership functions $\mu_i^L(z_i(x))$, $i = 1, \ldots, k$ and following the fuzzy decision by Bellman and Zadeh (1970), the original multiobjective linear programming problem can be interpreted as

$$\left. \begin{array}{l} \text{maximize} \quad \min_{i=1,\ldots,k} \mu_i^L(z_i(x)) \\ \text{subject to} \quad Ax \leq b, \ x \geq 0. \end{array} \right\} \quad (4.50)$$

By introducing an auxiliary variable λ, it can be reduced to the following conventional linear programming problem:

$$\left. \begin{array}{l} \text{maximize} \quad \lambda \\ \text{subject to} \quad \lambda \leq \mu_i^L(z_i(x)), \ i = 1, \ldots, k \\ \quad \quad \quad \quad Ax \leq b, \ x \geq 0. \end{array} \right\} \quad (4.51)$$

By assuming the existence of the optimal solution x^{io} of the individual objective function minimization problem under the constraints defined by

4.3 Fuzzy Multiobjective Linear Programming

$$\min_{x \in X} z_i(x), \quad i = 1, \ldots, k \tag{4.52}$$

where $X = \{x \in \mathbb{R}^n \mid Ax \leq b, x \geq 0\}$, Zimmermann (1978) suggested a way to determine the linear membership function $\mu_i^L(z_i(x))$. To be more specific, using the individual minimum

$$z_i^{\min} = z_i(x^{io}) = \min_{x \in X} z_i(x), \quad i = 1, \ldots, k, \tag{4.53}$$

together with

$$z_i^m = \max\{z_i(x^{1o}), \ldots, z_i(x^{i-1,o}), z_i(x^{i+1,o}), \ldots, z_i(x^{ko})\}, \quad i = 1, \ldots, k, \tag{4.54}$$

he determined the linear membership function as in (4.49) by choosing $z_i^1 = z_i^{\min}$ and $z_i^0 = z_i^m$. For this membership function, it can be easily shown that if the optimal solution of (4.50) or (4.51) is unique, it is also a Pareto optimal solution of the multiobjective linear programming problem.

In a case where not only fuzzy goals but also fuzzy constraints exist, using linear membership functions for the fuzzy constraints, similar discussion can be made. Zimmermann called the fuzzy decision the minimum operator, and for other aggregation methods than the minimum operator, he considered the product fuzzy decision. He called the product fuzzy decision the product operator. Employing the product operator, the problem to be solved becomes

$$\left. \begin{array}{ll} \text{maximize} & \prod_{i=1}^{k} \mu_i^L(z_i(x)) \\ \text{subject to} & Ax \leq b, x \geq 0. \end{array} \right\} \tag{4.55}$$

Unfortunately, with the product operator, even if we use the linear membership functions, the objective function of this problem becomes a nonlinear function, and hence, the linear programming method cannot be applied.

Example 4.5 (Two-objective production planning under the fuzzy environment). To illustrate the fuzzy multiobjective linear programming method proposed by Zimmermann, consider the following two-objective production planning problem as previously discussed in Example 3.2:

$$\begin{array}{ll} \text{minimize} & z_1 = -3x_1 - 8x_2 \\ \text{minimize} & z_2 = 5x_1 + 4x_2 \\ \text{subject to} & 2x_1 + 6x_2 \leq 27 \\ & 3x_1 + 2x_2 \leq 16 \\ & 4x_1 + x_2 \leq 18 \\ & x_1 \geq 0, \quad x_2 \geq 0. \end{array}$$

Note that these two-objective functions to be minimized represent the opposite of the total profit (z_1) and the amount of pollution (z_2).

The individual minimum and maximum of these objective functions are

$$z_1^{\min} = -37, \quad z_1^{\max} = 0, \quad z_2^{\min} = 0, \quad z_2^{\max} = 29.$$

Assume that the DM subjectively determines the corresponding linear membership functions $\mu_i^L(z_i)$, $i = 1, 2$ as follows:

$$\begin{cases} \text{fuzzy goal for } z_1!: & \mu_1(-35) = 0, \ \mu_1(-37) = 1 \\ \text{fuzzy goal for } z_2: & \mu_2(24) = 0, \ \mu_2(20) = 1. \end{cases}$$

To be more specific, the fuzzy goals of the DM are assumed to be expressed by the following membership functions:

$$\mu_1^L(z_1(x)) = \begin{cases} 0 & \text{if } z_1(x) \geq -35 \\ \dfrac{-3x_1 - 8x_2 + 35}{-2} & \text{if } -35 \geq z_1(x) \geq -37 \\ 1 & \text{if } z_1(x) \leq -37, \end{cases}$$

$$\mu_2^L(z_2(x)) = \begin{cases} 0 & \text{if } z_2(x) \geq 24 \\ \dfrac{5x_1 + 4x_2 - 24}{-4} & \text{if } 24 \geq z_2(x) \geq 20 \\ 1 & \text{if } z_2(x) \leq 20. \end{cases}$$

Then the equivalent linear programming problem is formulated as

$$\begin{array}{ll} \text{maximize} & \lambda \\ \text{subject to} & 1.5x_1 + 4x_2 - \lambda \geq 17.5 \\ & 1.25x_1 + x_2 + \lambda \leq 6 \\ & 2x_1 + 6x_2 \leq 27 \\ & 3x_1 + 2x_2 \leq 16 \\ & 4x_1 + x_2 \leq 18 \\ & x_1 \geq 0, \ x_2 \geq 0. \end{array}$$

Solving this problem by the simplex method of linear programming yields the optimal solution

$$x_1 = 0.92308, \quad x_2 = 4.19231, \quad \lambda = 0.65385.$$

This means that the overall satisfaction of the fuzzy goals of the DM is 0.65385, the corresponding total profit ($-z_1$) is 36.30769, and the amount of pollution (z_2)

is 21.38462. This point $(x_1, x_2) = (0.92308, 4.19231)$ or $(z_1, z_2) = (-36.30769, 21.38462)$ lies on the edge ED in Fig. 3.3 or Fig. 3.4. From this, it can be understood that it is also a Pareto optimal solution of the original multiobjective linear programming problem. ◇

4.4 Interactive Fuzzy Multiobjective Linear Programming

In the fuzzy approaches to multiobjective linear programming problems proposed by Zimmermann and his successors, it has been implicitly assumed that the fuzzy decision of Bellman and Zadeh (1970) is the proper representation of the fuzzy preferences of the DM. Therefore, these approaches are preferable only when the DM feels that the fuzzy decision is appropriate for combining the fuzzy goals and/or constraints. However, such situations seem to occur rarely in practice, and consequently it becomes evident that an interaction process with the DM is necessary.

In this section, assuming that the DM has a fuzzy goal for each of the objective functions in multiobjective linear programming problems, we present an interactive fuzzy multiobjective linear programming method incorporating the desirable features of the interactive approaches into the fuzzy approaches.

In general, the multiobjective linear programming problem is represented as the following vector-minimization problem:

$$\left.\begin{array}{ll} \text{minimize} & z(x) = (z_1(x), z_2(x), \ldots, z_k(x))^T \\ \text{subject to} & x \in X = \{x \in \mathbb{R}^n \mid Ax \leq b, x \geq 0\}, \end{array}\right\} \quad (4.56)$$

where x is an n-dimensional vector of decision variables, $z_1(x) = c_1 x, \ldots, z_k(x) = c_k x$ are k conflicting linear objective functions, and X is the feasible region of linearly constrained decisions.

Fundamental to the multiobjective linear programming is the concept of Pareto optimal solutions, also known as noninferior solutions. However, considering the imprecise nature inherent in human judgments in multiobjective linear programming problems, the DM may have a fuzzy goal expressed as "$z_i(x)$ should be substantially less than or equal to some value p_i" in a minimization problem. This type of statement can be quantified by eliciting a corresponding membership function. Figure 4.10 illustrates the possible shape of the membership function representing the fuzzy goal to achieve substantially less than or equal to p_i.

To elicit a membership function $\mu_i(z_i(x))$ from the DM for each of the objective functions $z_i(x)$, $i = 1, \ldots, k$, we first calculate the individual minimum $z_i^{\min} = \min_{x \in X} z_i(x)$ and maximum $z_i^{\max} = \max_{x \in X} z_i(x)$ of each objective function $z_i(x)$ under the given constraints. Taking into account the calculated individual minimum and maximum of each objective function together with the increase rate of satisfaction in terms of the membership degree, the DM must determine the

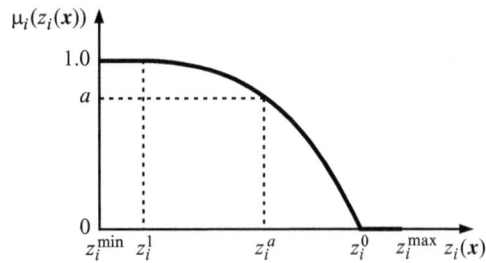

Fig. 4.10 Membership function of fuzzy goal for minimization problem

subjective membership function $\mu_i(z_i(x))$, which is a strictly monotone decreasing function with respect to $z_i(x)$. For the membership function shown in Fig. 4.10, it is assumed that $\mu_i(z_i(x)) = 0$ if $z_i(x) \geq z_i^0$ and $\mu_i(z_i(x)) = 1$ if $z_i(x) \leq z_i^1$, and z_i^a represents the value of $z_i(x)$ within $z_i^{\min}$ and $z_i^{\max}$ such that the value of membership function $\mu_i(z_i(x))$ is $a \in [0, 1]$.

So far, we have restricted ourselves to a minimization problem and consequently assumed that the DM has a fuzzy goal such as "$z_i(x)$ should be substantially less than or equal to p_i." In the fuzzy approaches, however, we can further treat a more general multiobjective linear programming problem in which the DM has two types of fuzzy goals expressed in words such as "$z_i(x)$ should be in the vicinity of r_i" (called fuzzy equal) and "$z_i(x)$ should be substantially less than or equal to p_i or greater than or equal to q_i" (called fuzzy min or fuzzy max).

Such a generalized multiobjective linear programming problem may now be expressed as

$$\left. \begin{array}{ll} \text{fuzzy min} & z_i(x), \quad i \in I_1 \\ \text{fuzzy max} & z_i(x), \quad i \in I_2 \\ \text{fuzzy equal} & z_i(x), \quad i \in I_3 \\ \text{subject to} & x \in X, \end{array} \right\} \quad (4.57)$$

where $I_1 \cup I_2 \cup I_3 = \{1, 2, \ldots, k\}$, $I_i \cap I_j = \emptyset, i, j = 1, 2, 3, i \neq j$.

In this formulation, "fuzzy min $z_i(x)$" or "fuzzy max $z_i(x)$" represents the fuzzy goal of the DM such as "$z_i(x)$ should be substantially less than or equal to p_i" or "$z_i(x)$ should be substantially greater than or equal to q_i," and "fuzzy equal $z_i(x)$" represents the fuzzy goal such as "$z_i(x)$ should be in the vicinity of r_i." Concerning the membership function for the fuzzy goal of the DM such as "$z_i(x)$ should be in the vicinity of r_i," it is obvious that a strictly monotone increasing function $d_{iL}(z_i)$, $i \in I_{3L}$ and a strictly monotone decreasing function $d_{iR}(z_i)$, $i \in I_{3R}$ corresponding to the left and right sides of r_i must be determined through interaction with the DM, where $I_3 = I_{3L} \cup I_{3R}$.

As an example, Fig. 4.11 illustrates a possible shape of the fuzzy equal membership functions where the left function is linear and the right function is exponential.

When the fuzzy equal is included in the fuzzy goals of the DM, it is desirable that $z_i(x)$ should be as close to r_i as possible. Consequently, the notion of Pareto

Fig. 4.11 Fuzzy equal membership function

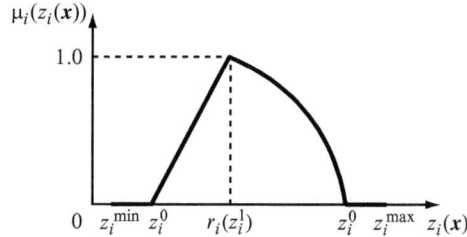

optimal solutions defined in terms of objective functions cannot be applied. For this reason, we introduce the concept of M-Pareto optimal solutions which is defined in terms of membership functions instead of objective functions, where M refers to membership.

Definition 4.5 (M-Pareto optimal solution). A point $x^* \in X$ is said to be an M-Pareto optimal solution to a generalized multiobjective linear programming problem if and only if there does not exist another $x \in X$ such that $\mu_i(z_i(x)) \geq \mu_i(z_i(x^*))$ for all i and $\mu_j(z_j(x)) \neq \mu_j(z_j(x^*))$ for at least one j.

Having elicited the membership functions $\mu_i(z_i(x))$, $i = 1, \ldots, k$ from the DM for each of the objective functions $z_i(x)$, $i = 1, \ldots, k$, the multiobjective linear programming and/or generalized multiobjective linear programming problem can be converted into the fuzzy multiobjective optimization problem of the membership functions defined by

$$\underset{x \in X}{\text{maximize}} \; (\mu_1(z_1(x)), \ldots, \mu_k(z_k(x))). \tag{4.58}$$

By introducing a general conjunctive function

$$\mu_D(\boldsymbol{\mu}(z(x))) = \mu_D(\mu_1(z_1(x)), \ldots, \mu_k(z_k(x))), \tag{4.59}$$

a general fuzzy multiobjective problem can be defined by

$$\underset{x \in X}{\text{maximize}} \; \mu_D(\boldsymbol{\mu}(z(x))). \tag{4.60}$$

Observe that the value of $\mu_D(\boldsymbol{\mu}(z(x)))$ can be interpreted as the overall degree of satisfaction with the DM's multiple fuzzy goals. From this viewpoint, the fuzzy decision or the minimum operator of Bellman and Zadeh (1970)

$$\min\{\mu_1(z_1(x)), \ldots, \mu_k(z_k(x))\} \tag{4.61}$$

can be viewed only as one special case of $\mu_D(\boldsymbol{\mu}(z(x)))$.

In the conventional fuzzy approaches discussed thus far, it has been implicitly assumed that the minimum operator is the proper representation of the DM's fuzzy preferences, and according to this idea, the general fuzzy multiobjective problem (4.60) has been interpreted as

$$\underset{x \in X}{\text{maximize}} \quad \min\{\mu_1(z_1(x)), \ldots, \mu_k(z_k(x))\}, \tag{4.62}$$

or equivalently

$$\left. \begin{array}{ll} \text{maximize} & v \\ \text{subject to} & v \le \mu_i(z_i(x)), \ i = 1, \ldots, k \\ & x \in X. \end{array} \right\} \tag{4.63}$$

However, it should be emphasized here that this approach is preferable only when the DM feels that the minimum operator is appropriate. In other words, in general decision situations, the DM does not always use the minimum operator for combining the fuzzy goals and/or constraints. Probably the most crucial problem in the general fuzzy multiobjective problem (4.60) is the identification of an appropriate aggregation function which well represents the DM's fuzzy preferences. If $\mu_D(\cdot)$ can be explicitly identified, then the general fuzzy multiobjective problem (4.60) reduces to a standard mathematical programming problem. However, this rarely happens, and as an alternative, an interaction process with the DM is necessary for finding the satisficing solution to the general fuzzy multiobjective problem (4.60).

In the interactive fuzzy multiobjective linear programming method proposed by Sakawa et al. (1987), the membership functions $\mu(z(x)) = (\mu_1(z_1(x)), \ldots, \mu_k(z_k(x)))^T$ for the objective functions $z(x) = (z_1(x), \ldots, z_k(x))^T$ are determined first. To generate a candidate for the satisficing solution which is also (M-)Pareto optimal, the DM is then asked to specify the aspiration levels of achievement for the values of all the membership functions, called the reference membership levels. The reference membership levels can be viewed as natural extensions of the reference point of Wierzbicki (1980) in the objective function space.

For the DM's reference membership levels $\hat{\mu} = (\hat{\mu}_1, \ldots, \hat{\mu}_k)^T$, the corresponding (M-)Pareto optimal solution, which is the nearest to $\hat{\mu}$ in the minimax sense or better than $\hat{\mu}$ if they are attainable, is obtained by solving the following minimax problem:

$$\left. \begin{array}{ll} \text{minimize} & \max_{i=1,\ldots,k} \{\hat{\mu}_i - \mu_i(z_i(x))\} \\ \text{subject to} & x \in X, \end{array} \right\} \tag{4.64}$$

or equivalently

$$\left. \begin{array}{ll} \text{minimize} & v \\ \text{subject to} & \hat{\mu}_i - \mu_i(z_i(x)) \le v, \ i = 1, \ldots, k \\ & x \in X. \end{array} \right\} \tag{4.65}$$

If all of the membership functions $\mu_i(z_i(x))$, $i = 1, \ldots, k$ are linear, the minimax problem becomes a linear programming problem, and hence, we can obtain an optimal solution by directly applying the simplex method of linear programming.

4.4 Interactive Fuzzy Multiobjective Linear Programming

However, with the strictly monotone decreasing or increasing membership functions, which may be nonlinear, the resulting minimax problem becomes a nonlinear programming problem. For notational convenience, denote the strictly monotone decreasing functions for the fuzzy min and the right function of the fuzzy equal by $d_{iR}(z_i)$, $i \in I_1 \cup I_{3R}$ and the strictly monotone increasing functions for the fuzzy max and the left function of the fuzzy equal by $d_{iL}(z_i)$, $i \in I_2 \cup I_{3L}$. Then in order to solve the formulated problem on the basis of the linear programming method, convert each constraint $\hat{\mu}_i - \mu_i(z_i(x)) \leq v$, $i = 1, \ldots, k$ of the minimax problem (4.65) into the following form using the strictly monotone property of $d_{iL}(\cdot)$ and $d_{iR}(\cdot)$:

$$\left.\begin{aligned}
&\text{minimize} \quad v \\
&\text{subject to} \quad z_i(x) \leq d_{iR}^{-1}(\hat{\mu}_i - v), \ i \in I_1 \cup I_{3R} \\
&\phantom{\text{subject to}} \quad z_i(x) \geq d_{iL}^{-1}(\hat{\mu}_i - v), \ i \in I_2 \cup I_{3L} \\
&\phantom{\text{subject to}} \quad x \in X.
\end{aligned}\right\} \quad (4.66)$$

It is important to note here that, if the value of v is fixed, the constraints of (4.66) can be reduced to a set of linear inequalities. Obtaining the optimal solution v^* to (4.66) is equivalent to determining the minimum of v so that there exists a feasible solution satisfying the constraints of (4.66). Since v satisfies $\hat{\mu}_{\max} - 1 \leq v \leq \hat{\mu}_{\min}$, where $\hat{\mu}_{\max}$ and $\hat{\mu}_{\min}$ denote the maximum and the minimum of $\hat{\mu}_i$, $i = 1, \ldots, k$, respectively, we have the following method for solving this problem by combined use of the bisection method and the simplex method of linear programming.

Step 1   Set $v_0 = \hat{\mu}_{\min}$ and test whether a feasible solution satisfying the constraints of (4.66) exists or not by making use of phase I of the simplex method. If a feasible solution exists, set $n = 0$ and proceed. Otherwise, the DM must reassess the membership function.

Step 2   Update the value of v for the bisection method as follows:

$$\begin{cases} v_{n+1} = v_n - 1/2^{n+1}\hat{\mu}_{\max} \text{ and } v^* = v_n & \text{if a feasible solution exists for } v_n, \\ v_{n+1} = v_n + 1/2^{n+1}\hat{\mu}_{\max} & \text{if no feasible solution exists for } v_n. \end{cases}$$

If the difference between v_n and v_{n+1} is sufficiently small, stop the procedure.

Step 3   For $v = v_{n+1}$, test whether a feasible solution to (4.66) exists or not using the sensitivity analysis technique for the changes in the right-hand side of the simplex method. Return to step 2.

In this way, we can determine the optimal value v^*. Then the DM selects an appropriate standing objective from among the objectives $z_i(x)$, $i = 1, \ldots, k$. For notational convenience in the following without loss of generality, let it be $z_1(x)$ and assume $1 \in I_1$. Then the following linear programming problem is solved for $v = v^*$:

$$\left.\begin{array}{ll}\text{minimize} & z_1(x) \\ \text{subject to} & z_i(x) \leq d_{iR}^{-1}(\hat{\mu}_i - v^*), \ i \,(\neq 1) \in I_1 \cup I_3 \\ & z_i(x) \geq d_{iL}^{-1}(\hat{\mu}_i - v^*), \ i \,(\neq 1) \in I_2 \cup I_3 \\ & x \in X. \end{array}\right\} \quad (4.67)$$

The relationships between the optimal solution of the minimax problem (4.64) and the (M-)Pareto optimal concept of the multiobjective linear programming problem can be characterized by the following theorems.

Theorem 4.2. *If $x^* \in X$ is a unique optimal solution to the minimax problem (4.65) for some $\hat{\mu}_i$, $i = 1, \ldots, k$, then x^* is a (M-)Pareto optimal solution to the (generalized) multiobjective linear programming problem.*

Theorem 4.3. *If $x^* \in X$ is a (M-)Pareto optimal solution to the (generalized) multiobjective linear programming problem, then there exist $\hat{\mu}_i$, $i = 1, \ldots, k$ such that x^* is an optimal solution to the minimax problem (4.65).*

The proof of this theorem follows directly from the definitions of optimality and (M-)Pareto optimality by making use of contradiction arguments.

It must be noted here that, for generating (M-)Pareto optimal solutions using this theorem, uniqueness of a solution must be verified. In order to test (M-)Pareto optimality of current optimal solution x^*, we solve the following Pareto optimality test problem as discussed in Theorem 3.7:

$$\left.\begin{array}{ll}\text{maximize} & \sum_{i=1}^{k} \varepsilon_i \\ \text{subject to} & \mu_i(z_i(x)) - \varepsilon_i = \mu_i(z_i(x^*)), \ i = 1, \ldots, k \\ & x \in X, \ \boldsymbol{\varepsilon} = (\varepsilon_1, \ldots, \varepsilon_k)^T \geq \mathbf{0}. \end{array}\right\} \quad (4.68)$$

For an optimal solution $(\bar{x}, \bar{\varepsilon})$ to this linear programming problem, as discussed in Theorem 3.7, (i) if $\bar{\varepsilon}_i = 0$ for all $i = 1, \ldots, k$, then x^* is a (M-)Pareto optimal solution to the (generalized) multiobjective linear programming problem, and (ii) if $\bar{\varepsilon}_i > 0$ for at least one i, not x^* but $\bar{x}$ is a (M-)Pareto optimal solution of the (generalized) multiobjective linear programming problem.

The DM must either be satisfied with the current (M-)Pareto optimal solution or act on this solution by updating the reference membership levels. In order to help the DM express a degree of preference, as was discussed in the previous subsection, the trade-off information between a standing membership function $\mu_1(z_1(x))$ and each of the other membership functions is very useful. Such trade-off information is easily obtainable since it is closely related to the simplex multipliers of problem (4.67).

Let the simplex multipliers corresponding to the constraints for $z_i(x), i = 2, \ldots, k$ of the linear programming problem (4.67) be denoted by $\pi_i^* = \pi_i(x^*), i = 2, \ldots, k$, where x^* is an optimal solution of (4.67). If x^* is a

4.4 Interactive Fuzzy Multiobjective Linear Programming

nondegenerate solution of (4.67) and all the constraints of (4.67) are active, then by using the results in Haimes and Chankong (1979), the trade-off rate between the objective functions can be represented by

$$-\frac{\partial z_1(x)}{\partial z_i(x)} = -\pi_i^*, \quad i = 2, \ldots, k. \tag{4.69}$$

Hence, by the chain rule, the trade-off rate between the membership functions is given by

$$-\frac{\partial \mu_1(z_1(x))}{\partial \mu_i(z_i(x))} = -\frac{\partial \mu_1(z_1(x))}{\partial z_1(x)} \frac{\partial z_1(x)}{\partial z_i(x)} \left\{ \frac{\partial \mu_i(z_i(x))}{\partial z_i(x)} \right\}^{-1}, \quad i = 2, \ldots, k. \tag{4.70}$$

Therefore, for each $i = 2, \ldots, k$, we have the following expression:

$$-\frac{\partial \mu_1(z_1(x))}{\partial \mu_i(z_i(x))} = \pi_i^* \frac{\partial \mu_1(z_1(x))/\partial z_1(x)}{\partial \mu_i(z_i(x))/\partial z_i(x)}, \quad i = 2, \ldots, k. \tag{4.71}$$

It should be stressed here that in order to obtain the trade-off rate from (4.71), all the constraints of problem (4.67) must be active. Therefore, if there are inactive constraints, it is necessary to replace $\hat{\mu}_i$ for inactive constraints with $d_{iR}(z_i(x^*))+v^*$ or $d_{iL}(z_i(x^*)) + v^*$ and solve the corresponding problem to obtain the simplex multipliers.

Observing that the trade-off rates $-\partial \mu_1/\partial \mu_i$, $i = 2, \ldots, k$ (4.71) indicate the decrement value of the membership function μ_1 with a unit increment of value of the membership function μ_i, the information of trade-off rates can be used to estimate the local shape of $(\mu_1(z_1(x^*)), \ldots, \mu_k(z_k(x^*)))$ around the current solution x^*.

We can now give the interactive algorithm in order to derive the satisficing solution for the DM from the (M-)Pareto optimal solution set where the steps marked with an asterisk involve interaction with the DM. This interactive fuzzy multiobjective programming method can also be interpreted as the fuzzy version of the reference point method with the trade-off information.

Procedure of Interactive Fuzzy Multiobjective Linear Programming

Step 0   Calculate the individual minimum and maximum of each objective function under the given constraints.
Step 1* Elicit a membership function of the fuzzy goal from the DM for each of the objective functions.
Step 2   Set the initial reference membership levels to 1s.
Step 3   For the reference membership levels, solve the corresponding minimax problem to obtain a (M-)Pareto optimal solution and its membership values together with the trade-off rates between the membership functions.
Step 4* If the DM is satisfied with the current membership values of the (M-)Pareto optimal solution, stop. Then the current (M-)Pareto optimal solution

is the satisficing solution of the DM. Otherwise, ask the DM to update the current reference membership levels by considering the values of the membership functions together with the trade-off rates between the membership functions and return to Step 3.

It should be stressed to the DM that any improvement of one membership function can be achieved only at the expense of at least one of the other membership functions.

Example 4.6 (Three-objective production planning problem under fuzzy environment). To demonstrate interactive fuzzy multiobjective programming, consider the production planning problem with environmental consideration in Example 4.5 again. In addition to the fuzzy goals of (i) maximization of the total profit and (ii) minimization of pollution, considering the profit contribution ratio of the two products, assume that the DM also has a fuzzy goal such as (iii) production of product P_2 should be 3/8 times as large as product P_1.

Then the corresponding generalized multiobjective linear programming problem can be formulated as follows:

$$\begin{aligned} \text{fuzzy max} \quad & 3x_1 + 8x_2 \\ \text{fuzzy min} \quad & 5x_1 + 4x_2 \\ \text{fuzzy equal} \quad & 3x_1 - 8x_2 \\ \text{subject to} \quad & x = (x_1, x_2)^T \in X, \end{aligned}$$

where X is the feasible region given in Example 3.1.

First, observe that the individual minimum and maximum of each objective function are

$$z_1^{\min} = 0, \quad z_1^{\max} = 37, \quad z_2^{\min} = 0, \quad z_2^{\max} = 29, \quad z_3^{\min} = -36, \quad z_3^{\max} = 13.5.$$

Considering these values, assume that the DM determines the following linear membership functions for the fuzzy goals:

$$\begin{cases} \text{fuzzy max} & \mu_1(18) = 0, \quad \mu_1(26) = 1 \\ \text{fuzzy min} & \mu_2(25) = 0, \quad \mu_2(18) = 1 \\ \text{fuzzy equal} & \mu_{3L}(-17) = 0, \, \mu_{3L}(0) = \mu_{3R}(0) = 1, \, \mu_{3R}(8) = 0, \end{cases}$$

where linear functions are assumed from $\mu_i = 0$ to $\mu_i = 1$ for $i = 1, 2, 3$.

Then, for the initial reference membership levels $\hat{\mu}_1 = \hat{\mu}_2 = \hat{\mu}_3 = 1$, solving the corresponding minimax problem yields an M-Pareto optimal solution

$$z_1 = 22.821, \quad z_2 = 20.782, \quad z_3 = -6.756 \quad (x_1 = 2.678, \, x_2 = 1.849),$$

the corresponding membership values

$$\mu_1 = 0.6026, \quad \mu_2 = 0.6026, \quad \mu_3 = 0.6026,$$

4.5 Interactive Fuzzy Linear Programming with Fuzzy Parameters

and the trade-off rates between the membership functions

$$-\frac{\partial \mu_1}{\partial \mu_2} = 0.8077, \quad -\frac{\partial \mu_1}{\partial \mu_3} = 1.1442.$$

On the basis of such information, suppose that the DM updates the reference membership levels to

$$\hat{\mu}_1 = 0.7, \quad \hat{\mu}_2 = 0.8, \quad \hat{\mu}_3 = 0.5$$

with the anticipation of improving the satisfaction levels for the total profit and the amount of pollution at the expense of the ratio of two products.

For the updated reference membership levels, the corresponding minimax problem yields an M-Pareto optimal solution

$$z_1 = 23.2221, \quad z_2 = 19.7306, \quad z_3 = -9.3029 \quad (x_1 = 2.3199, \; x_2 = 2.0328),$$

the membership values

$$\mu_1 = 0.6528, \quad \mu_2 = 0.7528, \quad \mu_3 = 0.4528,$$

and the trade-off rates

$$-\frac{\partial \mu_1}{\partial \mu_2} = 0.8077, \quad -\frac{\partial \mu_1}{\partial \mu_3} = 1.1442.$$

If the DM is satisfied with the current values of the membership functions, the procedure stops. Otherwise, a similar procedure continues until the satisficing solution for the DM is obtained. ◇

4.5 Interactive Fuzzy Linear Programming with Fuzzy Parameters

As discussed in Chap. 3, fundamental to multiobjective linear programming is the Pareto optimal concept, and for a Pareto optimal solution of the multiobjective linear programming problem, any improvement of one objective function can be achieved only at the expense of another.

In practice, however, it would certainly be more appropriate to consider that possible values of the parameters in the description of the objective functions and the constraints usually involve the ambiguity of the experts' understanding of the real system. For this reason, in this section, we consider the following multiobjective linear programming problem involving fuzzy parameters

$$\left.\begin{array}{l}\text{minimize} \quad (C_1 x, \ldots, C_k x)^T \\ \text{subject to} \quad x \in X(A, B) = \{x \in \mathbb{R}^n \mid A_j x \leq B_j, \ j = 1, \ldots, m \,;\, x \geq 0\},\end{array}\right\} \quad (4.72)$$

where $x = (x_1, \ldots, x_n)^T$ is an n-dimensional column vector of decision variable, $C_i = (C_{i1}, \ldots, C_{in})$ is an n-dimensional row vector of fuzzy parameters involved in the ith objective function, $C_i x$, $A_j = (A_{j1}, \ldots, A_{jn})$ is an n-dimensional row vector of fuzzy parameters involved in the jth constraint $A_j x \leq B_j$, and B_j is a fuzzy parameter of a right-hand side constant in the jth constraint $A_j x \leq B_j$.

These fuzzy parameters, reflecting the experts' ambiguous understanding of the nature of the parameters in the problem-formulation process, are assumed to be characterized as fuzzy numbers introduced by Dubois and Prade (1978, 1980). We now assume that all of the fuzzy parameters $C_{i1}, \ldots, C_{in}$, $i = 1, \ldots, k$, $A_{j1}, \ldots, A_{jn}$, $j = 1, \ldots, m$, and B_j, $j = 1, \ldots, m$ in the multiobjective linear programming problem with fuzzy parameters are fuzzy numbers whose membership functions are denoted by $\mu_{C_{i1}}(c_{i1}), \ldots, \mu_{C_{in}}(c_{in})$, $i = 1, \ldots, k$, $\mu_{A_{j1}}(a_{j1}), \ldots, \mu_{A_{jn}}(a_{jn})$, $j = 1, \ldots, m$, and $\mu_{B_j}(b_j)$, $j = 1, \ldots, m$, respectively. For simplicity in notation, define the following vectors:

$$c = (c_1, \ldots, c_k), \quad a = (a_1, \ldots, a_m), \quad b = (b_1, \ldots, b_m),$$

$$C = (C_1, \ldots, C_k), \quad A = (A_1, \ldots, A_m), \quad B = (B_1, \ldots, B_m).$$

Observing that the multiobjective linear programming problem with fuzzy parameters involves fuzzy numbers both in the objective functions and the constraints, it is evident that the notion of Pareto optimality defined for multiobjective linear programming problems cannot be applied directly. Thus, it seems essential to extend the notion of usual Pareto optimality in some sense. For that purpose, we first introduce the α-level set of the fuzzy numbers A_{jr}, B_j, and C_{ir}.

Definition 4.6 (α-level set). The α-level set of the fuzzy numbers A_{jr}, B_j, and C_{ir} is defined as an ordinary set $(A, B, C)_\alpha$ for which the degrees of the membership functions exceed the level α:

$$(A, B, C)_\alpha = \{ (a, b, c) \mid \mu_{A_{jr}}(a_{jr}) \geq \alpha, \ \mu_{B_j}(b_j) \geq \alpha, \ \mu_{C_{ir}}(c_{ir}) \geq \alpha; \\ i = 1, \ldots, k, \ j = 1, \ldots, m, \ r = 1, \ldots, n\}. \quad (4.73)$$

Now suppose that the DM decides that the degrees of all of the membership functions of the fuzzy numbers involved in the multiobjective linear programming problem with fuzzy parameters should be greater than or equal to some level α. Then for such a level α, by selecting a coefficient vector $(a, b, c) \in (A, B, C)_\alpha$, the multiobjective linear programming problem with fuzzy parameters can be reduced to the following nonfuzzy multiobjective linear programming problem which depends on the coefficient vector $(a, b, c) \in (A, B, C)_\alpha$:

4.5 Interactive Fuzzy Linear Programming with Fuzzy Parameters

$$\begin{aligned} & \text{minimize} \quad (c_1 x, \ldots, c_k x)^T \\ & \text{subject to} \quad x \in X(a, b) = \{x \in \mathbb{R}^n \mid a_j x \le b_j,\ j = 1, \ldots, m\,;\ x \ge 0\}. \end{aligned} \qquad (4.74)$$

Observe that there exists an infinite number of such multiobjective linear programming problems with fuzzy parameters depending on a coefficient vector $(a, b, c) \in (A, B, C)_\alpha$, and the values of (a, b, c) are arbitrarily chosen from the α-level set $(A, B, C)_\alpha$ in the sense that the degrees of all of the membership functions for the fuzzy numbers exceed the level α. However, if possible, it would be desirable for the DM to choose $(a, b, c) \in (A, B, C)_\alpha$ so as to minimize the objective functions under the constraints. From such a point of view, for a given multiobjective linear programming problem with fuzzy parameters and a certain level α, it seems to be quite natural to formulate the following nonfuzzy multiobjective linear programming problem:

$$\begin{aligned} & \text{minimize} \quad (c_1 x, \ldots, c_k x)^T \\ & \text{subject to} \quad x \in X(a, b) = \{x \in \mathbb{R}^n \mid a_j x \le b_j,\ j = 1, \ldots, m\,;\ x \ge 0\} \\ & \qquad\qquad (a, b, c) \in (A, B, C)_\alpha. \end{aligned} \qquad (4.75)$$

Let the problem (4.75) be denoted by the α-multiobjective linear programming problem. It should be emphasized here that, in such α-multiobjective linear programming problems, the parameters (a, b, c) are treated as decision variables rather than constants.

On the basis of the α-level sets of the fuzzy numbers, we can introduce the concept of an α-Pareto optimal solution to the α-multiobjective linear programming problem as a natural extension of the Pareto optimality concept for the multiobjective linear programming problem.

Definition 4.7 (α-Pareto optimal solution). A point $x^* \in X(a^*, b^*)$ is said to be an α-Pareto optimal solution to an α-multiobjective linear programming problem if and only if there does not exist another $x \in X(a, b)$ and $(a, b, c) \in (A, B, C)_\alpha$ such that $c_i x \le c_i^* x^*$, $i = 1, \ldots, k$ with strict inequality holding for at least one i. The corresponding values of parameters (a^*, b^*, c^*) are called the α-level optimal parameters.

Observe that the α-Pareto optimal solution and the α-level optimal parameters can be obtained through a direct application of the usual scalarizing methods for generating Pareto optimal solutions by regarding (x, a, b, c) as the decision variables in the α-multiobjective linear programming problem.

However, as can be immediately understood from Definition 4.7, in general, there are an infinite number of α-Pareto optimal solutions to the α-multiobjective linear programming problem, and then some kinds of subjective judgment should be added to the quantitative analyses by the DM. That is, the DM must select a compromise or satisficing solution from the α-Pareto optimal solution set based on a subjective value judgment. For that purpose, Sakawa and Yano (1986d, 1990) presented an interactive algorithm to derive the satisficing solution of the DM from the α-Pareto optimal solution set by updating the reference membership levels and/or the level α.

However, considering the imprecise nature of the DM's judgment, it is natural to assume that the DM may have imprecise or fuzzy goals for the objective functions in the α-multiobjective linear programming problem. In a minimization problem, a goal stated by the DM may be to achieve "substantially less than or equal to some value p_i." This type of statement can be quantified by eliciting a corresponding membership function for the fuzzy goal.

To elicit a membership function $\mu_i(c_i x)$ from the DM for each of the objective functions $c_i x$, $i = 1, \ldots, k$ in the α-multiobjective linear programming problem, we first calculate the individual minimum and maximum of each objective function under the given constraints for $\alpha = 0$ and $\alpha = 1$. By taking account of the calculated individual minimum and maximum of each objective function for $\alpha = 0$ and $\alpha = 1$ together with the rate of increase of membership satisfaction, the DM determines a membership function $\mu_i(c_i x)$ in a subjective manner which is a strictly monotone decreasing function with respect to $c_i x$.

So far we have restricted ourselves to a minimization problem and consequently assumed that the DM has a fuzzy goal such as "$c_i x$ should be substantially less than or equal to p_i." In the fuzzy approaches, as discussed in Sect. 4.4, we can further treat a more general case where the DM has two types of fuzzy goals, namely, fuzzy goals expressed in words such as "$c_i x$ should be in the vicinity of r_i" (called fuzzy equal) as well as "$c_i x$ should be substantially less than or equal to p_i or greater than or equal to q_i" (called fuzzy min or fuzzy max). Such a generalized α-multiobjective linear programming problem may now be expressed as

$$\left. \begin{array}{ll} \text{fuzzy min} & c_i x, \quad i \in I_1 \\ \text{fuzzy max} & c_i x, \quad i \in I_2 \\ \text{fuzzy equal} & c_i x, \quad i \in I_3 \\ \text{subject to} & x \in X(a,b) \\ & (a,b,c) \in (A,B,C)_\alpha, \end{array} \right\} \quad (4.76)$$

where $I_1 \cup I_2 \cup I_3 = \{1, 2, \ldots, k\}$, $I_i \cap I_j = \emptyset$, $i, j = 1, 2, 3, i \neq j$.

To elicit a membership function $\mu_i(c_i x)$ from the DM for a fuzzy goal like "$c_i x$ should be in the vicinity of r_i," it should be quite apparent that different functions can be utilized for the left and right sides of r_i.

Concerning the membership functions of the generalized α-multiobjective linear programming problem, it is reasonable to assume that $\mu_i(c_i x)$, $i \in I_1$ and the right side functions of $\mu_i(c_i x)$, $i \in I_3$ are strictly monotone increasing and continuous functions with respect to $c_i x$. To be more explicit, each membership function $\mu_i(c_i x)$ of the generalized α-multiobjective linear programming problem for $i \in I_1$, $i \in I_2$, or $i \in I_3$ is defined as follows:

4.5 Interactive Fuzzy Linear Programming with Fuzzy Parameters

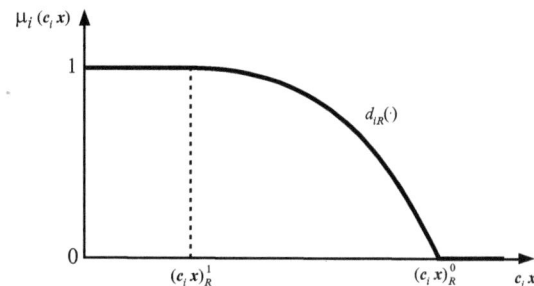

Fig. 4.12 Fuzzy min membership function

(i) For $i \in I_1$,

$$\mu_i(c_i x) = \begin{cases} 1 & \text{if } (c_i x)^1_R \geq c_i x \\ d_{iR}(c_i x) & \text{if } (c_i x)^1_R \leq c_i x \leq (c_i x)^0_R \\ 0 & \text{if } c_i x \geq (c_i x)^0_R. \end{cases} \quad (4.77)$$

(ii) For $i \in I_2$,

$$\mu_i(c_i x) = \begin{cases} 0 & \text{if } (c_i x)^0_L \geq c_i x \\ d_{iL}(c_i x) & \text{if } (c_i x)^0_L \leq c_i x \leq (c_i x)^1_L, \\ 1 & \text{if } c_i x \geq (c_i x)^1_L. \end{cases} \quad (4.78)$$

(iii) For $i \in I_3$,

$$\mu_i(c_i x) = \begin{cases} 0 & \text{if } c_i x \leq (c_i x)^0_L \\ d_{iL}(c_i x) & \text{if } (c_i x)^0_L \leq c_i x \leq (c_i x)^1_L \\ 1 & \text{if } (c_i x)^1_L \leq c_i x \leq (c_i x)^1_R \\ d_{iR}(c_x) & \text{if } (c_i x)^1_R \leq c_i x \leq (c_i x)^0_R \\ 0 & \text{if } (c_i x)^0_R \leq c_i x. \end{cases} \quad (4.79)$$

In the above definition, it is assumed that $d_{iR}(c_i x)$ is a strictly monotone decreasing continuous function with respect to $c_i x$ and $d_{iL}(c_i x)$ is a strictly monotone increasing continuous function with respect to $c_i x$. Both may be linear or nonlinear. The values $(c_i x)^0_L$ and $(c_i x)^0_R$ are best values of unacceptable levels for $c_i x$, and the values $(c_i x)^1_L$ and $(c_i x)^1_R$ are worst values of totally desirable levels for $c_i x$. Possible shapes of the membership functions for "fuzzy min," "fuzzy max," and "fuzzy equal" are depicted in Figs. 4.12–4.14.

The fuzzy equal is interpreted as the DM's fuzzy preference such as a fuzzy goal that $c_i x$ should be as close to r_i as possible. Consequently, the notion of α-Pareto optimal solutions defined in terms of objective functions cannot be applied. For this reason, we introduce the concept of M-α-Pareto optimal solutions which is defined in terms of membership functions instead of objective functions, where M refers to membership.

Fig. 4.13 Fuzzy max membership function

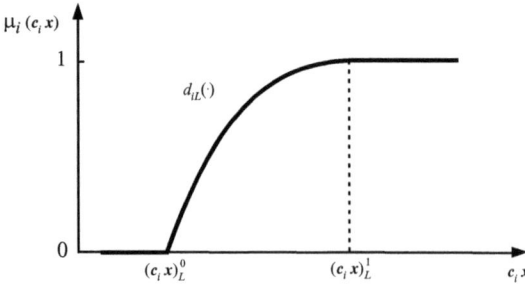

Fig. 4.14 Fuzzy equal membership function

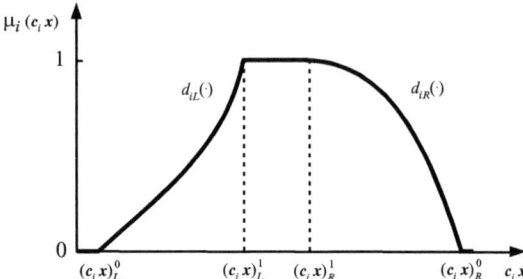

Definition 4.8 (M-α-Pareto optimal solution). A point $x^* \in X(a^*, b^*)$ is said to be an M-α-Pareto optimal solution to a generalized α-multiobjective linear programming problem if and only if there does not exist another $x \in X(a, b)$ and $(a, b, c) \in (A, B, C)_\alpha$ such that $\mu_i(c_i x) \geq \mu_i(c_i^* x^*)$, $i = 1, \ldots, k$ with strict inequality holding for at least one i. The corresponding values of parameters (a^*, b^*, c^*) are called the α-level optimal parameters.

Observe that the concept of the M-α-Pareto optimal solution defined in terms of membership functions is a natural extension to that of α-Pareto optimal solutions defined in terms of objective functions when "fuzzy equal" is included in the fuzzy goals of the DM.

Having elicited the membership functions $\mu_i(c_i x)$, $i = 1, \ldots, k$ from the DM for the objective functions $c_i x$, $i = 1, \ldots, k$ if we introduce a general aggregation function

$$\mu_D(\mu_1(c_1 x), \ldots, \mu_k(c_k x), \alpha), \tag{4.80}$$

a general fuzzy α-multiobjective problem can be defined by

$$\left. \begin{array}{ll} \text{maximize} & \mu_D(\mu_1(c_1 x), \ldots, \mu_k(c_k x), \alpha) \\ \text{subject to} & (x, a, b, c) \in P(\alpha), \end{array} \right\} \tag{4.81}$$

4.5 Interactive Fuzzy Linear Programming with Fuzzy Parameters

where $P(\alpha)$ is the set of M-α-Pareto optimal solutions and the corresponding α-level optimal parameters to the generalized α-multiobjective linear programming problem.

Probably the most crucial problem in the general fuzzy α-multiobjective problem (4.81) is the identification of an appropriate aggregation function which well represents the DM's fuzzy preference. If $\mu_D(\cdot)$ can be explicitly identified, then the general fuzzy α-multiobjective problem (4.81) reduces to a standard mathematical programming problem. However, this happens rarely, and then as discussed in the previous section, consider interactive decision making for the generalized α-multiobjective linear programming problem.

To generate a candidate for the satisficing solution, which is also M-α-Pareto optimal, the DM is asked to specify a level α of the α-level set and the reference membership levels. Observe that the idea of the reference membership levels, which first appeared in Sakawa et al. (1987), can be viewed as a natural extension of the idea of the reference point in Wierzbicki (1980).

Once the level α and the reference membership levels $\hat{\mu}_i$, $i = 1, \ldots, k$ are specified by the DM, the corresponding M-α-Pareto optimal solution, which is, in the minimax sense, nearest to $\hat{\mu} = (\hat{\mu}_1, \ldots, \hat{\mu}_k)$ or better than it if the reference membership levels are attainable, is obtained by solving the following minimax problem:

$$\left.\begin{array}{ll} \text{minimize} & \max_{i=1,\ldots,k} \{\hat{\mu}_i - \mu_i(c_i x)\} \\ \text{subject to} & x \in X(a, b) \\ & (a, b, c) \in (A, B, C)_\alpha \end{array}\right\} \quad (4.82)$$

or equivalently

$$\left.\begin{array}{ll} \text{minimize} & v \\ \text{subject to} & \hat{\mu}_i - \mu_i(c_i x) \leq v, \ i = 1, \ldots, k \\ & a_j x \leq b_j, \ j = 1, \ldots, m, \ x \geq 0 \\ & (a, b, c) \in (A, B, C)_\alpha. \end{array}\right\} \quad (4.83)$$

However, with the strictly monotone decreasing or increasing membership functions (4.77)–(4.79) which may be nonlinear, the resulting problem becomes a nonlinear programming problem. To solve the formulated problem on the basis of the linear programming techniques, we first convert each constraint $\hat{\mu}_i - \mu_i(c_i x) \leq v$, $i = 1, \ldots, k$ of the minimax problem (4.83) into the following form using the strictly monotone property of the membership functions of the fuzzy goals $d_{iL}(\cdot)$ and $d_{iR}(\cdot)$:

$$\left.\begin{array}{l} c_i x \leq d_{iR}^{-1}(\hat{\mu}_i - v), \ i \in I_1 \cup I_{3R} \\ c_i x \geq d_{iL}^{-1}(\hat{\mu}_i - v), \ i \in I_2 \cup I_{3L}. \end{array}\right\} \quad (4.84)$$

Now we introduce the following set-valued functions $S_{iR}(\cdot)$, $S_{iL}(\cdot)$, and $T_j(\cdot,\cdot)$:

$$\left.\begin{array}{l} S_{iR}(c_i) = \{(x,v) \mid c_i x \leq d_{iR}^{-1}(\hat{\mu}_i - v)\},\ i \in I_1 \cup I_{3R} \\ S_{iL}(c_i) = \{(x,v) \mid c_i x \geq d_{iL}^{-1}(\hat{\mu}_i - v)\},\ i \in I_2 \cup I_{3R} \\ T_j(a_j, b_j) = \{x \mid a_j x \leq b_j\},\ j = 1,\ldots,m. \end{array}\right\} \quad (4.85)$$

It can be verified that the following relations hold for $S_{iR}(\cdot)$, $S_{iL}(\cdot)$, and $T_j(\cdot,\cdot)$ when $x \geq 0$.

Proposition 4.1.

(i) *If $c_i^1 \leq c_i^2$, then $S_{iR}(c_i^1) \supseteq S_{iR}(c_i^2)$ and $S_{iL}(c_i^1) \subseteq S_{iL}(c_i^2)$.*
(ii) *If $a_j^1 \leq a_j^2$, then $T_j(a_j^1, b_j) \supseteq T_j(a_j^2, b_j)$.*
(iii) *If $b_j^1 \leq b_j^2$, then $T_j(a_j, b_j^1) \subseteq T_j(a_j, b_j^2)$.*

Using the properties of the α-level sets for the vectors C_i and A_j of the fuzzy numbers and the fuzzy numbers B_j, the feasible regions for c_i, a_j, and b_j can be denoted respectively by the closed intervals $[c_{i\alpha}^L, c_{i\alpha}^R]$, $[a_{j\alpha}^L, a_{j\alpha}^R]$, and $[b_{j\alpha}^L, b_{j\alpha}^R]$. From this fact, it should be noted that the values of parameters $(a_\alpha^L = (a_{1\alpha}^L, \ldots, a_{m\alpha}^L), b_\alpha^R = (b_{1\alpha}^R, \ldots, b_{m\alpha}^R), c_{i\alpha}^L, i \in I_1 \cup I_{3R}, c_{i\alpha}^R, i \in I_2 \cup I_{3L})$ are the α-level optimal parameters for any M-α-Pareto optimal solution. For convenience sake, we define the feasible region corresponding to the α-level optimal parameters as

$$X(a_\alpha^L, b_\alpha^R) = \left\{ x \in \mathbb{R}^n \mid a_{j\alpha}^L x \leq b_{j\alpha}^R,\ j = 1,\ldots,m,\ x \geq 0 \right\}. \quad (4.86)$$

Consequently, using of the results in Proposition 4.1, we can obtain an optimal solution to (4.83) by solving the following problem:

$$\left.\begin{array}{ll} \text{minimize} & v \\ \text{subject to} & c_{i\alpha}^L x \leq d_{iR}^{-1}(\hat{\mu}_i - v),\ i \in I_1 \cup I_{3R} \\ & c_{i\alpha}^R x \geq d_{iL}^{-1}(\hat{\mu}_i - v),\ i \in I_2 \cup I_{3L} \\ & x \in X(a_\alpha^L, b_\alpha^R). \end{array}\right\} \quad (4.87)$$

Since v satisfies $\hat{\mu}_{\max} - 1 \leq v \leq \hat{\mu}_{\min}$, where $\hat{\mu}_{\max}$ and $\hat{\mu}_{\min}$ denote the maximum and the minimum of $\hat{\mu}_i$, $i = 1,\ldots,k$, respectively, the problem (4.87) can be solved by the combined use of the bisection method and the simplex method of linear programming presented in the previous section.

After determining the optimal solution v^*, the DM selects an appropriate standing objective from among the objectives $c_i x$, $i = 1,\ldots,k$. For notational convenience in the following without loss of generality, let it be $c_1 x$ and assume that $1 \in I_1$. Then the following linear programming problem is solved for $v = v^*$:

4.5 Interactive Fuzzy Linear Programming with Fuzzy Parameters

$$\left.\begin{array}{ll} \text{minimize} & c_{1\alpha}^L x \\ \text{subject to} & c_{i\alpha}^L x \leq d_{iR}^{-1}(\hat{\mu}_i - v^*), \ i \in I_1 \cup I_{3R} \\ & c_{i\alpha}^R x \geq d_{iL}^{-1}(\hat{\mu}_i - v^*), \ i \in I_2 \cup I_{3L} \\ & x \in X(a_\alpha^L, b_\alpha^R). \end{array}\right\} \quad (4.88)$$

For convenience in our subsequent discussion, we assume that the optimal solution x^* to (4.88) satisfies the following conditions:

$$\left.\begin{array}{l} c_{i\alpha}^L x^* = d_{iR}^{-1}(\hat{\mu}_i - v^*), \ i \in I_1 \cup I_{3R} \\ c_{i\alpha}^R x^* = d_{iL}^{-1}(\hat{\mu}_i - v^*), \ i \in I_2 \cup I_{3L}, \end{array}\right\} \quad (4.89)$$

where $I_3 = I_{3L} \cup I_{3R}$ and $I_{3L} \cap I_{3R} = \emptyset$.

The relationships between the optimal solutions to (4.87) and the M-α-Pareto optimal concept of the generalized α-multiobjective linear programming problem can be characterized by the following theorems.

Theorem 4.4. *If a point $x^* \in X(a_\alpha^L, b_\alpha^R)$ is a unique optimal solution to (4.87), then x^* is an M-α-Pareto optimal solution to the generalized α-multiobjective linear programming problem.*

Theorem 4.5. *If a point $x^* \in X(a_\alpha^L, b_\alpha^R)$ is an M-α-Pareto optimal solution to the generalized α-multiobjective linear programming problem, then x^* is an optimal solution to (4.87) for some $\hat{\mu} = (\hat{\mu}_1, \ldots, \hat{\mu}_k)$.*

It must be observed here that for generating M-α-Pareto optimal solutions using Theorem 4.4, the uniqueness of the solution must be verified. In order to test M-α-Pareto optimality of a current optimal solution x^*, we formulate and solve the following linear programming problem:

$$\left.\begin{array}{ll} \text{maximize} & \sum_{i=1}^k \varepsilon_i \\ \text{subject to} & c_{i\alpha}^L x + \varepsilon_i = c_{i\alpha}^L x^*, \ \varepsilon_i \geq 0, \ i \in I_1 \cup I_{3R} \\ & c_{i\alpha}^R x - \varepsilon_i = c_{i\alpha}^R x^*, \ \varepsilon_i \geq 0, \ i \in I_2 \cup I_{3L} \\ & x \in X(a_\alpha^L, b_\alpha^R). \end{array}\right\} \quad (4.90)$$

Let $(\bar{x}, \bar{\varepsilon})$ be an optimal solution to this problem. If $\bar{\varepsilon}_i = 0$ for all $i = 1, \ldots, k$, then x^* is an M-α-Pareto optimal solution. If $\bar{\varepsilon}_i > 0$ for at least one i, as discussed in Chap. 3, it can be easily shown that $\bar{x}$ is an M-α-Pareto optimal solution.

Now given the M-α-Pareto optimal solution for the level α and the reference membership levels specified by the DM by solving the corresponding minimax problem, the DM must either be satisfied with the current M-α-Pareto optimal solution and α or update the reference membership levels and/or the level α. To help the DM specify the reference membership levels appropriately, trade-off rates between a standing membership function and each of the other membership functions as well as between the level α and the membership function are very useful. Such a trade-off rate is easily obtainable since it is closely related to the simplex multipliers of the problem (4.88).

To derive the trade-off information, define the following Lagrangian function L corresponding to the problem (4.88):

$$L = c_{1\alpha}^L x + \sum_{i \in I_1 \cup I_{3R}} \pi_{iR} \{c_{i\alpha}^L x - d_{iR}^{-1}(\hat{\mu}_i - v^*)\}$$
$$+ \sum_{i \in I_2 \cup I_{3L}} \pi_{iL} \{d_{iL}^{-1}(\hat{\mu}_i - v^*) - c_{i\alpha}^R x\} + \sum_{j=1}^m \lambda_j (a_{j\alpha}^L x - b_{j\alpha}^R), \quad (4.91)$$

where π_{iL}, π_{iR}, and λ_j are simplex multipliers corresponding to the constraints of (4.88).

Here we assume that the problem (4.88) has a unique and nondegenerate optimal solution satisfying the following conditions:

(i) $\pi_{iR} > 0$ for $i \in I_1 \cup I_{3R}, i \neq 1$,
(ii) $\pi_{iL} > 0$ for $i \in I_2 \cup I_{3L}$.

Then by using the results in Haimes and Chankong (1979), the following expression holds:

$$-\frac{\partial(c_{1\alpha}^L x)}{\partial(c_{i\alpha}^L x)} = \pi_{iR}, \ i \in I_1 \cup I_{3R}, \ i \neq 1, \quad (4.92)$$

$$-\frac{\partial(c_{1\alpha}^L x)}{\partial(c_{i\alpha}^R x)} = -\pi_{iL}, \ i \in I_2 \cup I_{3L}. \quad (4.93)$$

Furthermore, using the strictly monotone decreasing or increasing property of $d_{iR}(\cdot)$ or $d_{iL}(\cdot)$ together with the chain rule, if $d_{iR}(\cdot)$ and $d_{iL}(\cdot)$ are differentiable at the optimal solution to (4.88), it holds that

$$-\frac{\partial \mu_1(c_{1\alpha}^L x)}{\partial \mu_i(c_{i\alpha}^L x)} = \frac{d'_{1R}(c_{1\alpha}^L x)}{d'_{iR}(c_{i\alpha}^L x)} \pi_{iR}, \ i \in I_1 \cup I_{3R}, \ i \neq 1, \quad (4.94)$$

$$-\frac{\partial \mu_1(c_{1\alpha}^L x)}{\partial \mu_i(c_{i\alpha}^R x)} = -\frac{d'_{1R}(c_{1\alpha}^L x)}{d'_{iL}(c_{i\alpha}^R x)} \pi_{iL}, \ i \in I_2 \cup I_{3L}, \quad (4.95)$$

where $d'_{iR}(\cdot)$ and $d'_{iL}(\cdot)$ denote the differential coefficients of $d_{iR}(\cdot)$ and $d_{iL}(\cdot)$, respectively.

Regarding a trade-off rate between $\mu_1(c_{1\alpha}^L x)$ and α, the following relation holds based on the sensitivity theorem [for details, see, e.g., Luenberger (1973, 1984, 2008) or Fiacco (1983)]:

4.5 Interactive Fuzzy Linear Programming with Fuzzy Parameters

$$\frac{\partial \mu_1(c_{1\alpha}^L x)}{\partial \alpha} = d'_{1R}(c_{1\alpha}^L x) \left\{ \frac{\partial(c_{1\alpha}^L)}{\partial \alpha} x + \sum_{i \in I_1 \cup I_{3R}} \pi_{iR} \frac{\partial(c_{i\alpha}^L)}{\partial \alpha} x \right.$$

$$\left. - \sum_{i \in I_2 \cup I_{3L}} \pi_{iL} \frac{\partial(c_{i\alpha}^R)}{\partial \alpha} x + \sum_{j=1}^{m} \lambda_j \left\{ \frac{\partial(a_{j\alpha}^L)}{\partial \alpha} x - \frac{\partial(b_{j\alpha}^R)}{\partial \alpha} \right\} \right\}. \quad (4.96)$$

It should be noted that to obtain the trade-off rates in (4.94) and (4.95), all the constraints of the problem (4.88) must be active for the current optimal solution. Therefore, if there are inactive constraints, it is necessary to replace $\hat{\mu}_j$ for the inactive constraints with $d_{iR}(c_{i\alpha}^L x^*) + v^*$ or $d_{iL}(c_{i\alpha}^R x^*) + v^*$ and solve the corresponding problem (4.88) for obtaining the simplex multipliers.

Now, following the above discussions, we can present the interactive algorithm to derive the satisficing solution for the DM from the M-α-Pareto optimal solution set. The steps marked with an asterisk involve interaction with the DM.

Procedure of Interactive Fuzzy Multiobjective Linear Programming with Fuzzy Parameters

Step 0   Calculate the individual minimum and maximum of each objective function under the given constraints for $\alpha = 0$ and $\alpha = 1$.

Step 1*   Elicit a membership function $\mu_i(c_i x)$ from the DM for each of the objective functions.

Step 2*   Ask the DM to select the initial value of α ($0 \leq \alpha \leq 1$) and set the initial reference membership levels to $\hat{\mu}_i = 1, i = 1, \ldots, k$.

Step 3   For the level α and the reference membership levels specified by the DM, solve the minimax problem and perform the M-α-Pareto optimality test to obtain the M-α-Pareto optimal solution and the trade-off rates between the membership functions and the level α.

Step 4*   The DM is supplied with the corresponding M-α-Pareto optimal solution and the trade-off rates between the membership functions and the level α. If the DM is satisfied with the current membership function values of the M-α-Pareto optimal solution and α, stop. Otherwise, the DM must update the reference membership levels and/or the level α by considering the current values of the membership functions and the level α together with the trade-off rates between the membership functions and the level α and return to step 3.

It should be stressed to the DM that (i) any improvement of one membership function can be achieved only at the expense of at least one of the other membership functions for some fixed level α and (ii) the greater value of the degree α gives the worse values of the membership functions for some fixed reference membership levels.

Example 4.7. To clarify the concept of M-α-Pareto optimality as well as the presented method, consider the following three-objective linear programming problem with fuzzy parameters:

fuzzy max $z_1 = 3x_1 + C_{12}x_2$
fuzzy min $z_2 = C_{21}x_1 + 4x_2$
fuzzy equal $z_3 = C_{31}x_1 - 8x_2$
subject to $x \in X = \{(x_1, x_2) \mid 2x_1 + 6x_2 \leq 27,\ 3x_1 + 2x_2 \leq 16,$
             $\qquad\qquad 4x_1 + x_2 \leq 18,\ x_i \geq 0,\ i = 1, 2\}$

where C_{12}, C_{21}, and C_{31} are fuzzy numbers whose membership functions are given by

$$\mu_{C_{12}}(c_{12}) = \max(1 - 2|c_{12} - 8|, 0)$$
$$\mu_{C_{21}}(c_{21}) = \max(1 - |c_{21} - 5|, 0)$$
$$\mu_{C_{31}}(c_{31}) = \max(1 - 2|c_{31} - 3|, 0).$$

Now, for illustrative purposes, suppose that the DM specifies the following simple linear membership functions for the three-objective functions:

$$\mu_1(z_1) = \begin{cases} 1 & \text{if } z_1 \leq 18 \\ d_{1R}(z_1) = (z_1 - 18)/8 & \text{if } 18 \leq z_1 \leq 26 \\ 0 & \text{if } 26 \leq z_1 \end{cases}$$

$$\mu_2(z_2) = \begin{cases} 0 & \text{if } z_2 \leq 18 \\ d_{2L}(z_2) = (25 - z_2)/7 & \text{if } 18 \leq z_2 \leq 25 \\ 1 & \text{if } 25 \leq z_2 \end{cases}$$

$$\mu_3(z_3) = \begin{cases} 0 & \text{if } z_3 \leq -17 \\ d_{3L}(z_3) = (z_3 + 17)/17 & \text{if } -17 \leq z_3 \leq 0 \\ 1 & \text{if } z_3 = 0 \\ d_{3R}(z_3) = (8 - z_3)/8 & \text{if } 0 \leq z_3 \leq 8 \\ 0 & \text{if } 8 \leq z_3. \end{cases}$$

Also assume that the DM selects the initial value of the degree α to be 0.5 and the initial reference membership levels $(\hat{\mu}_1, \hat{\mu}_2, \hat{\mu}_3)$ to be $(1, 1, 1)$. The corresponding M-α-Pareto optimal solution can be obtained by solving the following problem:

$$\begin{aligned}
\text{minimize} \quad & v \\
\text{subject to} \quad & 3x_1 + 8.25x_2 \leq d_{1L}^{-1}(\hat{\mu}_1 - v) \\
& 4.5x_1 + 4x_2 \geq d_{2R}^{-1}(\hat{\mu}_2 - v) \\
& 3.25x_1 - 8x_2 \leq d_{3L}^{-1}(\hat{\mu}_3 - v) \\
& 2.75x_1 - 8x_2 \geq d_{3R}^{-1}(\hat{\mu}_3 - v) \\
& x \in X.
\end{aligned}$$

4.5 Interactive Fuzzy Linear Programming with Fuzzy Parameters

Solving this problem by the combined use of the bisection method and the simplex method of linear programming, we obtain the optimal value $v^* = 0.30590$. To obtain the corresponding optimal solution of the decision variable x^*, we solve the following linear programming problem for $v^* = 0.33774$:

$$\begin{aligned}
\text{minimize} \quad & 3x_1 + 8.25x_2 \\
\text{subject to} \quad & 4.5x_1 + 4x_2 \geq d_{2R}^{-1}(\hat{\mu}_2 - v^*) \\
& 3.25x_1 - 8x_2 \leq d_{3L}^{-1}(\hat{\mu}_3 - v^*) \\
& 2.75x_1 - 8x_2 \geq d_{3R}^{-1}(\hat{\mu}_3 - v^*) \\
& x \in X.
\end{aligned}$$

As a result, we get the following optimal solution and the related values x^*, $z_1^* = z_1(x^*, c_{1\alpha}^R)$, $z_2^* = z_2(x^*, c_{2\alpha}^L)$, $z_3^* = z_3(x^*, c_{3\alpha}^R)$, $\mu_i^* = \mu_i(z_i^*)$, $i = 1,\ldots,k$, and the simplex multipliers $(\pi_{2L}^*, \pi_{3L}^*, \pi_{3R}^*)$:

$$(x_1^*, x_2^*) = (2.86386, 1.81348),$$

$$(z_1^*, z_2^*, z_3^*) = (23.55280, 20.14130, -5.20030),$$

$$(\mu_1^*, \mu_2^*, \mu_3^*) = (0.6941, 0.6941, 0.6941),$$

$$(\pi_{2L}^*, \pi_{3L}^*, \pi_{3R}^*) = (1.036990, -0.512755, 0).$$

From (4.94) and (4.95), the trade-off rates between the membership functions are

$$-\frac{\partial \mu_1(z_1^*)}{\partial \mu_2(z_2^*)} = -\frac{d'_{1L}(z_1^*)}{d'_{2R}(z_2^*)} \pi_{2R}^* = -\frac{1/8}{-1/7} \cdot 1.036990 = 0.907366,$$

$$-\frac{\partial \mu_1(z_1^*)}{\partial \mu_3(z_3^*)} = -\frac{d'_{1L}(z_1^*)}{d'_{3L}(z_3^*)} \pi_{3L}^* = -\frac{1/8}{1/17} \cdot (-0.512755) = 1.089605.$$

Concerning the trade-off rate between $\mu_1(z_1)$ and α, from (4.96), we have

$$\frac{\partial \mu_1(z_1^*)}{\partial \alpha} = d'_{1L}(z_1^*) \left\{ \frac{\partial c_{1\alpha}^R}{\partial \alpha} x_2^* + \pi_{2R}^* \frac{\partial c_{2\alpha}^L}{\partial \alpha} x_1^* - \pi_{3L}^* \frac{\partial c_{3\alpha}^R}{\partial \alpha} x_1^* \right\}$$

$$= (1/8)\{(-2) \cdot 1.81348 + 1.036990 \cdot 1 \cdot 2.86386 - 0.512755 \cdot 2 \cdot 2.86386\}$$

$$= -0.449261.$$

Suppose that the DM updates the reference membership levels to $(\hat{\mu}_1, \hat{\mu}_2, \hat{\mu}_3) = (0.72, 0.72, 0.67)$ in order to improve the satisfaction levels for the total profit and the amount of pollution at the expense of the ratio of the two products. After solving similar problems, we get the following optimal solution and the related values:

$$(x_1^*, x_2^*) = (2.798925, 1.854721),$$

$$(z_1^*, z_2^*, z_3^*) = (23.698228, 20.014050, -5.741265),$$

$$(\mu_1^*, \mu_2^*, \mu_3^*) = (0.712279, 0.712279, 0.662279),$$

$$(\pi_{2R}^*, \pi_{3L}^*, \pi_{3R}^*) = (1.036990, -0.512755, 0).$$

Similarly, the trade-off rates can be calculated as

$$-\frac{\partial \mu_1(z_1^*)}{\partial \mu_2(z_2^*)} = -\frac{d'_{1L}(z_1^*)}{d'_{2R}(z_2^*)} \pi_{2R}^* = -\frac{(1/8)}{(-1/7)} 1.036990 = 0.907366,$$

$$-\frac{\partial \mu_1(z_1^*)}{\partial \mu_3(z_3^*)} = -\frac{d'_{1L}(z_1^*)}{d'_{3L}(z_3^*)} \pi_{3L}^* = -\frac{(1/8)}{(1/17)} (-0.512755) = 1.089605,$$

$$\frac{\partial \mu_1(z_1^*)}{\partial \alpha} = d'_{1L}(z_1^*) \left\{ \frac{\partial c_{1\alpha}^R}{\partial \alpha} x_2^* + \pi_{2R}^* \frac{\partial c_{2\alpha}^L}{\partial \alpha} x_1^* - \pi_{3L}^* \frac{\partial c_{3\alpha}^R}{\partial \alpha} x_1^* \right\}$$

$$= \frac{1}{8} \{(-2) \cdot 1.854721 + 1.036990 \cdot 1 \cdot 2.798925 - 0.512755 \cdot 2 \cdot 2.798925\}$$

$$= -0.459664.$$

As presented above, observe that the DM can obtain a satisficing solution from an M-α-Pareto optimal solution set by updating the reference membership levels and/or the level α on the basis of the current values of the membership functions and α together with the trade-off rates between the membership functions and the level α. ◊

It is significant to point out here that all the results presented in this chapter have already been extended to deal with multiobjective linear fractional (Sakawa and Yano 1985b, 1988) and nonlinear programming (Sakawa and Yano 1985a,c, 1986c,d, 1987, 1989) problems. For further details of multiobjective linear and nonlinear programming in a fuzzy environment, the readers might refer to Sakawa's 1993 book entitled "Fuzzy Sets and Interactive Multiobjective Optimization" (Sakawa 1993). In addition to this book, the books of Lai and Hwang (1992) and Carlsson and Fullér (2002) and the two books of Sakawa (2000, 2001) together with the edited volumes of Kacprzyk and Orlovski (1987), Verdegay and Delgado (1989), Słowìnski and Teghem (1990), Delgado et al. (1994), Słowìnski (1998), Ehrgott and Gandibleux (2002), and Kahraman (2008) provide helpful information for interested readers.

Problems

4.1 Prove that

$$(A \cap B)_\alpha = A_\alpha \cap B_\alpha$$

holds.

4.2 Prove that

$$\mu_G(x)\mu_C(x) \leq \min\{\mu_G(x), \mu_C(x)\} \leq \alpha\mu_G(x) + (1-\alpha)\mu_C(x), \ 0 \leq \alpha \leq 1$$

holds.

4.3 For the linear programming problem

$$\begin{aligned}
\text{minimize} \quad & z = -x_1 - 2x_2 \\
\text{subject to} \quad & 2x_1 + 6x_2 \leq 27 \\
& 8x_1 + 6x_2 \leq 45 \\
& 3x_1 + x_2 \leq 15 \\
& x_1 \geq 0, \ x_2 \geq 0,
\end{aligned}$$

suppose that the DM specifies linear membership functions for the fuzzy goal and constraints as follows:

fuzzy goal: $\mu_0(-9.5) = 0, \ \mu_0(-10.5) = 1$
fuzzy constraint 1: $\mu_1(30) = 0, \ \mu_1(27) = 1$
fuzzy constraint 2: $\mu_2(50) = 0, \ \mu_2(45) = 1$
fuzzy constraint 3: $\mu_3(17) = 0, \ \mu_3(15) = 1$.

Calculate an optimal solution and compare with the nonfuzzy solution.

4.4 For the linear programming problem

$$\begin{aligned}
\text{minimize} \quad & z = -5x_1 - 5x_2 \\
\text{subject to} \quad & 5x_1 + 7x_2 \leq 12 \\
& 9x_1 + x_2 \leq 10 \\
& -5x_1 + 3x_2 \leq 3 \\
& x_1 \geq 0, \ x_2 \geq 0,
\end{aligned}$$

suppose that the DM specifies linear membership functions for the fuzzy goal and constraints as follows:

fuzzy goal: $\mu_0(-8) = 0, \ \mu_0(-12) = 1$
fuzzy constraint 1: $\mu_1(15) = 0, \ \mu_1(12) = 1$
fuzzy constraint 2: $\mu_2(12) = 0, \ \mu_2(10) = 1$
fuzzy constraint 3: $\mu_3(4) = 0, \ \mu_3(3) = 1$.

Calculate an optimal solution and compare with the nonfuzzy solution.

4.5 In Example 4.5, when the linear membership function is determined by (4.53) and (4.54), obtain an optimal solution.

4.6 For the multiobjective linear programming problem

$$\begin{align}
\text{minimize} \quad & z_1 = -x_1 - 2x_2 \\
\text{minimize} \quad & z_2 = 3x_1 + 2x_2 \\
\text{subject to} \quad & 2x_1 + 6x_2 \leq 27 \\
& 8x_1 + 6x_2 \leq 45 \\
& 3x_1 + x_2 \leq 15 \\
& x_1 \geq 0, \; x_2 \geq 0,
\end{align}$$

suppose that the DM specifies linear membership functions for the fuzzy goals of the objective functions $z_1(x)$ and $z_2(x)$ as follows:

fuzzy goal of $z_1(x)$: $\mu_1(-8) = 0$, $\mu_1(-10) = 1$,
fuzzy goal of $z_2(x)$: $\mu_2(14) = 0$, $\mu_2(9) = 1$.

Calculate an optimal solution.

4.7 For the multiobjective linear programming problem

$$\begin{align}
\text{minimize} \quad & z_1 = -5x_1 - 5x_2 \\
\text{minimize} \quad & z_2 = 5x_1 + x_2 \\
\text{subject to} \quad & 5x_1 + 7x_2 \leq 12 \\
& 9x_1 + x_2 \leq 10 \\
& -5x_1 + 3x_2 \leq 3 \\
& x_1 \geq 0, \; x_2 \geq 0,
\end{align}$$

suppose that the DM specifies linear membership functions for the fuzzy goals of the objective functions $z_1(x)$ and $z_2(x)$ as follows:

fuzzy goal of $z_1(x)$: $\mu_1(-5) = 0$, $\mu_1(-10) = 1$,
fuzzy goal of $z_2(x)$: $\mu_2(5) = 0$, $\mu_2(1) = 1$.

Calculate an optimal solution.

4.8 Prove Theorems 4.2 and 4.3.

4.9 For the fuzzy multiobjective linear programming problem

$$\begin{align}
\text{fuzzy max} \quad & x_1 + 2x_2 \\
\text{fuzzy min} \quad & 3x_1 + 2x_2 \\
\text{fuzzy equal} \quad & x_1 - 2x_2 \\
\text{subject to} \quad & 2x_1 + 6x_2 \leq 27 \\
& 8x_1 + 6x_2 \leq 45 \\
& 3x_1 + x_2 \leq 15 \\
& x_1 \geq 0, \; x_2 \geq 0,
\end{align}$$

4.5 Interactive Fuzzy Linear Programming with Fuzzy Parameters

suppose that the DM specifies linear membership functions for the fuzzy goals of the three objective functions as follows:

fuzzy max $x_1 + 2x_2$: $\mu_1(5) = 0$, $\mu_1(8) = 1$,
fuzzy min $3x_1 + 2x_2$: $\mu_2(14) = 0$, $\mu_2(9) = 1$,
fuzzy equal $x_1 - 2x_2$: $\mu_{3L}(-7) = 0$, $\mu_{3L}(0) = \mu_{3R}(0) = 1$, $\mu_{3R}(4) = 0$.

Employ the interactive fuzzy multiobjective linear programming.

4.10 For the fuzzy multiobjective linear programming problem

$$
\begin{aligned}
\text{fuzzy max} \quad & 5x_1 + 5x_2 \\
\text{fuzzy min} \quad & 5x_1 + x_2 \\
\text{fuzzy equal} \quad & 3x_1 - 8x_2 \\
\text{subject to} \quad & 5x_1 + 7x_2 \leq 12 \\
& 9x_1 + x_2 \leq 10 \\
& -5x_1 + 3x_2 \leq 3 \\
& x_1 \geq 0, \ x_2 \geq 0,
\end{aligned}
$$

suppose that the DM specifies linear membership functions for the fuzzy goals of the three objective functions as follows:

fuzzy max $5x_1 + 5x_2$: $\mu_1(0) = 0$, $\mu_1(10) = 1$,
fuzzy min $5x_1 + x_2$: $\mu_2(6) = 0$, $\mu_2(1) = 1$,
fuzzy equal $3x_1 - 8x_2$: $\mu_{3L}(-10) = 0$, $\mu_{3L}(0) = \mu_{3R}(0) = 1$, $\mu_{3R}(3) = 0$.

Employ the interactive fuzzy multiobjective linear programming.

4.11 Prove Proposition 4.1.
4.12 Prove Theorems 4.4 and 4.5.
4.13 For the fuzzy multiobjective linear programming problem with fuzzy parameters

$$
\begin{aligned}
\text{fuzzy max} \quad & 5x_1 + C_{12}x_2 \\
\text{fuzzy min} \quad & C_{21}x_1 + x_2 \\
\text{fuzzy equal} \quad & C_{31}x_1 - 8x_2 \\
\text{subject to} \quad & 5x_1 + 7x_2 \leq 12 \\
& 9x_1 + x_2 \leq 10 \\
& -5x_1 + 3x_2 \leq 3 \\
& x_1 \geq 0, \ x_2 \geq 0,
\end{aligned}
$$

where C_{12}, C_{21}, and C_{31} are fuzzy numbers whose membership functions are given by

$$\mu_{C_{12}}(c_{12}) = \max\left(1 - |c_{12} - 5|, 0\right)$$
$$\mu_{C_{21}}(c_{21}) = \max\left(1 - |c_{21} - 5|, 0\right)$$
$$\mu_{C_{31}}(c_{31}) = \max\left(1 - 2|c_{31} - 3|, 0\right),$$

suppose that the DM specifies linear membership functions for the fuzzy goals of the three objective functions as follows:

fuzzy max $5x_1 + C_{12}x_2$: $\quad \mu_1(0) = 0, \quad \mu_1(10) = 1,$
fuzzy min $C_{21}x_1 + x_2$: $\quad \mu_2(6) = 0, \quad \mu_2(1) = 1,$
fuzzy equal $C_{31}x_1 - 8x_2$: $\quad \mu_{3L}(-10) = 0, \mu_{3L}(0) = \mu_{3R}(0) = 1, \mu_{3R}(3) = 0.$

Employ the interactive fuzzy multiobjective linear programming.

Chapter 5
Stochastic Linear Programming

In this chapter, after overviewing elementary probability, two-stage programming and chance constrained programming are explained in detail. In two-stage programming, a shortage or an excess arising from the violation of the constraints is penalized, and then the expectation of the amount of the penalties for the constraint violation is minimized. In contrast, chance constrained programming admits random data variations and permits constraint violations up to specified probability limits, and its formulation is somewhat variable, including the expectation model, the variance model, the probability model, and the fractile model.

5.1 Elementary Probability

Suppose we perform a random experiment, such as flipping coins or rolling dice, with an uncertain outcome. The sample space S is defined as the set of all possible outcomes. An event A is any subset of the possible outcomes.

We can derive new events from the union or intersection of the events A and B. The union $A \cup B$ of events A and B consists of all events in A or B or both. The intersection $A \cap B$ of events A and B consists of all events that are in both A and B.

If the intersection of two events is the null (empty) set, $A \cap B = \emptyset$, then the two events A and B are called mutually exclusive. The complement A^c of an event A consists of all events in the space S not A.

Many common occurrences are considered to be equally likely: the flip of an unbiased coin, the spin of a well-balanced roulette wheel, the roll of an unweighted die (fair die), and the selection of a card from a well-shuffled deck. This suggests the following definition of probability.

Definition 5.1 (Probability). If the sample space S of an experiment consists of finitely many outcomes that are equally likely, then the probability $P(A)$ of an event A is

$$P(A) = \frac{\text{number of outcomes in event } A}{\text{number of outcomes in space } S}. \tag{5.1}$$

Thus, in particular, $P(S) = 1$.

This quite simple definition can be extended to experiments in which equally likely outcomes are not available. Naturally, the extended definition should include Definition 5.1.

Definition 5.2 (Probability). Given a sample space S, with each event A of S (subset of S), there is associated a number $P(A)$, called the probability of A, such that the following three axioms of probability are satisfied.

Axiom 1 The probability of an event must be between 0 and 1, i.e., for every A in S,

$$0 \leq P(A) \leq 1. \tag{5.2}$$

Axiom 2 The probability of the sample space must equal 1, i.e., the entire sample space S has the probability

$$P(S) = 1. \tag{5.3}$$

Axiom 3 The probability of the union of mutually exclusive events is the sum of the event probabilities. For mutually exclusive events A and B, i.e., $A \cap B = \emptyset$,

$$P(A \cup B) = P(A) + P(B). \tag{5.4}$$

If S is infinite (has infinitely many events), Axiom 3 has to be replaced as follows. For any sequence of mutually exclusive events A_i, $i = 1, 2, \ldots$ such that $A_i \cap A_j = \emptyset$ for $i \neq j$,

$$P\left(\bigcup_{i=1}^{\infty} A_i\right) = \sum_{i=1}^{\infty} P(A_i). \tag{5.5}$$

The consequences of probability theory result from these axioms; we state several of the important results without proof.

The probability of the complementary event is one minus the probability of the event itself, i.e.,

$$P(A^c) = 1 - P(A). \tag{5.6}$$

The probability of the union of two events is the sum of the event probabilities less the probability of their intersection, i.e.,

5.1 Elementary Probability

$$P(A \cup B) = P(A) + P(B) - P(A \cap B). \tag{5.7}$$

It is often required to find the probability of an event A under the condition that an event B occurs. This probability, denoted by $P(A|B)$, is called the conditional probability of A given B. In this case B serves as a new (reduced) sample space, and that probability is the fraction of $P(B)$ which corresponds to $A \cap B$. Thus

$$P(A|B) = \frac{P(A \cap B)}{P(B)}. \tag{5.8}$$

This equation can be rewritten as

$$P(A \cap B) = P(A|B)P(B). \tag{5.9}$$

In words, the probability of the intersection of two events is the product of the conditional probability of one event given the conditioned event and the probability of the conditioned event.

Two events A and B are called independent when the conditional probability $P(A|B)$ is equal to $P(A)$ alone. In other word, if

$$P(A \cap B) = P(A)P(B), \tag{5.10}$$

two events A and B are called independent events. This means that the probability of A does not depend on the occurrence or nonoccurrence of B, and conversely. This justifies the term independent. Similarly, n events $A_1, A_2, \ldots, A_n$ are called independent if

$$P(A_1 \cap \cdots \cap A_n) = P(A_1) \cdots P(A_n). \tag{5.11}$$

The following theorem is well known as the total probability formula.

Theorem 5.1 (Total probability formula). *Given a sequence of events $A_1, A_2, \ldots, A_n$, if the events $A_1, A_2, \ldots, A_n$ are mutually exclusive ($A_i \cap A_j = \emptyset$ for all $i \neq j$) and their union is S ($A_1 \cup A_2 \cup \cdots \cup A_n = S$), then for any event B in S*

$$P(B) = \sum_{i=1}^{n} P(A_i)P(B|A_i), \tag{5.12}$$

where $P(A_i) > 0$, $i = 1, \ldots, n$.

This observation leads to Bayes' theorem, which is used to calculate conditional probabilities in many important situations.

Theorem 5.2 (Bayes' theorem). *Given a sequence of events $A_1, A_2, \ldots, A_n$, if the events $A_1, A_2, \ldots, A_n$ are mutually exclusive and their union is S, then for any event B in S*

$$P(A_i|B) = \frac{P(A_i)P(B|A_i)}{\sum_{i=1}^{n} P(A_i)P(B|A_i)}. \tag{5.13}$$

It is convenient to consider a real-valued function on a sample space. Such a real-valued function X defined on a sample space S is called a random variable.[1]

The distribution function of the random variable X, usually denoted by $F(x)$, is the function that gives the probability of X being less than or equal to any real number x. That is,

$$F(x) = P(X \leq x). \tag{5.14}$$

A random variable X is called discrete if its set of all possible values is countable. For discrete random variables, the probability mass function $p(x_i)$ is defined by

$$p(x_i) = P(X = x_i), \tag{5.15}$$

such that $p(x_i) \geq 0$, $i = 1, 2, \ldots$ and $\sum_{i=1}^{\infty} p(x_i) = 1$. The distribution function of the discrete random variable X is then given by

$$F(x) = P(X \leq x) = \sum_{y \leq x} p(y). \tag{5.16}$$

A random variable X called continuous if there exists a function $f(x)$, called the density function, such that

$$F(x) = P(X \leq x) = \int_{-\infty}^{x} f(y) dy, \tag{5.17}$$

where $\int_{-\infty}^{\infty} f(y) dy = 1$.

Differentiation of the above equation gives

$$\frac{dF(x)}{dx} = f(x) \tag{5.18}$$

which implies the density function is the derivative of the distribution function.

---

[1] Although the term random variable is somewhat confusing, it is well established and so it will be used in this book. However, a random function would be more appropriate, since a random variable is really a real-valued function whose domain is a sample space.

5.1 Elementary Probability

For a discrete random variable X, the expectation or mean is defined by

$$E[X] = \sum_i x_i p(x_i), \qquad (5.19)$$

provided the sum exists.

Similarly, for an continuous random variable X, it is defined by

$$E[X] = \int_{-\infty}^{\infty} x f(x) dx, \qquad (5.20)$$

provided the integral exists.

The variance of the random variable X is defined by

$$\begin{aligned} Var[X] &= E[(X - E[X])^2] \\ &= E[X^2] - E[2E[X]X] + (E[X])^2 \\ &= E[X^2] - (E[X])^2, \end{aligned} \qquad (5.21)$$

and the standard deviation of the random variable X is also defined by the positive root of the variance, $\sqrt{Var[X]}$.

The standardized random variable Y corresponding to X is defined by

$$Y = \frac{X - E[X]}{\sqrt{Var[X]}}, \qquad (5.22)$$

where $E[Y] = 0$ and $Var[Y] = 1$. The normal distribution, also known as Gauss distribution, is the most important continuous distribution and plays a central role in statistics. The normal distribution or Gauss distribution is defined as the probability distribution with the density function

$$f(x) = \frac{1}{\sqrt{2\pi}\sigma} e^{-\frac{(x-m)^2}{2\sigma^2}}, \qquad (5.23)$$

and its expectation and variance are

$$E[X] = \int_{-\infty}^{\infty} x f(x) dx = m \qquad (5.24)$$

and

$$Var[X] = E[X^2] - (E[X])^2 = \sigma^2, \qquad (5.25)$$

respectively. Such a normal distribution is frequently denoted by $X \sim N(m, \sigma^2)$. Since the exponent contains $(x - m)^2$, the curve of $f(x)$ is symmetric with respect to $x = m$ and bell-shaped.

From the density function $f(x)$, by integrating from $-\infty$ to x, the distribution function $F(x)$ of the normal distribution is given by

$$F(x) = \frac{1}{\sqrt{2\pi}\sigma} \int_{-\infty}^{x} e^{-\frac{(y-m)^2}{2\sigma^2}} dy. \qquad (5.26)$$

Associated with the normal random variable X, it is significant to introduce the standardized normal random variable Z given by

$$Z = \frac{X - E[X]}{\sqrt{Var[X]}} = \frac{X - m}{\sigma}, \qquad (5.27)$$

with $E[Z] = 0$ and $Var[Z] = 1$. Then it immediately follows that

$$\phi(z) = \frac{1}{\sqrt{2\pi}} e^{-\frac{z^2}{2}} \qquad (5.28)$$

is the density function and its integral

$$\Phi(z) = \int_{-\infty}^{z} \phi(y) dy = \frac{1}{\sqrt{2\pi}} \int_{-\infty}^{z} e^{-\frac{1}{2}y^2} dy \qquad (5.29)$$

is the distribution function of the standardized normal distribution $Z \sim N(0, 1)$. Values of the distribution function $\Phi(z)$ are well tabulated. Such tables can be found in most statistics textbooks.

The random variables considered thus far are functions that assign a real value to each event. If a two-dimensional real space is considered, the joint distribution of two random variables X and Y is defined by

$$F(x, y) = P(X \leq x, Y \leq y). \qquad (5.30)$$

The marginal distribution of the random variable X is defined by

$$F(x) = P(X \leq x) = \lim_{y \to \infty} P(X \leq x, Y \leq y) = F(x, \infty). \qquad (5.31)$$

The marginal distribution function of Y is similarly defined by $F(y) = F(\infty, y)$. The random variables X and Y are called independent if

$$F(x, y) = F(x)F(y) \text{ for all real } x \text{ and } y. \qquad (5.32)$$

If the two random variables are discrete, the joint probability mass function of the two random variables is defined by

$$p(x, y) = P(X = x, Y = y), \qquad (5.33)$$

5.1 Elementary Probability

and the corresponding joint distribution function is given by

$$F(x, y) = \sum_{a \leq x} \sum_{b \leq y} p(a, b). \tag{5.34}$$

In addition, if

$$p(x, y) = p(x)p(y) \tag{5.35}$$

for all x and y, the two discrete random variables X and Y are independent, where

$$p(x) = \sum_{y} p(x, y) \tag{5.36}$$

is the marginal probability mass function of X, and $p(y)$ is the marginal probability function of Y, which is similarly defined.

If the two random variables are continuous, the joint distribution function is given by

$$F(x, y) = \int_{-\infty}^{x} \int_{-\infty}^{y} f(u, v) du dv, \tag{5.37}$$

where $f(x, y)$ is called the joint density function of X and Y. Obviously, the joint density function

$$f(x, y) = \frac{\partial^2 F(x, y)}{\partial x \partial y} \tag{5.38}$$

can be derived by partial differentiations of x and y. In addition, if

$$f(x, y) = f(x) f(y) \tag{5.39}$$

for all x and y, then two continuous random variables X and Y are independent, where

$$f(x) = \int_{-\infty}^{\infty} f(x, y) dy \tag{5.40}$$

is the marginal density function of X, and $f(y)$ is the marginal density function of Y, which is similarly defined.

Let X and Y be the continuous random variables whose joint density function $f(x, y)$ and marginal density functions $f(x)$ and $f(y)$ are given. Then the conditional density function $f(x|y)$ of X given $Y = y$ is defined by

$$f(x|y) = \frac{f(x,y)}{f(y)}, \qquad (5.41)$$

provided $f(y) > 0$.

The conditional distribution function of X given $Y = y$ is defined by

$$F(x|y) = \int_{-\infty}^{x} \frac{f(x,y)}{f(y)} dx. \qquad (5.42)$$

The conditional expectation of X given $Y = y$ is defined by

$$E[X|Y=y] = \int_{-\infty}^{\infty} x \frac{f(x,y)}{f(y)} dx. \qquad (5.43)$$

From the total probability formula, it follows that

$$f(x) = \int_{-\infty}^{\infty} f(x|y) f(y) dy. \qquad (5.44)$$

Hence the expectation of X is given by

$$E[X] = \int_{-\infty}^{\infty} x f(x) dx$$
$$= \int_{-\infty}^{\infty} \int_{-\infty}^{\infty} x f(x|y) f(y) dx dy$$
$$= E[E[X|Y]] \qquad (5.45)$$

Consider the expectation or mean of $X + Y$. If the two random variables X and Y are discrete, then

$$E[X+Y] = \sum_x \sum_y (x+y) p(x,y) = \sum_x x p(x) + \sum_y y p(y)$$
$$= E[X] + E[Y]. \qquad (5.46)$$

This identity can be similarly verified if the two random variables are continuous, i.e.,

$$E[X+Y] = \int_{-\infty}^{\infty} \int_{-\infty}^{\infty} (x+y) f(x,y) dx dy = E[X] + E[Y]. \qquad (5.47)$$

More generally, if k_1 and k_2 are constants, then

$$E[k_1 X + k_2 Y] = k_1 E[X] + k_2 E[Y]. \qquad (5.48)$$

5.1 Elementary Probability

The variance of $X + Y$ is given by

$$\begin{aligned}
Var[X + Y] &= E[((X + Y) - E[X + Y])^2] \\
&= E[(X - E[X] + Y - E[Y])^2] \\
&= E[(X - E[X])^2] + E[(Y - E[Y])^2] + 2(E[XY] - E[X]E[Y]) \\
&= Var[X] + Var[Y] + 2Cov[X, Y],
\end{aligned} \quad (5.49)$$

where the quantity

$$\begin{aligned}
Cov[X, Y] &= E[(X - E[X])(Y - E[Y])] \\
&= E[XY] - E[X]E[Y]
\end{aligned} \quad (5.50)$$

is called the covariance of the two random variables X and Y. The correlation coefficient of X and Y, which is used as a measure of linear dependence between two random variables, is defined by

$$\rho[X, Y] = \frac{Cov[X, Y]}{\sqrt{Var[X]Var[Y]}}. \quad (5.51)$$

It directly follows that the correlation coefficient is always between -1 and $+1$, i.e., $-1 \leq \rho[X, Y] \leq 1$.

Two jointly distributed random variables X and Y are called uncorrelated if their covariance is zero. If X and Y are independent, $E[XY] = E[X]E[Y]$, then $Cov[X, Y] = 0$, meaning that X and Y are uncorrelated. However, note that the converse is not true.

The discussions thus far can be easily extended to n random variables, namely, for n random variables $X_1, X_2, \ldots, X_n$, the corresponding joint distribution function is defined by

$$F(x_1, x_2, \ldots, x_n) = P(X_1 \leq x_1, X_2 \leq x_2, \ldots, X_n \leq x_n). \quad (5.52)$$

The marginal distribution function of each X_i is defined by

$$F(x_i) = \lim_{\substack{x_j \to \infty \\ j \neq i}} F(x_1, x_2, \ldots, x_n) = F(\infty, \ldots, \infty, x_i, \infty, \ldots, \infty), \; i = 1, 2, \ldots, n. \quad (5.53)$$

The n random variables $X_1, X_2, \ldots, X_n$ are called mutually independent if

$$F(x_1, x_2, \ldots, x_n) = F(x_1)F(x_2) \cdots F(x_n) \text{ for all } x_1, x_2, \ldots, x_n. \quad (5.54)$$

The expectation or mean of the linear combination $k_1 X_1 + k_2 X_2 + \cdots + k_n X_n$ of the n random variables $X_1, X_2, \ldots, X_n$ is

$$E[k_1 X_1 + k_2 X_2 + \cdots + k_n X_n] = k_1 E[X_1] + k_2 E[X_2] + \cdots + k_n E[X_n], \quad (5.55)$$

where $k_1, k_2, \ldots, k_n$ are constants. The variance of $k_1 X_1 + k_2 X_2 + \cdots + k_n X_n$ is

$$Var[k_1 X_1 + k_2 X_2 + \cdots + k_n X_n] = \sum_i \sum_j k_i k_j Cov[X_i, X_j]$$

$$= \sum_{i=1}^{n} k_i^2 Var[X_i] + 2 \sum_{i=1}^{n} \sum_{j=i+1}^{n} k_i k_j Cov[X_i, X_j], \quad (5.56)$$

where $Cov[X_i, X_i] = Var[X_i]$.

Introducing an n-dimensional column vector $\boldsymbol{k} = (k_1, k_2, \ldots, k_n)^T$ of constants, an n-dimensional column vector $\boldsymbol{m} = (m_1, m_2, \ldots, m_n)^T$ of means, an n-dimensional column vector $\boldsymbol{X} = (X_1, X_2, \ldots, X_n)^T$ of random variables, and an $n \times n$ matrix Σ, these two formulae (5.55) and (5.56) can be written in more compact vector–matrix notation as follows:

$$E[\boldsymbol{k}^T \boldsymbol{X}] = \boldsymbol{k}^T \boldsymbol{m} \quad (5.57)$$

$$Var[\boldsymbol{k}^T \boldsymbol{X}] = \boldsymbol{k}^T \Sigma \boldsymbol{k}, \quad (5.58)$$

where the $n \times n$ matrix

$$\Sigma = \begin{bmatrix} Var[X_1] & Cov[X_1, X_2] & \cdots & Cov[X_1, X_n] \\ Cov[X_2, X_1] & Var[X_2] & \cdots & Cov[X_2, X_n] \\ \vdots & \vdots & \ddots & \vdots \\ Cov[X_n, X_1] & Cov[X_n, X_2] & \cdots & Var[X_n] \end{bmatrix} \quad (5.59)$$

is called a variance–covariance matrix. Associated with a variance–covariance matrix, an $n \times n$ matrix

$$R = \begin{bmatrix} 1 & \rho[X_1, X_2] & \cdots & \rho[X_1, X_n] \\ \rho[X_2, X_1] & 1 & \cdots & \rho[X_2, X_n] \\ \vdots & \vdots & \ddots & \vdots \\ \rho[X_n, X_1] & \rho[X_n, X_2] & \cdots & 1 \end{bmatrix} \quad (5.60)$$

is called a correlation matrix.

From the definition of correlation coefficients, using a correlation matrix together with an n-dimensional column vector $\boldsymbol{\sigma} = (\sigma_1, \sigma_2, \ldots, \sigma_n)^T$ of standard deviations of n random variables, a variance–covariance matrix is expressed by

$$\Sigma = \boldsymbol{\sigma}^T R \boldsymbol{\sigma}. \quad (5.61)$$

5.1 Elementary Probability

The multivariate normal distribution or Gauss distribution having the mean vector m and the variance–covariance matrix Σ is defined as the probability distribution with the density function

$$f(x) = \frac{1}{\sqrt{(2\pi)^n \det(\Sigma)}} e^{-\frac{1}{2}(x-m)^T \Sigma^{-1}(x-m)}, \tag{5.62}$$

where $x = (x_1, x_2, \ldots, x_n)^T$ and $\det(\Sigma)$ denote the determinant of Σ.

Some of the most important results in probability theory are in the form of limit theorems. The two most important theorems are presented below without proof.

Theorem 5.3 (Strong law of large numbers). *If $X_1, X_2, \ldots, X_n$ are independent and identically distributed random variables with mean $E[X_i] = m$, then with probability 1,*

$$\frac{X_1 + X_2 + \cdots + X_n}{n} \to m, \text{ as } n \to \infty. \tag{5.63}$$

The strong law of large numbers asserts that the sample mean $(X_1 + X_2 + \cdots + X_n)/n$ tends to the population mean with probability 1 as $n \to \infty$.

Theorem 5.4 (Central limit theorem). *If $X_1, X_2, \ldots, X_n$ are independent and identically distributed random variables with mean $E[X_i] = m$ and variance $Var[X_i] = \sigma^2$, then*

$$P\left(\frac{X_1 + X_2 + \cdots + X_n - nm}{\sigma\sqrt{n}} \leq k\right) \to \frac{1}{\sqrt{2\pi}} \int_{-\infty}^{k} e^{-\frac{x^2}{2}} dx, \text{ as } n \to \infty. \tag{5.64}$$

For the sample mean

$$Y_n = \frac{X_1 + X_2 + \cdots + X_n}{n}, \tag{5.65}$$

since $E[Y_n] = E[(X_1 + X_2 + \cdots + X_n)/n] = m$ and $Var[Y_n] = Var[(X_1 + X_2 + \cdots + X_n)/n] = \sigma^2/n$, each

$$Z_n = \frac{Y_n - m}{\sqrt{Var[Y_n]}} = (X_1 + X_2 + \cdots + X_n - nm)/(\sigma\sqrt{n}) \tag{5.66}$$

is the standardized random variable.

From this fact, the central limit theorem asserts that the sample mean $Y_n = (X_1 + X_2 + \cdots + X_n)/n$ tends toward the normal distribution $Y_n \sim N(m, \frac{\sigma}{\sqrt{n}})$ as $n \to \infty$, or equivalently, the standardized sample mean Z_n tends toward the standardized normal distribution $Z_n \sim N(0, 1)$ as $n \to \infty$.

Further details of probability theory with applications can be found in standard texts including Feller (1968, 1978), Cramér (1999), Chung (1974), Johnson and Kotz (1972), and Kotz et al. (2000).

In this book, henceforth to discriminate a random variable from the other variables or parameters, we attach a bar "–" to a character like $\bar{d}$.

5.2 Two-Stage Programming

In the real-world decision-making problems, some stochastic events may influence elements characterizing decision-making problems such as demands of products and the amount of available resources. When such a decision-making problem under uncertainty is formulated as a linear programming problem, it may be difficult that the constraints of the problem always hold completely. Normally, a shortage or an excess comes from the violation of the constraints, and the corresponding penalties are imposed as the occasion demands. From this point of view, two-stage programming had been investigated from the beginning of the development of linear programming. Two-stage programming, individually introduced by Beale (1955) and Dantzig (1955), was comprehensively discussed by Walkup and Wets (1967) and Wets (1974), and further extended by Beale et al. (1980) and Louveaux (1980) as multistage programming.

To understand the framework of two-stage programming, consider a decision problem of a manufacturing company. Let $x = (x_1, \ldots, x_n)^T$ denote activity levels in a production plant of the company, and then $w = Tx$ denotes the amount of products, where T is an $m_0 \times n$ matrix transforming the n kinds of activity levels into the m_0 types of products. Let $\bar{h}_i$ be the demand for the ith product which is only known in probability, and assume that the random variable $\bar{h}_i$ of the demand is indicated by the distribution function $F_i(h_i) = P(\bar{h}_i \leq h_i)$. Suppose that the decision maker (DM) selects the activity levels, say $x = \hat{x}$, and after the occurrence of the random event the demands $\bar{h} = (\bar{h}_1, \ldots, \bar{h}_{m_0})^T$ are fixed at $\hat{h}$. Then, if $\hat{y}_i^+ = \hat{h}_i - \hat{w}_i \geq 0$, the shortage of the ith product is $\hat{y}_i^+$, and if $\hat{y}_i^- = \hat{w}_i - \hat{h}_i \geq 0$, the excess of the ith product is $\hat{y}_i^-$. This situation can be formulated as

$$Tx + Iy^+ - Iy^- = \bar{h},$$

where y^+ and y^- represent the errors for estimating the demands, and I is the m-dimensional identity matrix. Let q^+ and q^- denote the penalty costs for making these errors, and let c be the original costs for the activities in the production plant. Then, the objective function to be minimized may be the expectation of

$$cx + q^+ y^+ + q^- y^-.$$

Adding the constraints

$$Ax \leq b$$

5.2 Two-Stage Programming

for the activity levels such as the capacity, budget, and technology, we can formulate the standard form of the two-stage programming problem as

$$\left.\begin{array}{rl} \text{minimize} & cx + E\left[q^+ y^+ + q^- y^-\right] \\ \text{subject to} & Tx + Iy^+ - Iy^- = \bar{h} \\ & Ax \leq b \\ & x \geq 0, \end{array}\right\} \quad (5.67)$$

where E means the expectation. In particular, a two-stage programming problem such that the coefficients of y^+ and y^- are identity matrices such as (5.67) is called a simple recourse problem.

For the selected x and the realized values $\hat{h}$, the following solution minimizes the objective function of (5.67) unless $q_i^+ + q_i^-$ is negative, and therefore we assume that $q_i^+ + q_i^- \geq 0, i = 1, \ldots, m_0$:

$$\left.\begin{array}{ll} y_i^+ = \hat{h}_i - \sum_{j=1}^n t_{ij} x_j, \quad y_i^- = 0 & \text{if } \hat{h}_i \geq \sum_{j=1}^n t_{ij} x_j \\ y_i^+ = 0, \qquad y_i^- = \sum_{j=1}^n t_{ij} x_j - \hat{h}_i & \text{if } \hat{h}_i < \sum_{j=1}^n t_{ij} x_j, \end{array}\right\} \quad (5.68)$$

where t_{ij} is the ij element of T.

Assume that the random variables $\bar{h}_i$ with the density functions $f_i(h_i)$, $i = 1, \ldots, m_0$ are mutually independent. From the independence of the random variables and (5.68), we can calculate the second term of the objective function of (5.67) as follows:

$$E\left[q^+ y^+ + q^- y^-\right]$$

$$= \sum_{i=1}^{m_0} q_i^+ \int_{\sum_{j=1}^n t_{ij} x_j}^{\infty} \left(h_i - \sum_{j=1}^n t_{ij} x_j\right) f_i(h_i) dh_i + \sum_{i=1}^{m_0} q_i^- \int_{-\infty}^{\sum_{j=1}^n t_{ij} x_j} \left(\sum_{j=1}^n t_{ij} x_j - h_i\right) f_i(h_i) dh_i$$

$$= \sum_{i=1}^{m_0} q_i^+ \left(E[\bar{h}_i] - \sum_{j=1}^n t_{ij} x_j\right) + \sum_{i=1}^{m_0} (q_i^+ + q_i^-)\left(\sum_{j=1}^n t_{ij} x_j\right) F_i\left(\sum_{j=1}^n t_{ij} x_j\right)$$

$$- \sum_{i=1}^{m_0} (q_i^+ + q_i^-) \int_{-\infty}^{\sum_{j=1}^n t_{ij} x_j} h_i f_i(h_i) dh_i. \quad (5.69)$$

Thus, (5.67) can be transformed into the problem

$$
\begin{aligned}
\text{minimize} \quad & cx + \sum_{i=1}^{m_0} q_i^+ \left(E[\bar{h}_i] - \sum_{j=1}^{n} t_{ij} x_j \right) \\
& + \sum_{i=1}^{m_0} (q_i^+ + q_i^-) \left\{ \left(\sum_{j=1}^{n} t_{ij} x_j \right) F_i \left(\sum_{j=1}^{n} t_{ij} x_j \right) - \int_{-\infty}^{\sum_{j=1}^{n} t_{ij} x_j} h_i f_i(h_i) dh_i \right\} \\
\text{subject to} \quad & Ax \leq b \\
& x \geq 0.
\end{aligned}
$$
(5.70)

Let z_i be the expression in the brace of the third term of the objective function of (5.70), i.e.,

$$z_i = \left(\sum_{j=1}^{n} t_{ij} x_j \right) F_i \left(\sum_{j=1}^{n} t_{ij} x_j \right) - \int_{-\infty}^{\sum_{j=1}^{n} t_{ij} x_j} h_i f_i(h_i) dh_i, \quad (5.71)$$

and then the partial differentiation of z_i for x_j and x_k can be calculated as

$$\frac{\partial z_i}{\partial x_j \partial x_k} = t_{ij} t_{ik} f_i \left(\sum_{j=1}^{n} t_{ij} x_j \right). \quad (5.72)$$

The Hessian matrix for z_i can be written as

$$\nabla^2 z_i(x) = f_i \left(\sum_{j=1}^{n} t_{ij} x_j \right) \begin{bmatrix} t_{i1}^2 & \cdots & t_{i1} t_{in} \\ \vdots & \ddots & \vdots \\ t_{in} t_{i1} & \cdots & t_{in}^2 \end{bmatrix}, \quad (5.73)$$

and since it is positive semidefinite, it follows that z_i is convex. From this fact and linearity of the first and the second terms of the objective function of (5.70), the problem (5.70) is a convex programming problem, and then it can be solved by a conventional convex programming techniques such as the sequential quadratic programming method (Fletcher 1980; Gill et al. 1981; Powell 1983).

Example 5.1. Consider the following production planning problem given in Example 1.1:

$$
\begin{aligned}
\text{minimize} \quad z = & -3x_1 - 8x_2 \\
\text{subject to} \quad & 2x_1 + 6x_2 \leq 27 \\
& 3x_1 + 2x_2 \leq 16 \\
& 4x_1 + 1x_2 \leq 18 \\
& x_1 \geq 0, \ x_2 \geq 0.
\end{aligned}
$$

5.2 Two-Stage Programming

In Example 1.1, we assume that produced goods do not remain unsold, namely, all the products can be sold. From a practical standpoint, however, the demands of products which mean the volumes of sales cannot be estimated exactly, and they might be only known as random variables estimated statistically. Moreover, the lost earnings for the shortage of the products or the holding costs for the excess of them should be taken into account, and then we can formulate the production planning problem in a form of two-stage programming.

Let $\bar{h}_1$ and $\bar{h}_2$ be random variables for the demands of products P_1 and P_2, respectively. Assume that the random variables $\bar{h}_1$ and $\bar{h}_2$ are uniform distributions on the interval $[0, 5]$. Thus, their density functions are $f_1(h_1) = \frac{1}{5}$ and $f_2(h_2) = \frac{1}{5}$, and their distribution functions are $F_1(h_1) = \frac{1}{5}h_1$ and $F_2(h_2) = \frac{1}{5}h_2$. Introducing variables y_i^+ and y_i^- representing the shortage and the excess of product P_i, the following relations hold:

$$x_i + y_i^+ - y_i^- = \bar{h}_i, \ i = 1, 2.$$

Suppose that the lost earnings for one unit of the shortage of products P_1 and P_2 are 5 and 7, respectively, and that the holding costs for one unit of the excess of products P_1 and P_2 are 5 and 4, respectively. Taking the volumes y_i^- of unsold products into account, the total profit becomes $3(x_1 - y_1^-) + 8(x_2 - y_2^-)$. Setting $q_1^+ = 5, q_2^+ = 7$, $q_1^- = 5 + 3 = 8$, and $q_2^- = 4 + 8 = 12$, we can formulate the following two-stage programming problem:

$$\begin{aligned}
\text{minimize} \quad & -3x_1 - 8x_2 + E[5y_1^+ + 7y_2^+ + 8y_1^- + 12y_2^-] \\
\text{subject to} \quad & x_1 + y_1^+ - y_1^- = \bar{h}_1 \\
& x_2 + y_2^+ - y_2^- = \bar{h}_2 \\
& 2x_1 + 6x_2 \leq 27 \\
& 3x_1 + 2x_2 \leq 16 \\
& 4x_1 + 1x_2 \leq 18 \\
& x_1 \geq 0, \ x_2 \geq 0.
\end{aligned}$$

From (5.70), the above two-stage programming problem can be reduced to the following convex programming problem:

$$\begin{aligned}
\text{minimize} \quad & \frac{13}{10}x_1^2 + \frac{19}{10}x_2^2 - 8x_1 - 15x_2 + 30 \\
\text{subject to} \quad & 2x_1 + 6x_2 \leq 27 \\
& 3x_1 + 2x_2 \leq 16 \\
& 4x_1 + 1x_2 \leq 18 \\
& x_1 \geq 0, \ x_2 \geq 0.
\end{aligned}$$

By using the solver in Excel (the Excel solver), we can obtain an optimal solution of this problem. The optimal solution is $(x_1^*, x_2^*) = (2.87868, 3.54044)$, and the optimal value is -11.54724. The total profit $3x_1^* + 8x_2^*$ is 36.95956, and the expectation $E[5y_1^+ + 7y_2^+ + 8y_1^- + 12y_2^-]$ of the objective function is calculated as

$$-11.54724 - (-36.95956) = 25.41232.$$

5.3 Chance Constrained Programming

As discussed in the previous section, in two-stage programming, some penalties are imposed for constraint violations. For decision problems under probabilistic uncertainty, from a different viewpoint, Charnes and Cooper (1959, 1963) proposed chance constrained programming which admits random data variations and permits constraint violations up to specified probability limits.

For a better understanding of chance constrained programming, in the remainder of this section, assuming that some or all of the coefficients in objective functions and constants are random variables, we deal with the stochastic linear programming problem formulated as

$$\left. \begin{array}{l} \text{minimize} \quad \bar{c}x \\ \text{subject to} \quad \bar{A}x \leq \bar{b} \\ \quad x \geq 0, \end{array} \right\} \quad (5.74)$$

where $\bar{c} = (\bar{c}_1, \ldots, \bar{c}_n)$ is an n-dimensional row vector of random variables, $x = (x_1, \ldots, x_n)^T$ is an n-dimensional column vector of decision variables, $\bar{A} = [\bar{a}_{ij}]$ is an $m \times n$ matrix of random variables, and $\bar{b} = (\bar{b}_1, \ldots, \bar{b}_m)^T$ is an m-dimensional column vector of random variables.

Considering that the constraints $\bar{A}x \leq \bar{b}$ of the stochastic linear programming problem (5.74) need not hold almost surely, but they can instead hold with given probabilities, Charnes and Cooper (1959) initiated chance constrained programming. To be more precise, the original m constraints

$$\sum_{j=1}^{n} \bar{a}_{ij} x_j \leq \bar{b}_i, \ i = 1, \ldots, m \quad (5.75)$$

of (5.74) are interpreted as

$$P\left(\sum_{j=1}^{n} \bar{a}_{ij} x_j \leq \bar{b}_i\right) \geq \beta_i, \ i = 1, \ldots, m, \quad (5.76)$$

5.3 Chance Constrained Programming

where P means probability, and $\beta_1, \ldots, \beta_m$ are given probabilities of the extents to which constraint violations are admitted. We refer to β_i as the satisficing probability level in this book. The inequalities (5.76) are called chance constraints meaning that the ith constraint may be violated, but at most $1 - \beta_i$ proportion of the time.

For finding deterministic constraints which are equivalent to the chance constraints, first, assume that only $\bar{b}_i$ in the right-hand side of the chance constraints (5.76) is a random variable and $\bar{a}_{ij}$ is a constant. Hence the notation: $\bar{a}_{ij} = a_{ij}$ is used. Let $F_i(\tau)$ denote the distribution function of $\bar{b}_i$. Since

$$P\left(\sum_{j=1}^{n} a_{ij} x_j \leq \bar{b}_i\right) = 1 - F_i\left(\sum_{j=1}^{n} a_{ij} x_j\right), \quad i = 1, \ldots, m,$$

the chance constraints (5.76) can be rewritten as

$$F_i\left(\sum_{j=1}^{n} a_{ij} x_j\right) \leq 1 - \beta_i, \quad i = 1, \ldots, m. \tag{5.77}$$

Let $K_{1-\beta_i}$ denote the maximum of τ such that $\tau = F_i^{-1}(1 - \beta_i)$, and then the inequality (5.77) can be simply expressed as

$$\sum_{j=1}^{n} a_{ij} x_j \leq K_{1-\beta_i}, \quad i = 1, \ldots, m. \tag{5.78}$$

In particular, if $\bar{b}_i$ is a normal random variable with the mean $m_{\bar{b}_i}$ and the variance $\sigma_{\bar{b}_i}^2$, it follows that

$$P\left(\sum_{j=1}^{n} a_{ij} x_j \leq \bar{b}_i\right) = P\left(\frac{\bar{b}_i - m_{\bar{b}_i}}{\sigma_{\bar{b}_i}} \geq \frac{\sum_{j=1}^{n} a_{ij} x_j - m_{\bar{b}_i}}{\sigma_{\bar{b}_i}}\right)$$

$$= 1 - \Phi\left(\frac{\sum_{j=1}^{n} a_{ij} x_j - m_{\bar{b}_i}}{\sigma_{\bar{b}_i}}\right), \tag{5.79}$$

where Φ is the distribution function of the standard normal distribution $N(0, 1)$. Thus, the chance constraints (5.76) can be transformed into

$$\sum_{j=1}^{n} a_{ij} x_j \leq m_{\bar{b}_i} + \sigma_{\bar{b}_i} \Phi^{-1}(1 - \beta_i), \quad i = 1, \ldots, m, \tag{5.80}$$

where Φ^{-1} is the inverse function of Φ.

Example 5.2. Consider the following constraints, where each of the right-hand sides is a normal random variable:

$$2x_1 + 6x_2 \leq \overline{27}$$
$$3x_1 + 2x_2 \leq \overline{16}$$
$$4x_1 + x_2 \leq \overline{18}.$$

For the above normal random variables, assume that $\overline{27} \sim N(27, 3^2)$, $\overline{16} \sim N(16, 2^2)$, and $\overline{18} \sim N(18, 2^2)$. If the DM sets the corresponding satisficing probability levels as $\beta_i = 0.8$, $i = 1, 2, 3$, from (5.80), the above chance constraints can be expressed by the following ordinary linear inequalities without random variables:

$$2x_1 + 6x_2 \leq 27 + 3\Phi^{-1}(0.2) = 24.475$$
$$3x_1 + 2x_2 \leq 16 + 2\Phi^{-1}(0.2) = 14.317$$
$$4x_1 + x_2 \leq 18 + 2\Phi^{-1}(0.2) = 16.317,$$

where Φ^{-1} is the inverse function of the distribution function Φ of the standard normal distribution $N(0, 1)$. ◇

Second, consider a more general case where not only $\bar{b}_i$ but also $\bar{a}_{ij}$ in the left-hand side of (5.75) are random variables, and specifically we assume that $\bar{b}_i$ and $\bar{a}_{ij}$ are normal random variables. Let $m_{\bar{b}_i}$ and $\sigma^2_{\bar{b}_i}$ be the mean and the variance of $\bar{b}_i$, respectively. Also let $m_{\bar{a}_{ij}}$ be the mean of $\bar{a}_{ij}$ and $V_{\bar{a}_i}$ be the variance–covariance matrix of the vector $\bar{a}_i = (\bar{a}_{i1}, \ldots, \bar{a}_{in})$. Moreover, assume that $\bar{b}_i$ and $\bar{a}_{ij}$ are independent of each other.

Since the random variable

$$\frac{\bar{b}_i - \sum_{j=1}^n \bar{a}_{ij} x_j - \left(m_{\bar{b}_i} - \sum_{j=1}^n m_{\bar{a}_{ij}} x_j \right)}{\sqrt{\sigma^2_{\bar{b}_i} + x^T V_{\bar{a}_i} x}}, \quad i = 1, \ldots, m \qquad (5.81)$$

is the standard normal random variable $N(0, 1)$ with mean 0 and variance 1^2, it follows that

$$P\left(\sum_{j=1}^n \bar{a}_{ij} x_j \leq \bar{b}_i \right)$$

5.3 Chance Constrained Programming

$$= P\left(\frac{\bar{b}_i - \sum_{j=1}^{n}\bar{a}_{ij}x_j - \left(m_{\bar{b}_i} - \sum_{j=1}^{n}m_{\bar{a}_{ij}}x_j\right)}{\sqrt{\sigma_{\bar{b}_i}^2 + x^T V_{\bar{a}_i}x}} \geq \frac{-\left(m_{\bar{b}_i} - \sum_{j=1}^{n}m_{\bar{a}_{ij}}x_j\right)}{\sqrt{\sigma_{\bar{b}_i}^2 + x^T V_{\bar{a}_i}x}}\right)$$

$$= 1 - \Phi\left(\frac{\sum_{j=1}^{n}m_{\bar{a}_{ij}}x_j - m_{\bar{b}_i}}{\sqrt{\sigma_{\bar{b}_i}^2 + x^T V_{\bar{a}_i}x}}\right), \tag{5.82}$$

where Φ is the distribution function of the standard normal distribution $N(0, 1)$. Hence, the chance constraints (5.76) can be transformed into

$$\sum_{j=1}^{n} m_{\bar{a}_{ij}} x_j - \Phi^{-1}(1-\beta_i)\sqrt{\sigma_{\bar{b}_i}^2 + x^T V_{\bar{a}_i} x} \leq m_{\bar{b}_i}, \quad i = 1, \ldots, m. \tag{5.83}$$

Example 5.3. Consider the following constraints, where not only the right-hand sides but also the coefficients of the left-hand sides are normal random variables:

$$\bar{a}_1 x = \bar{2}x_1 + \bar{6}x_2 \leq \overline{27}$$
$$\bar{a}_2 x = \bar{3}x_1 + \bar{2}x_2 \leq \overline{16}$$
$$\bar{a}_3 x = \bar{4}x_1 + \bar{1}x_2 \leq \overline{18}.$$

For the above normal random variables, assume that $\overline{27} \sim N(27, 3^2)$, $\overline{16} \sim N(16, 2^2)$, $\overline{18} \sim N(18, 2^2)$, $E[\bar{a}_1] = E[(\bar{2}, \bar{6})] = (2, 6)$, $E[\bar{a}_2] = E[(\bar{3}, \bar{2})] = (3, 2)$, $E[\bar{a}_3] = E[(\bar{4}, \bar{1})] = (4, 1)$, $V_{\bar{a}_1} = \begin{bmatrix} 0.5 & 0.2 \\ 0.2 & 1 \end{bmatrix}$, $V_{\bar{a}_2} = \begin{bmatrix} 0.5 & 0.1 \\ 0.1 & 0.5 \end{bmatrix}$, and $V_{\bar{a}_3} = \begin{bmatrix} 1 & 0.1 \\ 0.1 & 0.25 \end{bmatrix}$.

If the DM sets the corresponding satisficing probability levels as $\beta_i = 0.8$, $i = 1, 2, 3$, from (5.83), the chance constraints (5.76) can be expressed by the following nonlinear inequalities without random variables:

$$2x_1 + 6x_2 - \Phi^{-1}(0.2)\sqrt{3^2 + (x_1, x_2)\begin{bmatrix} 0.5 & 0.2 \\ 0.2 & 1 \end{bmatrix}\begin{pmatrix} x_1 \\ x_2 \end{pmatrix}} \leq 27$$

$$3x_1 + 2x_2 - \Phi^{-1}(0.2)\sqrt{2^2 + (x_1, x_2)\begin{bmatrix} 0.5 & 0.1 \\ 0.1 & 0.5 \end{bmatrix}\begin{pmatrix} x_1 \\ x_2 \end{pmatrix}} \leq 16$$

$$4x_1 + x_2 - \Phi^{-1}(0.2)\sqrt{2^2 + (x_1, x_2)\begin{bmatrix} 1 & 0.1 \\ 0.1 & 0.25 \end{bmatrix}\begin{pmatrix} x_1 \\ x_2 \end{pmatrix}} \leq 18.$$

◇

Charnes and Cooper (1963) also considered three types of decision rules for optimizing objective functions with random variables: (i) the minimum or maximum expected value model, (ii) the minimum variance model, and (iii) the maximum probability model, which are referred to as the expectation model, the variance model, and the probability model, respectively. Moreover, Kataoka (1963) and Geoffrion (1967) individually proposed the fractile model.

5.3.1 Expectation Model

Recalling that some or all of coefficients of $\bar{c} = (\bar{c}_1, \ldots, \bar{c}_n)$ in the stochastic linear programming problem (5.74) are random variables, the objective function in the expectation model is represented as

$$E[\bar{c}x] = E\left[\sum_{j=1}^{n} \bar{c}_j x_j\right], \quad (5.84)$$

where E means the expectation. Let $m_{\bar{c}_j}$ denote the mean of $\bar{c}_j$, and then the objective function of the expectation model is simply written as

$$E[\bar{c}x] = E\left[\sum_{j=1}^{n} \bar{c}_j x_j\right] = \sum_{j=1}^{n} E[\bar{c}_j]x_j = E[\bar{c}]x = \boldsymbol{m}_{\bar{c}}x, \quad (5.85)$$

where $\boldsymbol{m}_{\bar{c}} = (m_{\bar{c}_1}, \ldots, m_{\bar{c}_n})$.

Considering the constraints in the stochastic linear programming problem (5.74) as chance constraints introduced in (5.76), the expectation model with the chance constraints where the expectation of the objective function is minimized is formulated as

$$\left.\begin{aligned} \text{minimize} \quad & E[\bar{c}x] = \boldsymbol{m}_{\bar{c}}x \\ \text{subject to} \quad & P\left(\sum_{j=1}^{n} \bar{a}_{ij}x_j \leq \bar{b}_i\right) \geq \beta_i, \ i = 1, \ldots, m \\ & x \geq \boldsymbol{0}, \end{aligned}\right\} \quad (5.86)$$

where β_i, $i = 1, \ldots, m$ are the satisficing probability levels.

5.3 Chance Constrained Programming

Example 5.4. Consider the following production planning problem with random variable coefficients where the constraints are the same as those of Example 5.3.

$$\begin{aligned}
\text{minimize} \quad & -\bar{3}x_1 - \bar{8}x_2 \\
\text{subject to} \quad & \bar{2}x_1 + \bar{6}x_2 \leq \overline{27} \\
& \bar{3}x_1 + \bar{2}x_2 \leq \overline{16} \\
& \bar{4}x_1 + \bar{1}x_2 \leq \overline{18} \\
& x_1 \geq 0, \ x_2 \geq 0,
\end{aligned}$$

where $\overline{27} \sim N(27, 3^2)$, $\overline{16} \sim N(16, 2^2)$, $\overline{18} \sim N(18, 2^2)$, $E[\bar{a}_1] = E[(\bar{2}, \bar{6})] = (2, 6)$, $E[\bar{a}_2] = E[(\bar{3}, \bar{2})] = (3, 2)$, $E[\bar{a}_3] = E[(\bar{4}, \bar{1})] = (4, 1)$, $V_{\bar{a}_1} = \begin{bmatrix} 0.5 & 0.2 \\ 0.2 & 1 \end{bmatrix}$, $V_{\bar{a}_2} = \begin{bmatrix} 0.5 & 0.1 \\ 0.1 & 0.5 \end{bmatrix}$, and $V_{\bar{a}_3} = \begin{bmatrix} 1 & 0.1 \\ 0.1 & 0.25 \end{bmatrix}$.

Assume that the coefficients $(-\bar{3}, -\bar{8})$ of the objective function are random variables and their means are $E[-\bar{3}] = -3$ and $E[-\bar{8}] = -8$. Moreover, suppose that the DM specifies the satisficing probability levels as $\beta_i = 0.8$, $i = 1, 2, 3$, and then the expectation model (5.86) with the chance constraints is formulated as follows:

$$\begin{aligned}
\text{minimize} \quad & -3x_1 - 8x_2 \\
\text{subject to} \quad & 2x_1 + 6x_2 - \Phi^{-1}(0.2)\sqrt{3^2 + (x_1, x_2)\begin{bmatrix} 0.5 & 0.2 \\ 0.2 & 1 \end{bmatrix}\begin{pmatrix} x_1 \\ x_2 \end{pmatrix}} \leq 27 \\
& 3x_1 + 2x_2 - \Phi^{-1}(0.2)\sqrt{2^2 + (x_1, x_2)\begin{bmatrix} 0.5 & 0.1 \\ 0.1 & 0.5 \end{bmatrix}\begin{pmatrix} x_1 \\ x_2 \end{pmatrix}} \leq 16 \\
& 4x_1 + x_2 - \Phi^{-1}(0.2)\sqrt{2^2 + (x_1, x_2)\begin{bmatrix} 1 & 0.1 \\ 0.1 & 0.25 \end{bmatrix}\begin{pmatrix} x_1 \\ x_2 \end{pmatrix}} \leq 18 \\
& x_1 \geq 0, \ x_2 \geq 0.
\end{aligned}$$

By using the Excel solver, we can obtain the following optimal solution and the optimal value:

$$(x_1^*, x_2^*) = (2.3006, 3.0495),$$
$$E[\bar{c}x^*] = -3x_1^* - 8x_2^* = -31.2979.$$

◊

If only $\bar{b}_i$ in the right-hand side of the chance constraints (5.76) are random variables with the probability distribution functions $F_i(\tau)$, the expectation model (5.86) with the chance constraints can be equivalently transformed to the linear programming problem

$$\begin{aligned}
\text{minimize} \quad & E[\bar{c}x] = m_{\bar{c}}x \\
\text{subject to} \quad & \sum_{j=1}^{n} a_{ij}x_j \leq K_{1-\beta_i}, \ i = 1, \ldots, m \\
& x \geq 0,
\end{aligned} \qquad (5.87)$$

where $K_{1-\beta_i}$ denotes the maximum of τ such that $\tau = F_i^{-1}(1-\beta_i)$.

Example 5.5. Consider the following production planning problem with random variable coefficients where the constraints are the same as those of Example 5.2:

$$\begin{aligned}
\text{minimize} \quad & -\bar{3}x_1 - \bar{8}x_2 \\
\text{subject to} \quad & 2x_1 + 6x_2 \leq \overline{27} \\
& 3x_1 + 2x_2 \leq \overline{16} \\
& 4x_1 + 1x_2 \leq \overline{18} \\
& x_1 \geq 0, \ x_2 \geq 0.
\end{aligned}$$

For the above normal random variables in the constraints, we assume that $\overline{27} \sim N(27, 3^2)$, $\overline{16} \sim N(16, 2^2)$, $\overline{18} \sim N(18, 2^2)$. Moreover, assume that the coefficients $(-\bar{3}, -\bar{8})$ of the objective function are random variables and their means are $E[-\bar{3}] = -3$ and $E[-\bar{8}] = -8$.

Suppose that the DM specifies the satisficing probability levels as $\beta_i = 0.8$, $i = 1, 2, 3$, and then the expectation model (5.87) with the chance constraints is formulated as the following linear programming problem:

$$\begin{aligned}
\text{minimize} \quad & -3x_1 - 8x_2 \\
\text{subject to} \quad & 2x_1 + 6x_2 \leq 27 + 3\Phi^{-1}(0.2) = 24.475 \\
& 3x_1 + 2x_2 \leq 16 + 2\Phi^{-1}(0.2) = 14.317 \\
& 4x_1 + \ x_2 \leq 18 + 2\Phi^{-1}(0.2) = 16.317 \\
& x_1 \geq 0, \ x_2 \geq 0.
\end{aligned}$$

By using the Excel solver, we can obtain the following optimal solution and the optimal value:

$$(x_1^*, x_2^*) = (2.6393, 3.1994),$$
$$E[\bar{c}x^*] = -3x_1^* - 8x_2^* = -33.5133.$$

If the DM specifies the satisficing probability levels a little higher, say $\beta_i = 0.9$, $i = 1, 2, 3$, the problem (5.87) is formulated as

5.3 Chance Constrained Programming

$$\begin{array}{ll} \text{minimize} & -3x_1 - 8x_2 \\ \text{subject to} & 2x_1 + 6x_2 \leq 27 + 3\Phi^{-1}(0.1) = 23.155 \\ & 3x_1 + 2x_2 \leq 16 + 2\Phi^{-1}(0.1) = 13.437 \\ & 4x_1 + x_2 \leq 18 + 2\Phi^{-1}(0.1) = 15.437 \\ & x_1 \geq 0,\ x_2 \geq 0, \end{array}$$

and an optimal solution is

$$(x_1^*, x_2^*) = (2.4508, 3.0423).$$

Since the feasible region of this problem diminishes, the optimal value increases as

$$E[\bar{c}x^*] = -3x_1^* - 8x_2^* = -31.6907.$$

◇

5.3.2 Variance Model

The realization value of the objective function may vary quite widely even if the expected value of the objective function is minimized. In such a case, it may be suspicious if a plan based on the solution of the expectation model would work well due to the large variation of the objective function value. Some DMs would prefer plans with lower uncertainty. To meet this demand, the objective function in the variance model is formulated as

$$Var[\bar{c}x] = Var\left[\sum_{j=1}^{n} \bar{c}_j x_j\right], \qquad (5.88)$$

where Var means variance. Let $V_{\bar{c}}$ denote an $n \times n$ variance-covariance matrix for the vector $\bar{c}$ of the random variables, and then the objective function of the variance model can be calculated as

$$Var[\bar{c}x] = Var\left[\sum_{j=1}^{n} \bar{c}_j x_j\right] = x^T V_{\bar{c}} x. \qquad (5.89)$$

Considering the constraints in the stochastic linear programming problem (5.74) as chance constraints introduced in (5.76), the variance model with the chance constraints where the variance of the objective function is minimized is formulated as

$$\left.\begin{array}{ll}\text{minimize} & Var[\bar{c}x] = x^T V_{\bar{c}} x \\ \text{subject to} & P\left(\sum_{j=1}^{n} \bar{a}_{ij} x_j \leq \bar{b}_i\right) \geq \beta_i, \ i = 1, \ldots, m \\ & x \geq 0, \end{array}\right\} \quad (5.90)$$

where β_i, $i = 1, \ldots, m$ are the satisficing probability levels.

Since the expectation of the objective function value is not included in the formulation (5.90), the optimal solution to (5.90) may result in very poor performance of the objective function value. To avoid such a problem, the constraint on the expectation of the objective function value is often incorporated as follows:

$$\left.\begin{array}{ll}\text{minimize} & Var[\bar{c}x] = x^T V_{\bar{c}} x \\ \text{subject to} & P\left(\sum_{j=1}^{n} \bar{a}_{ij} x_j \leq \bar{b}_i\right) \geq \beta_i, \ i = 1, \ldots, m \\ & E[\bar{c}x] = m_{\bar{c}} x \leq \gamma \\ & x \geq 0. \end{array}\right\} \quad (5.91)$$

By adding the constraint $E[\bar{c}x] \leq \gamma$ in the formulation, we can guarantee a specified level γ to the expectation of the objective function value.

Example 5.6. Consider the following production planning problem with random variable coefficients where the constraints are the same as those of Example 5.3.

$$\begin{array}{ll}\text{minimize} & -\bar{3}x_1 - \bar{8}x_2 \\ \text{subject to} & \bar{2}x_1 + \bar{6}x_2 \leq \overline{27} \\ & \bar{3}x_1 + \bar{2}x_2 \leq \overline{16} \\ & \bar{4}x_1 + \bar{1}x_2 \leq \overline{18} \\ & x_1 \geq 0, \ x_2 \geq 0, \end{array}$$

where $\overline{27} \sim N(27, 3^2)$, $\overline{16} \sim N(16, 2^2)$, $\overline{18} \sim N(18, 2^2)$, $E[\bar{a}_1] = E[(\bar{2}, \bar{6})] = (2, 6)$, $E[\bar{a}_2] = E[(\bar{3}, \bar{2})] = (3, 2)$, $E[\bar{a}_3] = E[(\bar{4}, \bar{1})] = (4, 1)$, $V_{\bar{a}_1} = \begin{bmatrix} 0.5 & 0.2 \\ 0.2 & 1 \end{bmatrix}$, $V_{\bar{a}_2} = \begin{bmatrix} 0.5 & 0.1 \\ 0.1 & 0.5 \end{bmatrix}$, and $V_{\bar{a}_3} = \begin{bmatrix} 1 & 0.1 \\ 0.1 & 0.25 \end{bmatrix}$. Assume that the means and the variance-covariance matrix of the coefficients $-\bar{3}$ and $-\bar{8}$ of the objective function which are random variables are $E[-\bar{3}] = -3$, $E[-\bar{8}] = -8$ and $V_{\bar{c}} = \begin{bmatrix} 1 & 0.5 \\ 0.5 & 2 \end{bmatrix}$.

Moreover, suppose that the DM specifies the satisficing probability levels as $\beta_i = 0.8$, $i = 1, 2, 3$ and provides the expectation level as $\gamma = -20$. The variance model (5.87) with the chance constraints and the expectation level is formulated as

5.3 Chance Constrained Programming

$$\text{minimize} \quad (x_1, x_2) \begin{bmatrix} 1 & 0.5 \\ 0.5 & 2 \end{bmatrix} \begin{pmatrix} x_1 \\ x_2 \end{pmatrix}$$

$$\text{subject to} \quad 2x_1 + 6x_2 - \Phi^{-1}(0.2) \sqrt{3^2 + (x_1, x_2) \begin{bmatrix} 0.5 & 0.2 \\ 0.2 & 1 \end{bmatrix} \begin{pmatrix} x_1 \\ x_2 \end{pmatrix}} \leq 27$$

$$3x_1 + 2x_2 - \Phi^{-1}(0.2) \sqrt{2^2 + (x_1, x_2) \begin{bmatrix} 0.5 & 0.1 \\ 0.1 & 0.5 \end{bmatrix} \begin{pmatrix} x_1 \\ x_2 \end{pmatrix}} \leq 16$$

$$4x_1 + x_2 - \Phi^{-1}(0.2) \sqrt{2^2 + (x_1, x_2) \begin{bmatrix} 1 & 0.1 \\ 0.1 & 0.25 \end{bmatrix} \begin{pmatrix} x_1 \\ x_2 \end{pmatrix}} \leq 18$$

$$E[\bar{c}x] = -3x_1 - 8x_2 \leq -20$$

$$x_1 \geq 0, \ x_2 \geq 0.$$

By using the Excel solver, we can obtain the following optimal solution and the related values:

$$(x_1^*, x_2^*) = (0.68966, 2.24138),$$

$$x^{*T} V_{\bar{c}} x^* = 12.06897,$$

$$E[\bar{c}x^*] = -3x_1^* - 8x_2^* = -20.$$

In this example, without the constraint of the expectation level, the optimal solution is $(x_1^*, x_2^*) = (0, 0)$ because the objective function $(x_1, x_2) \begin{bmatrix} 1 & 0.5 \\ 0.5 & 2 \end{bmatrix} \begin{pmatrix} x_1 \\ x_2 \end{pmatrix}$ is not negative for all $x \in \mathbb{R}_+^2 = \{x \in \mathbb{R}^2 \mid x_1 \geq 0, x_2 \geq 0\}$. ◊

If only $\bar{b}_i$ in the right-hand side of the chance constraints (5.76) are random variables with the probability distribution functions $F_i(\tau)$, the variance model (5.87) with the chance constraints and the expectation level can be equivalently transformed to the convex quadratic programming problem

$$\left.\begin{aligned}
\text{minimize} \quad & Var[\bar{c}x] = x^T V_{\bar{c}} x \\
\text{subject to} \quad & \sum_{j=1}^{n} a_{ij} x_j \leq K_{1-\beta_i}, \ i = 1, \ldots, m \\
& E[\bar{c}x] = m_{\bar{c}} x \leq \gamma \\
& x \geq 0,
\end{aligned}\right\} \quad (5.92)$$

where $K_{1-\beta_i}$ denotes the maximum of τ such that $\tau = F_i^{-1}(1 - \beta_i)$.

Example 5.7. Consider the following production planning problem with random variable coefficients where the constraints are the same as those of Example 5.5:

$$\begin{aligned}\text{minimize} \quad & -\bar{3}x_1 - \bar{8}x_2 \\ \text{subject to} \quad & 2x_1 + 6x_2 \leq \overline{27} \\ & 3x_1 + 2x_2 \leq \overline{16} \\ & 4x_1 + 1x_2 \leq \overline{18} \\ & x_1 \geq 0, \ x_2 \geq 0,\end{aligned}$$

where $\overline{27} \sim N(27, 3^2)$, $\overline{16} \sim N(16, 2^2)$, and $\overline{18} \sim N(18, 2^2)$. Also assume that the means and the variance-covariance matrix of the coefficients $-\bar{3}$ and $-\bar{8}$ of the objective function which are random variables are $E[-\bar{3}] = -3$, $E[-\bar{8}] = -8$ and
$$V_{\bar{c}} = \begin{bmatrix} 1 & 0.5 \\ 0.5 & 2 \end{bmatrix}.$$

Moreover, as in Example 5.6, assume that the DM specifies the satisficing probability levels as $\beta_i = 0.8$, $i = 1, 2, 3$ and provides the expectation level as $\gamma = -20$. The variance model (5.92) with the chance constraints and the expectation level is formulated as

$$\begin{aligned}\text{minimize} \quad & (x_1, x_2) \begin{bmatrix} 1 & 0.5 \\ 0.5 & 2 \end{bmatrix} \begin{pmatrix} x_1 \\ x_2 \end{pmatrix} \\ \text{subject to} \quad & 2x_1 + 6x_2 \leq 27 + 3\Phi^{-1}(0.2) = 24.475 \\ & 3x_1 + 2x_2 \leq 16 + 2\Phi^{-1}(0.2) = 14.317 \\ & 4x_1 + \ x_2 \leq 18 + 2\Phi^{-1}(0.2) = 16.317 \\ & E[\bar{c}x] = -3x_1 - 8x_2 \leq -20 \\ & x_1 \geq 0, \ x_2 \geq 0.\end{aligned}$$

By using the Excel solver, we can obtain the following optimal solution and the related values:

$$(x_1^*, x_2^*) = (0.68966, 2.24138),$$
$$x^{*T} V_{\bar{c}} x^* = 12.06897,$$
$$E[\bar{c}x^*] = -3x_1^* - 8x_2^* = -20.$$

It should be noted that the optimal solution of this problem is the same as that of Example 5.6 because all the chance constraints are not active and only the constraint for the expectation level is active in both the problems. ◇

5.3.3 Probability Model

In the probability model, the probability that the objective function value is smaller than a certain target value is maximized, and then the objective function of the probability model is represented as

5.3 Chance Constrained Programming

$$P\left(\bar{c}x \leq f_0\right), \tag{5.93}$$

where f_0 is a given target value for the objective function.

Considering the constraints in the stochastic linear programming problem (5.74) as chance constraints introduced in (5.76), the probability model maximizing the probability that the objective function value is smaller than a certain target value is formulated as

$$\left.\begin{array}{ll} \text{maximize} & P\left(\bar{c}x \leq f_0\right) \\ \text{subject to} & P\left(\sum_{j=1}^{n} \bar{a}_{ij} x_j \leq \bar{b}_i\right) \geq \beta_i, \ i = 1, \ldots, m \\ & x \geq 0, \end{array}\right\} \tag{5.94}$$

where $\beta_i, \ i = 1, \ldots, m$ are the satisficing probability levels.

Let $\bar{c} = (\bar{c}_1, \ldots, \bar{c}_n)$ be a multivariate normal random variable with a mean vector $m_{\bar{c}} = (m_{\bar{c}_1}, \ldots, m_{\bar{c}_n})$ and an $n \times n$ variance-covariance matrix $V_{\bar{c}}$. Assuming $x \neq 0$, the random variable

$$\frac{\bar{c}x - m_{\bar{c}}x}{\sqrt{x^T V_{\bar{c}} x}} \tag{5.95}$$

is the standard normal random variable $N(0, 1)$. Using (5.95), it follows that

$$P\left(\bar{c}x \leq f_0\right) = P\left(\frac{\bar{c}x - m_{\bar{c}}x}{\sqrt{x^T V_{\bar{c}} x}} \leq \frac{f_0 - m_{\bar{c}}x}{\sqrt{x^T V_{\bar{c}} x}}\right) = \Phi\left(\frac{f_0 - m_{\bar{c}}x}{\sqrt{x^T V_{\bar{c}} x}}\right), \tag{5.96}$$

where Φ is the distribution function of the standard normal random variable.

Hence, the problem (5.94) can be equivalently transformed to

$$\left.\begin{array}{ll} \text{maximize} & \Phi\left(\dfrac{f_0 - m_{\bar{c}}x}{\sqrt{x^T V_{\bar{c}} x}}\right) \\ \text{subject to} & P\left(\sum_{j=1}^{n} \bar{a}_{ij} x_j \leq \bar{b}_i\right) \geq \beta_i, \ i = 1, \ldots, m \\ & x \geq 0. \end{array}\right\} \tag{5.97}$$

The problem (5.97) can be solved by using the bisection method (Yano 2012) or the Dinkelbach algorithm (Dinkelbach 1967). First, consider the bisection method. By introducing the auxiliary variable v, (5.97) can be also equivalently transformed as

$$\left.\begin{array}{l}\text{maximize}\quad v\\ \text{subject to}\quad \Phi^{-1}(v)\sqrt{x^T V_{\bar{c}} x} \leq f_0 - m_{\bar{c}} x\\ \qquad\qquad P\left(\sum_{j=1}^{n} \bar{a}_{ij} x_j \leq \bar{b}_i\right) \geq \beta_i,\ i = 1,\ldots,m\\ \qquad\qquad x \geq 0,\end{array}\right\} \qquad (5.98)$$

from the fact that

$$\Phi\left(\frac{f_0 - m_{\bar{c}} x}{\sqrt{x^T V_{\bar{c}} x}}\right) \geq v \qquad (5.99)$$

is equivalent to

$$\Phi^{-1}(v)\sqrt{x^T V_{\bar{c}} x} \leq f_0 - m_{\bar{c}} x, \qquad (5.100)$$

and from (5.99) the maximum of v is searched in the interval $(0, 1)$. For notational convenience, we define the following function:

$$g_0(x; v) = \Phi^{-1}(v)\sqrt{x^T V_{\bar{c}} x} - (f_0 - m_{\bar{c}} x). \qquad (5.101)$$

Here, we assume that, for an optimal solution (x^*, v^*) to the problem (5.98), $v^* > 0.5$ holds. It should be noted that if the DM specifies the target value f_0 for the objective function to be substantially large, this assumption is satisfied. From this assumption, we have $\Phi^{-1}(v^*) > 0$, and then for a fixed value of v^*, it follows that $g_0(x; v^*)$ is convex.

By using the following bisection method, we can find an optimal solution to the problem (5.98).

Step 1 Initialize the upper and lower bounds of the value v as

$$v_0 = 0.5 + \epsilon,$$
$$v_1 = 1 - \epsilon,$$
$$v = \frac{v_0 + v_1}{2},$$

where ϵ is a sufficiently small positive number.

Step 2 For the given parameter v, solve the convex programming problem

$$\left.\begin{array}{l}\text{minimize}\quad g_0(x; v)\\ \text{subject to}\quad P\left(\sum_{j=1}^{n} \bar{a}_{ij} x_j \leq \bar{b}_i\right) \geq \beta_i,\ i = 1,\ldots,m\\ \qquad\qquad x \geq 0,\end{array}\right\} \qquad (5.102)$$

and let x^* denote an optimal solution of the problem.

5.3 Chance Constrained Programming

Step 3 If $v_1 - v_0 < \epsilon$, then go to Step 4. Otherwise, if the optimal value is positive, i.e., $g_0(x^*; v) > 0$, then update the upper bound of the value v as

$$v_1 = v.$$

If the optimal value is negative, i.e., $g_0(x^*; v) < 0$, then update the lower bound of the value v as

$$v_0 = v.$$

After updating the value of v as $v = \frac{v_0 + v_1}{2}$, return to Step 2. If the optimal value is zero, i.e., $g_0(x^*; v) = 0$, then go to Step 4.

Step 4 Set $v^* = v$. Let (x^*, v^*) be the optimal solution to the problem (5.98), and stop.

Although we assume that, for the optimal solution (x^*, v^*) to the problem (5.98), $v^* > 0.5$ holds, we give an index for specifying an appropriate level of the target value f_0 such that this assumption is satisfied.

Let $\hat{x}$ denote an optimal solution to the following problem where the expected value of the objective function is minimized:

$$\left. \begin{array}{ll} \text{minimize} & \boldsymbol{m_{\bar{c}}} x \\ \text{subject to} & P\left(\sum_{j=1}^{n} \bar{a}_{ij} x_j \leq \bar{b}_i \right) \geq \beta_i, \ i = 1, \ldots, m \\ & x \geq \mathbf{0}. \end{array} \right\} \quad (5.103)$$

Let

$$\hat{f}_0 = \boldsymbol{m_{\bar{c}}} \hat{x}, \quad (5.104)$$

and then the following proposition holds.

Proposition 5.1. *Let (x^*, v^*) be an optimal solution to the problem (5.98) with a target value f_0 larger than $\hat{f}_0$, i.e., $f_0 > \hat{f}_0$. Then, $v^* > 0.5$ holds.*

Proof. From the condition $f_0 > \hat{f}_0$ of the proposition,

$$f_0 - \boldsymbol{m_{\bar{c}}} \hat{x} > 0,$$

and then we have

$$\Phi\left(\frac{f_0 - \boldsymbol{m_{\bar{c}}} \hat{x}}{\sqrt{\hat{x}^T V_{\bar{c}} \hat{x}}} \right) > 0.5.$$

Letting

$$\hat{v} = \Phi\left(\frac{f_0 - m_{\tilde{c}}\hat{x}}{\sqrt{\hat{x}^T V_{\tilde{c}} \hat{x}}}\right),$$

we have

$$\Phi^{-1}(\hat{v})\sqrt{\hat{x}^T V_{\tilde{c}} \hat{x}} = f_0 - m_{\tilde{c}}\hat{x}.$$

Therefore, $(\hat{x}, \hat{v})$ is a feasible solution of the problem (5.98) with the target value f_0. Since (x^*, v^*) is the optimal solution of (5.98), $v^* \geq \hat{v} > 0.5$ holds. □

Next, we present the Dinkelbach algorithm (Dinkelbach 1967) for solving the problem (5.98). In the Dinkelbach algorithm, the following problem equivalent to (5.98) is solved:

$$\left.\begin{array}{l} \text{maximize} \quad \dfrac{f_0 - m_{\tilde{c}}x}{\sqrt{x^T V_{\tilde{c}} x}} \\ \text{subject to} \quad P\left(\sum_{j=1}^{n} \bar{a}_{ij} x_j \leq \bar{b}_i\right) \geq \beta_i, \; i = 1, \ldots, m \\ \quad x \geq 0, \end{array}\right\} \quad (5.105)$$

where the target value f_0 is also assumed to be larger than $\hat{f}_0$, i.e., $f_0 > \hat{f}_0$, as in the bisection method.

From $f_0 > \hat{f}_0$, there is a feasible solution of (5.105) such that $f_0 - m_{\tilde{c}}x \geq 0$, and let $\bar{X}$ denote the set of all of such feasible solutions. Then, an optimal solution x^* to (5.105) also belongs to $\bar{X}$. The Dinkelbach algorithm is summarized as follows:

Step 1   Let $N(x) = f_0 - m_{\tilde{c}}x$, and let $D(x) = \sqrt{x^T V_{\tilde{c}} x}$. For all $x \in \bar{X}$, it follows that $N(x) \geq 0$ and $D(x) > 0$. Let x^1 denote the initial solution in $\bar{X}$. Set

$$q^1 = \frac{N(x^1)}{D(x^1)},$$

and $l = 1$.

Step 2   By using a conventional convex programming method, solve the problem

$$\left.\begin{array}{l} \text{maximize} \quad \lambda \\ \text{subject to} \quad N(x) - q^l D(x) \geq \lambda \\ \quad x \in \bar{X}, \end{array}\right\} \quad (5.106)$$

and let (x^{l+1}, λ^{l+1}) denote an optimal solution of the problem (5.106).

5.3 Chance Constrained Programming

Step 3 Let ϵ be a sufficiently small positive number. If $0 \leq \lambda^{l+1} < \epsilon$, then let x^{l+1} be the optimal solution to the problem (5.105), and stop. Otherwise, after setting

$$q^{l+1} = \frac{N(x^{l+1})}{D(x^{l+1})}$$

and $l = l + 1$, return to Step 2.

Example 5.8. Consider the following production planning problem given in Example 5.6:

$$\begin{aligned}
\text{minimize} \quad & -\bar{3}x_1 - \bar{8}x_2 \\
\text{subject to} \quad & \bar{2}x_1 + \bar{6}x_2 \leq \overline{27} \\
& \bar{3}x_1 + \bar{2}x_2 \leq \overline{16} \\
& \bar{4}x_1 + \bar{1}x_2 \leq \overline{18} \\
& x_1 \geq 0, \ x_2 \geq 0,
\end{aligned}$$

where $\overline{27} \sim N(27, 3^2)$, $\overline{16} \sim N(16, 2^2)$, $\overline{18} \sim N(18, 2^2)$, $E[\bar{a}_1] = E[(\bar{2}, \bar{6})] = (2, 6)$, $E[\bar{a}_2] = E[(\bar{3}, \bar{2})] = (3, 2)$, $E[\bar{a}_3] = E[(\bar{4}, \bar{1})] = (4, 1)$, $V_{\bar{a}_1} = \begin{bmatrix} 0.5 & 0.2 \\ 0.2 & 1 \end{bmatrix}$, $V_{\bar{a}_2} = \begin{bmatrix} 0.5 & 0.1 \\ 0.1 & 0.5 \end{bmatrix}$, $V_{\bar{a}_3} = \begin{bmatrix} 1 & 0.1 \\ 0.1 & 0.25 \end{bmatrix}$, $E[\bar{-3}] = -3$, $E[\bar{-8}] = -8$, and $V_{\bar{c}} = \begin{bmatrix} 1 & 0.5 \\ 0.5 & 2 \end{bmatrix}$.

Assume that the DM specifies the satisficing probability levels as $\beta_i = 0.8$, $i = 1, 2, 3$ and provides the target value as $f_0 = -25$. The probability model (5.97) with the chance constraints is formulated as follows:

$$\text{maximize} \quad \frac{(-25) - (-3x_1 - 8x_2)}{\sqrt{(x_1, x_2) \begin{bmatrix} 1 & 0.5 \\ 0.5 & 2 \end{bmatrix} \begin{pmatrix} x_1 \\ x_2 \end{pmatrix}}}$$

$$\text{subject to} \quad 2x_1 + 6x_2 - \Phi^{-1}(0.2)\sqrt{3^2 + (x_1, x_2) \begin{bmatrix} 0.5 & 0.2 \\ 0.2 & 1 \end{bmatrix} \begin{pmatrix} x_1 \\ x_2 \end{pmatrix}} \leq 27$$

$$3x_1 + 2x_2 - \Phi^{-1}(0.2)\sqrt{2^2 + (x_1, x_2) \begin{bmatrix} 0.5 & 0.1 \\ 0.1 & 0.5 \end{bmatrix} \begin{pmatrix} x_1 \\ x_2 \end{pmatrix}} \leq 16$$

$$4x_1 + x_2 - \Phi^{-1}(0.2)\sqrt{2^2 + (x_1, x_2) \begin{bmatrix} 1 & 0.1 \\ 0.1 & 0.25 \end{bmatrix} \begin{pmatrix} x_1 \\ x_2 \end{pmatrix}} \leq 18$$

$$x_1 \geq 0, \ x_2 \geq 0.$$

Since the above problem is not a convex programming problem, it is difficult to solve through the direct use of the Excel solver. However, a simple VBA[2] program with the Excel solver based on the bisection method or the Dinkelbach algorithm allows to solve it. Developing such a program, we obtain the following optimal solution and the related values:

$$(x_1^*, x_2^*) = (2.10462, 3.11762),$$

$$P(\bar{c}x \leq f_0) = \Phi \left(\frac{(-25) - (-3x_1^* - 8x_2^*)}{\sqrt{(x_1^*, x_2^*) \begin{bmatrix} 1 & 0.5 \\ 0.5 & 2 \end{bmatrix} \begin{pmatrix} x_1^* \\ x_2^* \end{pmatrix}}} \right) = 0.87158,$$

$$m_{\bar{c}} x^* = -3x_1^* - 8x_2^* = -31.25485.$$

That is, the probability that the objective function value $-3x_1 - 8x_2$ is smaller than -25 is 0.87, and the mean of the objective function value is -31.25. ◊

If only $\bar{b}_i$ in the right-hand side of the chance constraints (5.76) are random variables with the probability distribution functions $F_i(\tau)$, the probability model (5.97) with the chance constraints can be equivalently transformed to the fractional programming problem

$$\left. \begin{array}{ll} \text{maximize} & \dfrac{f_0 - m_{\bar{c}} x}{\sqrt{x^T V_{\bar{c}} x}} \\ \text{subject to} & \displaystyle\sum_{j=1}^{n} a_{ij} x_j \leq K_{1-\beta_i}, \ i = 1, \ldots, m \\ & x \geq 0, \end{array} \right\} \quad (5.107)$$

where $K_{1-\beta_i}$ denotes the maximum of τ such that $\tau = F_i^{-1}(1 - \beta_i)$.

Example 5.9. Consider the following production planning problem given in Example 5.7:

$$\begin{array}{ll} \text{minimize} & -\bar{3}x_1 - \bar{8}x_2 \\ \text{subject to} & 2x_1 + 6x_2 \leq \overline{27} \\ & 3x_1 + 2x_2 \leq \overline{16} \\ & 4x_1 + 1x_2 \leq \overline{18} \\ & x_1 \geq 0, \ x_2 \geq 0, \end{array}$$

---

[2] VBA (visual basic for applications) is a programming language for Excel, and then one can code a procedure in Excel.

5.3 Chance Constrained Programming

where $\overline{27} \sim N(27, 3^2)$, $\overline{16} \sim N(16, 2^2)$, $\overline{18} \sim N(18, 2^2)$, $E[\overline{-3}] = -3$, $E[\overline{-8}] = -8$, and $V_{\bar{c}} = \begin{bmatrix} 1 & 0.5 \\ 0.5 & 2 \end{bmatrix}$.

Moreover, as in Example 5.8, we assume that the DM specifies the satisficing probability levels as $\beta_i = 0.8$, $i = 1, 2, 3$ and provides the target value as $f_0 = -25$. The probability model (5.98) with the chance constraints is formulated as

$$\text{maximize} \quad \frac{(-25) - (-3x_1 - 8x_2)}{\sqrt{(x_1, x_2) \begin{bmatrix} 1 & 0.5 \\ 0.5 & 2 \end{bmatrix} \begin{pmatrix} x_1 \\ x_2 \end{pmatrix}}}$$

$$\begin{aligned}
\text{subject to} \quad & 2x_1 + 6x_2 \leq 27 + 3\Phi^{-1}(0.2) = 24.475 \\
& 3x_1 + 2x_2 \leq 16 + 2\Phi^{-1}(0.2) = 14.317 \\
& 4x_1 + x_2 \leq 18 + 2\Phi^{-1}(0.2) = 16.317 \\
& E[\bar{c}x] = -3x_1 - 8x_2 \leq -20 \\
& x_1 \geq 0, \; x_2 \geq 0.
\end{aligned}$$

By using the VBA program with the Excel solver based on the bisection method or the Dinkelbach algorithm, we obtain the following optimal solution and the related values:

$$(x_1^*, x_2^*) = (2.30353, 3.31135),$$

$$P(\bar{c}x \leq f_0) = \Phi \left(\frac{(-25) - (-3x_1^* - 8x_2^*)}{\sqrt{(x_1^*, x_2^*) \begin{bmatrix} 1 & 0.5 \\ 0.5 & 2 \end{bmatrix} \begin{pmatrix} x_1^* \\ x_2^* \end{pmatrix}}} \right) = 0.92261,$$

$$m_{\bar{c}}x^* = -3x_1^* - 8x_2^* = -33.40136.$$

5.3.4 Fractile Model

The fractile model is considered as complementary to the probability model; the target variable to the objective function is minimized, provided that the probability that the objective function value is smaller than the target variable is guaranteed to be larger than a given assured level. Then, the objective function of the fractile model is represented as

$$f \quad \text{subject to} \quad P(\bar{c}x \leq f) \geq \theta, \tag{5.108}$$

where f and θ are the target variable to the objective function and the given assured level for the probability that the objective function value is smaller than or equal to the target variable. It should be emphasized here that f is a decision variable, while θ is a parameter specified by the DM.

Considering the constraints in the stochastic linear programming problem (5.74) as chance constraints introduced in (5.76), the fractile model minimizing the target variable f under the probabilistic constraints with the assured probability level θ for the objective function and the satisficing probability levels β_i, $i = 1, \ldots, m$ for the original constraints is formulated as

$$\left. \begin{aligned} \text{minimize} \quad & f \\ \text{subject to} \quad & P\left(\bar{c}x \leq f\right) \geq \theta \\ & P\left(\sum_{j=1}^{n} \bar{a}_{ij} x_j \leq \bar{b}_i\right) \geq \beta_i, \ i = 1, \ldots, m \\ & x \geq 0. \end{aligned} \right\} \quad (5.109)$$

Assuming again that $\bar{c}$ is a multivariate normal random variable with a mean vector $m_{\bar{c}}$ and an $n \times n$ variance-covariance matrix $V_{\bar{c}}$ with a similar transformation to (5.96), the probabilistic constraint

$$P\left(\bar{c}x \leq f\right) \geq \theta \quad (5.110)$$

for the objective function can be transformed as

$$\Phi\left(\frac{f - m_{\bar{c}}x}{\sqrt{x^T V_{\bar{c}} x}}\right) \geq \theta, \quad (5.111)$$

where Φ is the distribution function of the standard normal distribution. Let Φ^{-1} be the inverse of Φ, and then (5.111) is also equivalent to

$$f \geq m_{\bar{c}}x + \Phi^{-1}(\theta)\sqrt{x^T V_{\bar{c}} x}. \quad (5.112)$$

Since minimizing f is equivalent to minimizing the right-hand side of (5.112), the fractile model with the chance constraints (5.109) can be equivalently transformed to

$$\left. \begin{aligned} \text{minimize} \quad & m_{\bar{c}}x + \Phi^{-1}(\theta)\sqrt{x^T V_{\bar{c}} x} \\ \text{subject to} \quad & P\left(\sum_{j=1}^{n} \bar{a}_{ij} x_j \leq \bar{b}_i\right) \geq \beta_i, \ i = 1, \ldots, m \\ & x \geq 0. \end{aligned} \right\} \quad (5.113)$$

5.3 Chance Constrained Programming

Example 5.10. Consider the following production planning problem given in Example 5.6:

$$\begin{aligned}
\text{minimize} \quad & -\bar{3}x_1 - \bar{8}x_2 \\
\text{subject to} \quad & \bar{2}x_1 + \bar{6}x_2 \leq \overline{27} \\
& \bar{3}x_1 + \bar{2}x_2 \leq \overline{16} \\
& \bar{4}x_1 + \bar{1}x_2 \leq \overline{18} \\
& x_1 \geq 0, \ x_2 \geq 0,
\end{aligned}$$

where $\overline{27} \sim N(27, 3^2)$, $\overline{16} \sim N(16, 2^2)$, $\overline{18} \sim N(18, 2^2)$, $E[\bar{a}_1] = E[(\bar{2}, \bar{6})] = (2, 6)$, $E[\bar{a}_2] = E[(\bar{3}, \bar{2})] = (3, 2)$, $E[\bar{a}_3] = E[(\bar{4}, \bar{1})] = (4, 1)$, $V_{\bar{a}_1} = \begin{bmatrix} 0.5 & 0.2 \\ 0.2 & 1 \end{bmatrix}$, $V_{\bar{a}_2} = \begin{bmatrix} 0.5 & 0.1 \\ 0.1 & 0.5 \end{bmatrix}$, $V_{\bar{a}_3} = \begin{bmatrix} 1 & 0.1 \\ 0.1 & 0.25 \end{bmatrix}$, $E[\bar{-3}] = -3$, $E[\bar{-8}] = -8$, and $V_{\bar{c}} = \begin{bmatrix} 1 & 0.5 \\ 0.5 & 2 \end{bmatrix}$.

Assume that the DM specifies the satisficing probability levels as $\beta_i = 0.8$, $i = 1, 2, 3$ and provides the assured probability level for the probabilistic constraint with respect to the objective function as $\theta = 0.8$. The fractile model (5.113) with the chance constraints is formulated as

$$\text{minimize} \quad (-3x_1 - 8x_2) + \Phi^{-1}(0.8) \sqrt{(x_1, x_2) \begin{bmatrix} 1 & 0.5 \\ 0.5 & 2 \end{bmatrix} \begin{pmatrix} x_1 \\ x_2 \end{pmatrix}}$$

$$\text{subject to} \quad 2x_1 + 6x_2 - \Phi^{-1}(0.2) \sqrt{3^2 + (x_1, x_2) \begin{bmatrix} 0.5 & 0.2 \\ 0.2 & 1 \end{bmatrix} \begin{pmatrix} x_1 \\ x_2 \end{pmatrix}} \leq 27$$

$$3x_1 + 2x_2 - \Phi^{-1}(0.2) \sqrt{2^2 + (x_1, x_2) \begin{bmatrix} 0.5 & 0.1 \\ 0.1 & 0.5 \end{bmatrix} \begin{pmatrix} x_1 \\ x_2 \end{pmatrix}} \leq 16$$

$$4x_1 + x_2 - \Phi^{-1}(0.2) \sqrt{2^2 + (x_1, x_2) \begin{bmatrix} 1 & 0.1 \\ 0.1 & 0.25 \end{bmatrix} \begin{pmatrix} x_1 \\ x_2 \end{pmatrix}} \leq 18$$

$$x_1 \geq 0, \ x_2 \geq 0.$$

By using the Excel solver, we can obtain the following optimal solution and the related values:

$$(x_1^*, x_2^*) = (2.30059, 3.04951),$$

$$f^* = (-3x_1^* - 8x_2^*) + \Phi^{-1}(0.8) \sqrt{(x_1^*, x_2^*) \begin{bmatrix} 1 & 0.5 \\ 0.5 & 2 \end{bmatrix} \begin{pmatrix} x_1^* \\ x_2^* \end{pmatrix}} = -26.61893,$$

$$m_{\bar{c}} x^* = -3x_1^* - 8x_2^* = -31.29788.$$

The followings are the implication of this solution. Under the condition that the probability that the objective function value $-\bar{3}x_1 - \bar{8}x_2$ is smaller than the target variable f is larger than the probability level $\theta = 0.8$ specified by the DM, the target variable f meaning the sum of the mean and the probabilistic variation of the objective function is minimized to -26.62, and the mean of the objective function $-\bar{3}x_1 - \bar{8}x_2$ is -31.30. ◊

If only $\bar{b}_i$ in the right-hand side of the chance constraints (5.76) are random variables with the probability distribution functions $F_i(\tau)$, the fractile model (5.113) with the chance constraints can be equivalently transformed to

$$\left. \begin{aligned} \text{minimize} \quad & m_{\bar{c}}x + \Phi^{-1}(\theta)\sqrt{x^T V_{\bar{c}} x} \\ \text{subject to} \quad & \sum_{j=1}^{n} a_{ij}x_j \leq K_{1-\beta_i}, \ i = 1,\ldots,m \\ & x \geq 0. \end{aligned} \right\} \quad (5.114)$$

It should be noted that the problem (5.114) is a convex programming problem because the objective function is convex due to the fact that $\Phi^{-1}(\theta_l) \geq 0$ for any $\theta_l \in [1/2, 1)$.

Example 5.11. Consider the following production planning problem given in Example 5.7:

$$\begin{aligned} \text{minimize} \quad & -\bar{3}x_1 - \bar{8}x_2 \\ \text{subject to} \quad & 2x_1 + 6x_2 \leq \overline{27} \\ & 3x_1 + 2x_2 \leq \overline{16} \\ & 4x_1 + 1x_2 \leq \overline{18} \\ & x_1 \geq 0, \ x_2 \geq 0, \end{aligned}$$

where $\overline{27} \sim N(27, 3^2)$, $\overline{16} \sim N(16, 2^2)$, $\overline{18} \sim N(18, 2^2)$, $E[-\bar{3}] = -3$, $E[-\bar{8}] = -8$, and $V_{\bar{c}} = \begin{bmatrix} 1 & 0.5 \\ 0.5 & 2 \end{bmatrix}$.

Assume that the DM specifies the satisficing probability levels as $\beta_i = 0.8$, $i = 1, 2, 3$ and provides the assured probability level for the probabilistic constraint with respect to the objective function as $\theta = 0.8$. The fractile model (5.114) with the chance constraints is formulated as

$$\begin{aligned} \text{minimize} \quad & (-3x_1 - 8x_2) + \Phi^{-1}(0.8)\sqrt{(x_1, x_2)\begin{bmatrix} 1 & 0.5 \\ 0.5 & 2 \end{bmatrix}\begin{pmatrix} x_1 \\ x_2 \end{pmatrix}} \\ \text{subject to} \quad & 2x_1 + 6x_2 \leq 27 + 3\Phi^{-1}(0.2) = 24.475 \\ & 3x_1 + 2x_2 \leq 16 + 2\Phi^{-1}(0.2) = 14.317 \\ & 4x_1 + x_2 \leq 18 + 2\Phi^{-1}(0.2) = 16.317 \\ & x_1 \geq 0, \ x_2 \geq 0. \end{aligned}$$

5.3 Chance Constrained Programming

By using the Excel solver, we can obtain the following optimal solution and the related values:

$$(x_1^*, x_2^*) = (2.63931, 3.19942),$$

$$f^* = (-3x_1^* - 8x_2^*) + \Phi^{-1}(0.8)\sqrt{(x_1^*, x_2^*)\begin{bmatrix} 1 & 0.5 \\ 0.5 & 2 \end{bmatrix}\begin{pmatrix} x_1^* \\ x_2^* \end{pmatrix}} = -28.47178,$$

$$m_{\bar{c}}x^* = -3x_1^* - 8x_2^* = -33.51328.$$

◇

Now, it would be appropriate to point out here that there exist some similarities between certain two-stage programming and chance constrained programming. Williams (1965) was the first to show the mathematical equivalence between stochastic programming problems with simple recourse and those with chance constraints. Symonds (1968) showed the relations between some stochastic programming problems and the corresponding chance constrained problems. Further equivalences were discussed in Gartska and Wets (1974) and Gartska (1980a,b). However, it should be noted that all of these equivalences are somewhat weak because they require a priori knowledge of the optimal solution to one of the problems.

Consider a chance constrained problem for a stochastic linear programming problem with random variables $\bar{b}_i$, $i = 1, \ldots, m$ only in the right-hand side of the constraints

$$\left. \begin{aligned} \text{minimize} \quad & cx \\ \text{subject to} \quad & P\left(\sum_{j=1}^{n} a_{ij} x_j \leq \bar{b}_i\right) \geq \beta_i, \ i = 1, \ldots, m \\ & x \geq 0, \end{aligned} \right\} \quad (5.115)$$

where β_i, $i = 1, \ldots, m$ are the satisficing probability levels, and let f_i and F_i be the density and the distribution functions of $\bar{b}_i$. From its purpose, the satisficing probability level satisfies $0 < \beta_i < 1$. As we showed before it is transformed into

$$\left. \begin{aligned} \text{minimize} \quad & cx \\ \text{subject to} \quad & \sum_{j=1}^{n} a_{ij} x_j \leq F_i^{-1}(1 - \beta_i), \ i = 1, \ldots, m \\ & x \geq 0. \end{aligned} \right\} \quad (5.116)$$

Let x^c be an optimal solution to (5.116), and π_i^c, $i = 1, \ldots, m$ and λ_j^c, $j = 1, \ldots, n$ be the simplex multipliers of the m chance constraints and the n nonnegativity constraints, respectively.

The Lagrangian function associated with (5.116) is defined as

$$L^C(\boldsymbol{x}, \boldsymbol{\pi}, \boldsymbol{\lambda}) = \boldsymbol{c}\boldsymbol{x} + \sum_{i=1}^{m} \pi_i \left(\sum_{j=1}^{n} a_{ij} x_j - F_i^{-1}(1 - \beta_i) \right) - \sum_{j=1}^{n} \lambda_j x_j, \quad (5.117)$$

where π_i, $i = 1, \ldots, m$ and λ_j, $j = 1, \ldots, n$ are the Lagrange multipliers.

Since (5.116) is a convex programming problem, the following Kuhn–Tucker necessary conditions, which are given in Theorem B.2 in Appendix B, are also sufficient.

$$\frac{\partial L^C}{\partial x_j}(\boldsymbol{x}^c, \boldsymbol{\pi}^c, \boldsymbol{\lambda}^c) = c_j + \sum_{i=1}^{m} \pi_i^c a_{ij} - \lambda_j^c = 0, \; j = 1, \ldots, n, \quad (5.118)$$

$$\pi_i^c \left(\sum_{j=1}^{n} a_{ij} x_j^c - F_i^{-1}(1 - \beta_i) \right) = 0, \; i = 1, \ldots, m, \quad (5.119)$$

$$\lambda_j^c x_j^c = 0, \; j = 1, \ldots, n, \quad (5.120)$$

$$\pi_i^c \geq 0, \; i = 1, \ldots, m, \quad (5.121)$$

$$\lambda_j^c \geq 0, \; j = 1, \ldots, n, \quad (5.122)$$

$$\sum_{j=1}^{n} a_{ij} x_j^c \leq F_i^{-1}(1 - \beta_i), \; i = 1, \ldots, m, \quad (5.123)$$

$$\boldsymbol{x}^c \geq \boldsymbol{0}. \quad (5.124)$$

As a counterpart, the corresponding simple recourse problem is formulated as

$$\left. \begin{array}{ll} \text{minimize} & \boldsymbol{c}\boldsymbol{x} + E\left[\sum_{i=1}^{m} q_i^- y_i^- \right] \\ \text{subject to} & \sum_{j=1}^{n} a_{ij} x_j + y_i^+ - y_i^- = \bar{b}_i, \; i = 1, \ldots, m \\ & y_i^+ \geq 0, \; y_i^- \geq 0, \; i = 1, \ldots, m \\ & \boldsymbol{x} \geq \boldsymbol{0}, \end{array} \right\} \quad (5.125)$$

and it can be transformed into

$$\left. \begin{array}{ll} \text{minimize} & \boldsymbol{c}\boldsymbol{x} + \sum_{i=1}^{m} q_i^- \left[\left(\sum_{j=1}^{n} a_{ij} x_j \right) F_i \left(\sum_{j=1}^{n} a_{ij} x_j \right) - \int_{-\infty}^{\sum_{j=1}^{n} a_{ij} x_j} b_i f_i(b_i) db_i \right] \\ \text{subject to} & \boldsymbol{x} \geq \boldsymbol{0}. \end{array} \right\}$$

$$(5.126)$$

5.3 Chance Constrained Programming

By introducing the variables $z = (z_1, \ldots, z_m)^T$ and letting

$$z_i = \sum_{j=1}^{n} a_{ij} x_j, \quad i = 1, \ldots, m, \tag{5.127}$$

the problem (5.126) is further rewritten as

$$\left. \begin{array}{ll} \text{minimize} & cx + \sum_{i=1}^{m} q_i^- \left[z_i F_i(z_i) - \int_{-\infty}^{z_i} b_i f_i(b_i) db_i \right] \\ \text{subject to} & z_i = \sum_{j=1}^{n} a_{ij} x_j, \quad i = 1, \ldots, m \\ & x \geq 0. \end{array} \right\} \tag{5.128}$$

Let (x^r, z^r) be an optimal solution to (5.128), and π_i^r, $i = 1, \ldots, m$ and λ_j^r, $j = 1, \ldots, n$ be the simplex multipliers of the m equality constraints and the n nonnegativity constraints, respectively.

The Lagrangian function associated with (5.128) is defined as

$$L^R(x, z, \pi, \lambda) = cx + \sum_{i=1}^{m} q_i^- \left[z_i F_i(z_i) - \int_{-\infty}^{z_i} b_i f_i(b_i) db_i \right]$$

$$+ \sum_{i=1}^{m} \pi_i \left(\sum_{j=1}^{n} a_{ij} x_j - z_i \right) - \sum_{j=1}^{n} \lambda_j x_j. \tag{5.129}$$

As we discussed in Sect. 5.2, the problem (5.128) is a convex programming problem, and therefore the following Kuhn–Tucker conditions are necessary and sufficient conditions for optimality.

$$\frac{\partial L^R}{\partial x_j}(x^r, z^r, \pi^r, \lambda^r) = c_j + \sum_{i=1}^{m} \pi_i^r a_{ij} - \lambda_j^r = 0, \quad j = 1, \ldots, m, \tag{5.130}$$

$$\frac{\partial L^R}{\partial z_i}(x^r, z^r, \pi^r, \lambda^r) = q_i^- F_i(z_i^r) - \pi_i^r = 0, \quad i = 1, \ldots, n, \tag{5.131}$$

$$\lambda_j^r x_j^r = 0, \quad j = 1, \ldots, n, \tag{5.132}$$

$$\lambda_j^r \geq 0, \quad j = 1, \ldots, n, \tag{5.133}$$

$$z_i^r = \sum_{j=1}^{n} a_{ij} x_j^r, \quad i = 1, \ldots, m, \tag{5.134}$$

$$x^r \geq 0. \tag{5.135}$$

In the following theorem, it is shown that the optimal solution of the chance constrained problem (5.116) coincides that of the simple recourse problem (5.128) when the satisficing probability levels β_i, $i = 1, \ldots, m$ are specified in connection with the penalty costs q_i^-, $i = 1, \ldots, m$ to violating the constraints.

Theorem 5.5. *Let*

$$\beta_i = 1 - \frac{\pi_i^c}{q_i^-}, \quad i = 1, \ldots, m. \tag{5.136}$$

Then, an optimal solution x^c and the simplex multipliers π_i^c, $i = 1, \ldots, m$ and λ_j^c, $j = 1, \ldots, n$ of the chance constrained problem (5.116) with the parameters (5.136) correspond to those of the simple recourse problem (5.128).

Proof. From the optimality of the chance constrained problem (5.116), the Kuhn–Tucker conditions (5.118)–(5.124) are satisfied.

Since (5.118), (5.130) holds. If $\pi_i^c > 0$ for (5.119), from (5.127) and (5.136), it follows

$$\sum_{j=1}^{n} a_{ij} x_j^c = F_i^{-1}(1 - \beta_i)$$

$$F_i\left(\sum_{j=1}^{n} a_{ij} x_j^c\right) = 1 - \beta_i$$

$$F_i(z_i) = \frac{\pi_i^c}{q_i^-},$$

and then (5.131) holds. If $\pi_i^c = 0$, from (5.136) $q_i^- = 0$, and then (5.131) also holds. From the definition (5.127) of z^c, $i = 1, \ldots, m$, (5.134) holds. The conditions (5.120) and (5.122) are the same as (5.132) and (5.133). □

The opposite result is given in the following theorem, namely, it is shown that the optimal solution of the simple recourse problem (5.128) coincides that of the chance constrained problem (5.116) when the penalty costs q_i^-, $i = 1, \ldots, m$ to violating the constraints are specified in connection with the satisficing probability levels β_i, $i = 1, \ldots, m$.

Theorem 5.6. *Let*

$$q_i^- = \frac{\pi_i^r}{1 - \beta_i}, \quad i = 1, \ldots, m. \tag{5.137}$$

Then, an optimal solution (x^r, z^r) and the simplex multipliers π_i^r, $i = 1, \ldots, m$ and λ_j^r, $j = 1, \ldots, n$ of the simple recourse problem (5.128) with the parameters (5.137) correspond to those of the chance constrained problem (5.116).

Proof. From the optimality of the simple recourse problem (5.128), the Kuhn–Tucker conditions (5.130)–(5.135) are satisfied.

5.3 Chance Constrained Programming

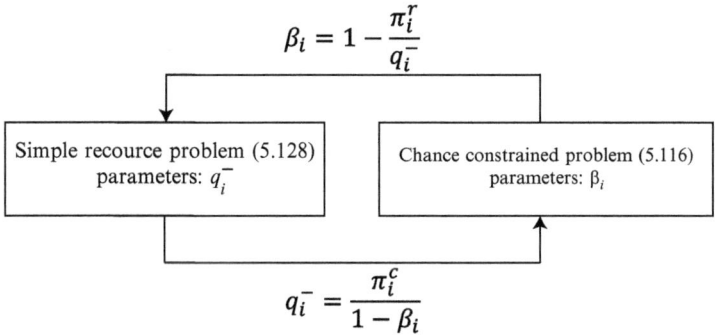

Fig. 5.1 Relationship between the simple recourse problem and the chance constrained problem

Since (5.130), (5.118) holds. From (5.131), (5.134) and (5.137), it follows

$$F_i(z_i^r) = \frac{\pi_i^r}{\bar{q}_i}$$

$$F_i\left(\sum_{j=1}^n a_{ij} x_j^r\right) = 1 - \beta_i$$

$$\sum_{j=1}^n a_{ij} x_j^r = F_i^{-1}(1 - \beta_i),$$

and then (5.119) and (5.123) hold. The conditions (5.132) and (5.133) are the same as (5.120) and (5.122). From (5.137), π_i^r must be nonnegative, and then (5.121) holds. □

The relation between the simple recourse problem (5.128) and the chance constrained problem (5.116) is also shown in Fig. 5.1.

The above two theorems give the corresponding relationship between stochastic programming problems with simple recourse and those with chance constraints, and then to make this relation more easily understood, we illustrate it with a simple numerical example.

Example 5.12. Consider the following stochastic production planning problem with the same constraints as those of Example 5.2:

$$\begin{aligned} \text{minimize} \quad & -3x_1 - 8x_2 \\ \text{subject to} \quad & 2x_1 + 6x_2 \leq \overline{27} \\ & 3x_1 + 2x_2 \leq \overline{16} \\ & 4x_1 + x_2 \leq \overline{18} \\ & x_1 \geq 0, \; x_2 \geq 0, \end{aligned}$$

where $\overline{27} \sim N(27, 3^2)$, $\overline{16} \sim N(16, 2^2)$, and $\overline{18} \sim N(18, 2^2)$.

Suppose that the DM specifies the satisficing probability levels as $\beta_i = 0.8$, $i = 1, 2, 3$, and then the chance constrained problem is formulated as

$$\begin{aligned} \text{minimize} \quad & -3x_1 - 8x_2 \\ \text{subject to} \quad & 2x_1 + 6x_2 \le 27 + 3\Phi^{-1}(0.2) = 24.475 \\ & 3x_1 + 2x_2 \le 16 + 2\Phi^{-1}(0.2) = 14.317 \\ & 4x_1 + x_2 \le 18 + 2\Phi^{-1}(0.2) = 16.317 \\ & x_1 \ge 0, \ x_2 \ge 0. \end{aligned}$$

By using the Excel solver, we can obtain the following optimal solution and the simplex multipliers:

$$(x_1^*, x_2^*) = (2.63931, 3.19942),$$
$$(\pi_1^*, \pi_2^*, \pi_3^*) = (1.28571, 0.14286, 0.0).$$

From (5.70), the simple recourse problem (5.125) can be transformed into

$$\underset{x \ge 0}{\text{minimize}} \ cx + \sum_{i=1}^{m} q_i^{-} \left\{ \left(\sum_{j=1}^{n} a_{ij} x_j \right) F_i \left(\sum_{j=1}^{n} a_{ij} x_j \right) - \int_{-\infty}^{\sum_{j=1}^{n} a_{ij} x_j} b_i f_i(b_i) db_i \right\}.$$

Thus, the objective function of this problem is calculated as

$$(-3x_1 - 8x_2) + \frac{\pi_1^*}{1 - \beta_1} \{(2x_1 + 6x_2) F(2x_1 + 6x_2; \mu = 27, \sigma = 3)$$
$$+ 3\phi((2x_1 + 6x_2 - 27)/3) - 27\Phi((2x_1 + 6x_2 - 27)/3)\}$$
$$+ \frac{\pi_2^*}{1 - \beta_2} \{(3x_1 + 2x_2) F(3x_1 + 2x_2; \mu = 16, \sigma = 2)$$
$$+ 2\phi((3x_1 + 2x_2 - 16)/2) - 16\Phi((3x_1 + 2x_2 - 16)/2)\},$$

where $F(\cdot; \mu, \sigma)$, ϕ, and Φ are the distribution function of the normal distribution with the mean μ and the variance σ^2, the density function and the distribution function of the standardized normal distribution, respectively.

Setting $\pi_1^* = 1.28571$, $\pi_2^* = 0.14286$, $\beta_1 = \beta_2 = 0.8$, by using the Excel solver, we can obtain the following optimal solution:

$$(x_1^*, x_2^*) = (2.63931, 3.19942).$$

One finds that the above optimal solution of the simple recourse problem is almost the same as that of the chance constrained problem. ◊

5.3 Chance Constrained Programming

So far, we have restricted ourselves to two different approaches to stochastic linear programming: two-stage programming with simple recourse and chance constrained programming. A brief and unified survey by Stancu-Minasian and Wets (1976), which covers major approaches to stochastic programming proposed before 1975, would be very useful for interested readers. Further details concerning the models, theory, algorithms, extensions, and applications can be found in the textbooks or monographs of Birge and Louveaux (1997), Kall (1976), Kall and Mayer (2005), Sengupta (1972), Stancu-Minasian (1984), and Vajda (1972).

Problems

5.1 Consider the following production planning problem:

$$\begin{aligned}
\text{minimize} \quad z = & -5x_1 - 5x_2 \\
\text{subject to} \quad & 5x_1 + 7x_2 \leq 12 \\
& 9x_1 + 1x_2 \leq 10 \\
& -5x_1 + 3x_2 \leq 3 \\
& x_1 \geq 0, \; x_2 \geq 0.
\end{aligned}$$

In the above problem, x_1 and x_2 denote the amounts of production for products P_1 and P_2, respectively. Suppose that the demands of products P_1 and P_2 are represented by random variables which are the same uniform distribution on the interval $[0, 2]$. Moreover, suppose that the lost earnings for one unit of the shortage of products P_1 and P_2 are 5 and 7, respectively, and that the holding costs for one unit of the excess of products P_1 and P_2 are 5 and 4, respectively. Formulate the corresponding two-stage programming problem, and solve it.

5.2 Show that, for $i = 1, \ldots, m$,

$$P\left(\sum_{j=1}^{n} a_{ij} x_j \geq \bar{b}_i\right) \geq \beta_i \quad \text{is equivalent to} \quad \sum_{j=1}^{n} a_{ij} x_j \geq K_{\beta_i},$$

where K_{β_i} is the minimum of τ such that $\tau = F_i^{-1}(\beta_i)$.

5.3 As discussed in Miller and Wagner (1965), instead of the individual chance constraints (5.76), consider the joint chance constraint

$$P\left(\sum_{j=1}^{n} a_{1j} x_j \leq \bar{b}_1, \ldots, \sum_{j=1}^{n} a_{mj} x_j \leq \bar{b}_m,\right) \geq \beta.$$

Assuming $\bar{b}_i, i = 1, \ldots, m$ are independent random variables with distribution functions $F_i(\tau_i), i = 1, \ldots, m$, derive the corresponding equivalent deterministic constraint.

5.4 For the constraints including normal random variables in the right-hand side

$$5x_1 + 7x_2 \leq \overline{12}$$
$$9x_1 + x_2 \leq \overline{10}$$
$$-5x_1 + 3x_2 \leq \bar{3},$$

where $\overline{12} \sim N(12, 2^2)$, $\overline{10} \sim N(10, 1^2)$, and $\bar{3} \sim N(3, 0.5^2)$, suppose that the DM specifies the satisficing probability levels as $\beta_i = 0.8, i = 1, 2, 3$.
Formulate the chance constraints.

5.5 For the constraints including normal random variables not only in the right-hand side but also in the left-hand side

$$\bar{a}_1 x = \bar{5}x_1 + \bar{7}x_2 \leq \overline{12}$$
$$\bar{a}_2 x = \bar{9}x_1 + \bar{1}x_2 \leq \overline{10}$$
$$\bar{a}_3 x = -\bar{5}x_1 + \bar{3}x_2 \leq \bar{3},$$

where $\overline{12} \sim N(12, 2^2)$, $\overline{10} \sim N(10, 1^2)$, $\bar{3} \sim N(3, 0.5^2)$, $E[\bar{a}_1] = E[(\bar{5}, \bar{7})] = (5, 7)$, $E[\bar{a}_2] = E[(\bar{9}, \bar{1})] = (9, 1)$, $E[\bar{a}_3] = E[(-\bar{5}, \bar{3})] = (-5, 3)$, $V_{\bar{a}_1} = \begin{bmatrix} 0.5 & 0.2 \\ 0.2 & 1 \end{bmatrix}$, $V_{\bar{a}_2} = \begin{bmatrix} 1 & 0.1 \\ 0.1 & 0.25 \end{bmatrix}$, and $V_{\bar{a}_3} = \begin{bmatrix} 0.5 & 0.1 \\ 0.1 & 0.25 \end{bmatrix}$, suppose that the DM specifies the satisficing probability levels as $\beta_i = 0.8, i = 1, 2, 3$.
Formulate the chance constraints.

5.6 In the expectation model (5.86), assume that the DM desires to suppress the variance $x^T V_{\bar{c}} x$ of the objective function to lower than some acceptable level γ, where $V_{\bar{c}}$ is a variance-covariance matrix of the random variable coefficient vector $\bar{c}$ of the objective function. Formulate the corresponding stochastic programming problem.

5.7 Consider a portfolio selection problem with n risky assets. An investor must choose a portfolio $x = (x_1, x_2, \ldots, x_n)^T$, where x_j is the proportion of the assets allocated to the jth security. Let $\bar{r} = (\bar{r}_1, \ldots, \bar{r}_n)$ be the vector of random returns of the assets. The random return on the portfolio has the mean $m_r x$ and the variance $x^T V_r x$, where m_r is the vector denoting the mean returns and V_r is the variance-covariance matrix of the returns. Formulate the problem of minimizing the variance of the return on the portfolio under prescribing the minimum acceptable level a of the expected return.

5.8 For the linear programming problem with random variable coefficients

5.3 Chance Constrained Programming

$$\begin{aligned}
\text{minimize} \quad & -\bar{5}x_1 - \bar{5}x_2 \\
\text{subject to} \quad & \bar{5}x_1 + \bar{7}x_2 \leq \overline{12} \\
& \bar{9}x_1 + \bar{1}x_2 \leq \overline{10} \\
& -\bar{5}x_1 + \bar{3}x_2 \leq \bar{3} \\
& x_1 \geq 0, \; x_2 \geq 0,
\end{aligned}$$

where $\overline{12} \sim N(12, 2^2)$, $\overline{10} \sim N(10, 1^2)$, $\bar{3} \sim N(3, 0.5^2)$, $E[\bar{a}_1] = E[(\bar{5}, \bar{7})] = (5, 7)$, $E[\bar{a}_2] = E[(\bar{9}, \bar{1})] = (9, 1)$, $E[\bar{a}_3] = E[(-\bar{5}, \bar{3})] = (-5, 3)$, $V_{\bar{a}_1} = \begin{bmatrix} 0.5 & 0.2 \\ 0.2 & 1 \end{bmatrix}$, $V_{\bar{a}_2} = \begin{bmatrix} 1 & 0.1 \\ 0.1 & 0.25 \end{bmatrix}$, $V_{\bar{a}_3} = \begin{bmatrix} 0.5 & 0.1 \\ 0.1 & 0.25 \end{bmatrix}$, and $E[\bar{c}] = E[(-\bar{5}, -\bar{5})] = (-5, -5)$, suppose that the DM employs the expectation model with the chance constraints and specifies the satisficing probability levels as $\beta_i = 0.8, i = 1, 2, 3$.

Solve the formulated problem based on the expectation model with the chance constraints.

5.9 For the linear programming problem with random variable coefficients only in the right-hand side of the constraints and the objective function

$$\begin{aligned}
\text{minimize} \quad & -\bar{5}x_1 - \bar{5}x_2 \\
\text{subject to} \quad & 5x_1 + 7x_2 \leq \overline{12} \\
& 9x_1 + x_2 \leq \overline{10} \\
& -5x_1 + 3x_2 \leq \bar{3} \\
& x_1 \geq 0, \; x_2 \geq 0,
\end{aligned}$$

where $\overline{12} \sim N(12, 2^2)$, $\overline{10} \sim N(10, 1^2)$, $\bar{3} \sim N(3, 0.5^2)$, and $E[\bar{c}] = E[(-\bar{5}, -\bar{5})] = (-5, -5)$, suppose that the DM employs the expectation model with the chance constraints and specifies the satisficing probability levels as $\beta_i = 0.8, i = 1, 2, 3$.

Solve the formulated problem based on the expectation model with the chance constraints.

5.10 For the linear programming problem with random variable coefficients

$$\begin{aligned}
\text{minimize} \quad & -\bar{5}x_1 - \bar{5}x_2 \\
\text{subject to} \quad & \bar{5}x_1 + \bar{7}x_2 \leq \overline{12} \\
& \bar{9}x_1 + \bar{1}x_2 \leq \overline{10} \\
& -\bar{5}x_1 + \bar{3}x_2 \leq \bar{3} \\
& x_1 \geq 0, \; x_2 \geq 0,
\end{aligned}$$

where $\overline{12} \sim N(12, 2^2)$, $\overline{10} \sim N(10, 1^2)$, $\bar{3} \sim N(3, 0.5^2)$, $E[\bar{a}_1] = E[(\bar{5}, \bar{7})] = (5, 7)$, $E[\bar{a}_2] = E[(\bar{9}, \bar{1})] = (9, 1)$, $E[\bar{a}_3] = E[(-\bar{5}, \bar{3})] = (-5, 3)$, $V_{\bar{a}_1} = \begin{bmatrix} 0.5 & 0.2 \\ 0.2 & 1 \end{bmatrix}$, $V_{\bar{a}_2} = \begin{bmatrix} 1 & 0.1 \\ 0.1 & 0.25 \end{bmatrix}$, $V_{\bar{a}_3} = \begin{bmatrix} 0.5 & 0.1 \\ 0.1 & 0.25 \end{bmatrix}$, $E[\bar{c}] = E[(-\bar{5}, -\bar{5})] =$

$(-5, -5)$, and $V_{\bar{c}} = \begin{bmatrix} 1 & 0.5 \\ 0.5 & 1 \end{bmatrix}$, suppose that the DM employs the variance model with the chance constraints and the expectation level and specifies the satisficing probability levels as $\beta_i = 0.8, i = 1, 2, 3$ and the expectation level for the objective function as $E[\bar{c}x] \leq -8$.

Solve the formulated problem based on the variance model with the chance constraints and the expectation level.

5.11 For the linear programming problem with random variable coefficients only in the right-hand side of the constraints and the objective function

$$\begin{aligned}
\text{minimize} \quad & -\bar{5}x_1 - \bar{5}x_2 \\
\text{subject to} \quad & 5x_1 + 7x_2 \leq \overline{12} \\
& 9x_1 + x_2 \leq \overline{10} \\
& -5x_1 + 3x_2 \leq \bar{3} \\
& x_1 \geq 0, \; x_2 \geq 0,
\end{aligned}$$

where $\overline{12} \sim N(12, 2^2)$, $\overline{10} \sim N(10, 1^2)$, $\bar{3} \sim N(3, 0.5^2)$, $E[\bar{c}] = E[(-\bar{5}, -\bar{5})] = (-5, -5)$, and $V_{\bar{c}} = \begin{bmatrix} 1 & 0.5 \\ 0.5 & 1 \end{bmatrix}$, suppose that the DM employs the variance model with the chance constraints and the expectation level and specifies the satisficing probability levels as $\beta_i = 0.8, i = 1, 2, 3$ and the expectation level for the objective function as $E[\bar{c}x] \leq -8$.

Solve the formulated problem based on the variance model with the chance constraints and the expectation level.

5.12 In the probability model (5.94), assume that the random variable coefficient vector $\bar{c}$ of the objective function is expressed by $\bar{c} = c^0 + \bar{t}c^1$, $c^1 > 0$, where $\bar{t}$ is a random variable with a continuous and strictly increasing probability distribution function T. Show that the problem (5.94) can be equivalently transformed to the fractional programming problem

$$\begin{aligned}
\text{maximize} \quad & \frac{f_0 - c^0 x}{c^1 x} \\
\text{subject to} \quad & P\left(\sum_{j=1}^{n} \bar{a}_{ij} x_j \leq \bar{b}_i\right) \geq \beta_i, \; i = 1, \ldots, m \\
& x \geq 0.
\end{aligned}$$

5.13 For the linear programming problem with random variable coefficients

$$\begin{aligned}
\text{minimize} \quad & -\bar{5}x_1 - \bar{5}x_2 \\
\text{subject to} \quad & \bar{5}x_1 + \bar{7}x_2 \leq \overline{12} \\
& \bar{9}x_1 + \bar{1}x_2 \leq \overline{10} \\
& -\bar{5}x_1 + \bar{3}x_2 \leq \bar{3} \\
& x_1 \geq 0, \; x_2 \geq 0,
\end{aligned}$$

5.3 Chance Constrained Programming

where $\overline{12} \sim N(12, 2^2)$, $\overline{10} \sim N(10, 1^2)$, $\overline{3} \sim N(3, 0.5^2)$, $E[\bar{a}_1] = E[(\bar{5}, \bar{7})] = (5, 7)$, $E[\bar{a}_2] = E[(\bar{9}, \bar{1})] = (9, 1)$, $E[\bar{a}_3] = E[(-\bar{5}, \bar{3})] = (-5, 3)$, $V_{\bar{a}_1} = \begin{bmatrix} 0.5 & 0.2 \\ 0.2 & 1 \end{bmatrix}$, $V_{\bar{a}_2} = \begin{bmatrix} 1 & 0.1 \\ 0.1 & 0.25 \end{bmatrix}$, $V_{\bar{a}_3} = \begin{bmatrix} 0.5 & 0.1 \\ 0.1 & 0.25 \end{bmatrix}$, $E[\bar{c}] = E[(-\bar{5}, -\bar{5})] = (-5, -5)$, and $V_{\bar{c}} = \begin{bmatrix} 1 & 0.5 \\ 0.5 & 1 \end{bmatrix}$, suppose that the DM employs the probability model with the chance constraints and specifies the satisficing probability levels as $\beta_i = 0.8$, $i = 1, 2, 3$ and the target value for the objective function as $f_0 = -8$.
Solve the formulated problem based on the probability model with the chance constraints.

5.14 For the linear programming problem with random variable coefficients only in the right-hand side of the constraints and the objective function

$$\begin{aligned}
\text{minimize} \quad & -\bar{5}x_1 - \bar{5}x_2 \\
\text{subject to} \quad & 5x_1 + 7x_2 \leq \overline{12} \\
& 9x_1 + x_2 \leq \overline{10} \\
& -5x_1 + 3x_2 \leq \bar{3} \\
& x_1 \geq 0, \ x_2 \geq 0,
\end{aligned}$$

where $\overline{12} \sim N(12, 2^2)$, $\overline{10} \sim N(10, 1^2)$, $\bar{3} \sim N(3, 0.5^2)$, $E[\bar{c}] = E[(-\bar{5}, -\bar{5})] = (-5, -5)$, and $V_{\bar{c}} = \begin{bmatrix} 1 & 0.5 \\ 0.5 & 1 \end{bmatrix}$, suppose that the DM employs the probability model with the chance constraints and specifies the satisficing probability levels as $\beta_i = 0.8$, $i = 1, 2, 3$ and the target value for the objective function as $f_0 = -8$.
Solve the formulated problem based on the probability model with the chance constraints.

5.15 For the linear programming problem with random variable coefficients

$$\begin{aligned}
\text{minimize} \quad & -\bar{5}x_1 - \bar{5}x_2 \\
\text{subject to} \quad & \bar{5}x_1 + \bar{7}x_2 \leq \overline{12} \\
& \bar{9}x_1 + \bar{1}x_2 \leq \overline{10} \\
& -\bar{5}x_1 + \bar{3}x_2 \leq \bar{3} \\
& x_1 \geq 0, \ x_2 \geq 0,
\end{aligned}$$

where $\overline{12} \sim N(12, 2^2)$, $\overline{10} \sim N(10, 1^2)$, $\bar{3} \sim N(3, 0.5^2)$, $E[\bar{a}_1] = E[(\bar{5}, \bar{7})] = (5, 7)$, $E[\bar{a}_2] = E[(\bar{9}, \bar{1})] = (9, 1)$, $E[\bar{a}_3] = E[(-\bar{5}, \bar{3})] = (-5, 3)$, $V_{\bar{a}_1} = \begin{bmatrix} 0.5 & 0.2 \\ 0.2 & 1 \end{bmatrix}$, $V_{\bar{a}_2} = \begin{bmatrix} 1 & 0.1 \\ 0.1 & 0.25 \end{bmatrix}$, $V_{\bar{a}_3} = \begin{bmatrix} 0.5 & 0.1 \\ 0.1 & 0.25 \end{bmatrix}$, $E[\bar{c}] = E[(-\bar{5}, -\bar{5})] = (-5, -5)$, and $V_{\bar{c}} = \begin{bmatrix} 1 & 0.5 \\ 0.5 & 1 \end{bmatrix}$, suppose that the DM employs the fractile model with the chance constraints and specifies the satisficing probability levels as

$\beta_i = 0.8, i = 1, 2, 3$ and the assured probability level for the objective function as $\theta = 0.6$.

Solve the formulated problem based on the fractile model with the chance constraints.

5.16 For the linear programming problem with random variable coefficients only in the right-hand side of the constraints and the objective function

$$\begin{aligned}
\text{minimize} \quad & -\bar{5}x_1 - \bar{5}x_2 \\
\text{subject to} \quad & 5x_1 + 7x_2 \leq \overline{12} \\
& 9x_1 + x_2 \leq \overline{10} \\
& -5x_1 + 3x_2 \leq \bar{3} \\
& x_1 \geq 0, \ x_2 \geq 0,
\end{aligned}$$

where $\overline{12} \sim N(12, 2^2)$, $\overline{10} \sim N(10, 1^2)$, $\bar{3} \sim N(3, 0.5^2)$, $E[\bar{c}] = E[(-\bar{5}, -\bar{5})] = (-5, -5)$, and $V_{\bar{c}} = \begin{bmatrix} 1 & 0.5 \\ 0.5 & 1 \end{bmatrix}$, suppose that the DM employs the fractile model with the chance constraints and specifies the satisficing probability levels as $\beta_i = 0.8, i = 1, 2, 3$ and the assured probability level for the objective function as $\theta = 0.6$.

Solve the formulated problem based on the fractile model with the chance constraints.

5.17 For the linear programming problem with random variable coefficients only in the right-hand side of the constraints

$$\begin{aligned}
\text{minimize} \quad & -5x_1 - 5x_2 \\
\text{subject to} \quad & 5x_1 + 7x_2 \leq \overline{12} \\
& 9x_1 + x_2 \leq \overline{10} \\
& -5x_1 + 3x_2 \leq \bar{3} \\
& x_1 \geq 0, \ x_2 \geq 0,
\end{aligned}$$

where $\overline{12} \sim N(12, 2^2)$, $\overline{10} \sim N(10, 1^2)$, $\bar{3} \sim N(3, 0.5^2)$, suppose that the DM specifies the satisficing probability levels as $\beta_i = 0.8, i = 1, 2, 3$.

Verify that the optimal solution of the chance constrained problem is almost the same as that of the corresponding simple recourse problem with the penalty parameters (5.136).

Chapter 6
Interactive Fuzzy Multiobjective Stochastic Linear Programming

This chapter is devoted to several interactive fuzzy programming approaches to multiobjective stochastic programming problems. Through the use of the several stochastic models including the expectation model, the variance model, the probability model, the fractile model, and the simple recourse model together with chance constrained programming, the formulated multiobjective stochastic programming problems are transformed into deterministic ones. Assuming that the decision maker has the fuzzy goal for each of the objective functions, several interactive fuzzy satisficing methods are presented for deriving a satisficing solution for the decision maker by updating the reference membership levels.

6.1 Multiobjective Chance Constrained Programming

Assuming that the coefficients in the objective functions and the right-hand side constants of the constraints are random variables, we deal with multiobjective linear programming problems formulated as

$$\left.\begin{array}{ll} \text{minimize} & z_1(x) = \bar{c}_1 x \\ & \cdots \\ \text{minimize} & z_k(x) = \bar{c}_k x \\ \text{subject to} & Ax \leq \bar{b} \\ & x \geq 0, \end{array}\right\} \quad (6.1)$$

where x is an n-dimensional decision variable column vector and A is an $m \times n$ coefficient matrix, $\bar{c}_l$, $l = 1, \ldots, k$ are n-dimensional random variable row vectors with finite means $E[\bar{c}_l]$ and $n \times n$ positive-definite variance–covariance matrices $V_l = [v^l_{jh}] = [Cov\{c_{lj}, c_{lh}\}]$, $l = 1, \ldots, k$, and $\bar{b}$ is an n-dimensional random variable column vector whose elements are mutually independent.

Multiobjective linear programming problems with random variable coefficients are referred to as multiobjective stochastic linear programming problems, which are often appropriate to model actual decision-making situations mathematically. For example, consider a production planning problem to optimize the gross profit and the production cost simultaneously under the condition that unit profits of the products, unit production costs of them, and the maximal available amounts of the resources depend on seasonal factors or market prices. Such a production planning problem can be formulated as a multiobjective stochastic programming problem expressed by (6.1).

Observing that (6.1) contains random variable coefficients, the definitions and solution methods for ordinary mathematical programming problems not depending on stochastic events cannot be directly applied. Realizing such difficulty, we deal with the constraints in (6.1) as chance constrained conditions (Charnes and Cooper 1959) which permit constraint violations up to specified probability limits. Let β_i, $i = 1, \ldots, m$ denote a minimum probability level that the ith constraint should be satisfied, and we call it the satisficing probability level. Then, replacing the constraints in (6.1) with the chance constrained conditions with the satisficing probability levels β_i, $i = 1, \ldots, m$, we can reformulate (6.1) as the chance constrained problem

$$\left.\begin{aligned}
\text{minimize} \quad & z_1(x) = \bar{c}_1 x \\
& \cdots\cdots \\
\text{minimize} \quad & z_k(x) = \bar{c}_k x \\
\text{subject to} \quad & P(a_1 x \leq \bar{b}_1) \geq \beta_1 \\
& \cdots\cdots\cdots \\
& P(a_m x \leq \bar{b}_m) \geq \beta_m \\
& x \geq 0,
\end{aligned}\right\} \quad (6.2)$$

where a_i is the ith row vector of A and $\bar{b}_i$ is the ith element of $\bar{b}$.

Assume that the random variable $\bar{b}_i$, $i = 1, \ldots, m$ has a continuous distribution function $F_i(r) = P(\bar{b}_i \leq r)$, and then the ith constraint in (6.2) can be rewritten as

$$\begin{aligned}
P(a_i x \leq \bar{b}_i) \geq \beta_i &\Leftrightarrow 1 - P(\bar{b}_i \leq a_i x) \geq \beta_i \\
&\Leftrightarrow 1 - F_i(a_i x) \geq \beta_i \\
&\Leftrightarrow a_i x \leq F_i^{-1}(1 - \beta_i).
\end{aligned} \quad (6.3)$$

From (6.3), for a vector of the satisficing probability levels $\boldsymbol{\beta} = (\beta_1, \ldots, \beta_m)^T$, (6.2) can be equivalently transformed to

6.1 Multiobjective Chance Constrained Programming

$$\left.\begin{array}{ll} \text{minimize} & z_1(x) = \bar{c}_1 x \\ \quad \cdots\cdots \\ \text{minimize} & z_k(x) = \bar{c}_k x \\ \text{subject to} & x \in X(\boldsymbol{\beta}), \end{array}\right\} \quad (6.4)$$

where

$$X(\boldsymbol{\beta}) = \left\{x \mid a_i x \le F_i^{-1}(1 - \beta_i),\ i = 1, \ldots, m,\ x \ge 0 \right\}. \quad (6.5)$$

As we already considered in Chap. 5, for a more general case where not only $\bar{b}_i$ but also $\bar{a}_{ij}$ in the left-hand side of the constraints are random variables, we assumed that $\bar{b}_i$ and $\bar{a}_{ij}$ are normal random variables. Let $m_{\bar{b}_i}$ and $\sigma^2_{\bar{b}_i}$ be the mean and the variance of $\bar{b}_i$, respectively. Also let $m_{\bar{a}_{ij}}$ be the mean of $\bar{a}_{ij}$ and $V_{\bar{a}_i}$ be the variance–covariance matrix of the vector $\bar{a}_i = (\bar{a}_{i1}, \ldots, \bar{a}_{in})$. Moreover, assume that $\bar{b}_i$ and $\bar{a}_{ij}$ are independent of each other. Then, we already showed that the feasible region of the chance constraints is given as

$$X(\boldsymbol{\beta}) = \left\{ x \ \Big|\ \sum_{j=1}^n m_{\bar{a}_{ij}} x_j - \Phi^{-1}(1 - \beta_i) \sqrt{\sigma^2_{\bar{b}_i} + x^T V_{\bar{a}_i} x} \le m_{\bar{b}_i},\ i = 1, \ldots, m,\ x \ge 0 \right\}, \quad (6.6)$$

where Φ^{-1} is the inverse function of the distribution function of the standard normal distribution $N(0, 1)$.

For the sake of simplicity, we deal with the constraints with random variables only in the right-hand side in the rest of this chapter.

6.1.1 Expectation Model

As a first attempt to deal with multiobjective stochastic programming problems, assuming that the decision maker (DM) desires to simply minimize the expected values of the objective functions, we introduce the expectation model (Sakawa and Kato 2002; Sakawa et al. 2003). By replacing the objective functions $z_l(x) = \bar{c}_l x$, $l = 1, \ldots, k$ in (6.4) with their expectations, the multiobjective stochastic programming problem can be reformulated as

$$\left.\begin{array}{ll} \text{minimize} & z_1^E(x) = E[z_1(x)] = E[\bar{c}_1] x \\ \quad \cdots\cdots \\ \text{minimize} & z_k^E(x) = E[z_k(x)] = E[\bar{c}_k] x \\ \text{subject to} & x \in X(\boldsymbol{\beta}), \end{array}\right\} \quad (6.7)$$

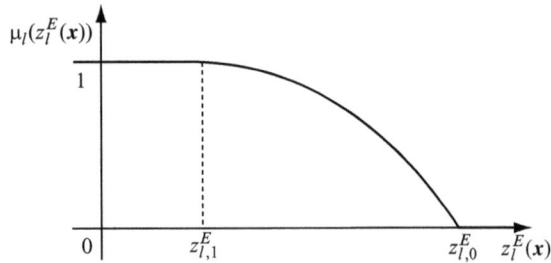

Fig. 6.1 Example of the membership function $\mu_l(z_l^E(x))$

where $E[z_l(x)]$ denotes the expectation of $z_l(x)$, and $E[\bar{c}_l] = (E[\bar{c}_{l1}], \ldots, E[\bar{c}_{ln}])$.

In order to consider the imprecise nature of the DM's judgments for each expectation of the objective function $z_l^E(x) = E[\bar{c}_l]x$, $l = 1, \ldots, k$ in (6.7), if we introduce the fuzzy goals such as "$z_l^E(x)$ should be substantially less than or equal to a certain value," (6.7) can be interpreted as

$$\underset{x \in X(\beta)}{\text{maximize}} \left\{ \mu_1\left(z_1^E(x)\right), \ldots, \mu_k\left(z_k^E(x)\right) \right\}, \tag{6.8}$$

where μ_l is a membership function to quantify the fuzzy goal for the lth objective function in (6.7). To be more specific, if the DM feels that $z_l^E(x)$ should be smaller than or equal to at most $z_{l,0}^E$ and satisfies $z_l^E(x)$ smaller than or equal to $z_{l,1}^E (< z_{l,0}^E)$, the shape of a typical membership function can be shown as in Fig. 6.1.

As a possible way to help the DM determine the values of $z_{l,0}^E$ and $z_{l,1}^E$, it is recommended to calculate the individual minima and maxima of $E[\bar{c}_l]x$, $l = 1, \ldots, k$ obtained by solving the linear programming problems

$$\left. \begin{array}{l} \underset{x \in X(\beta)}{\text{minimize}} \quad z_l^E(x) = E[\bar{c}_l]x, \ l = 1, \ldots, k, \\ \underset{x \in X(\beta)}{\text{maximize}} \quad z_l^E(x) = E[\bar{c}_l]x, \ l = 1, \ldots, k. \end{array} \right\} \tag{6.9}$$

It should be noted here that (6.8) is regarded as a multiobjective decision-making problem and that there rarely exists a complete optimal solution that simultaneously optimizes all the objective functions. As discussed in Chap. 4, by directly extending Pareto optimality in ordinary multiobjective programming problems, Sakawa and his colleagues defined M-Pareto optimality on the basis of membership function values as a reasonable solution concept for the fuzzy multiobjective decision-making problems (Sakawa and Yano 1985c; Sakawa et al. 1987; Sakawa and Yano 1990; Sakawa 1993).

Introducing an aggregation function $\mu_D(x)$ for the k membership functions in (6.8), the fuzzy multiobjective programming problem can be rewritten as

$$\underset{x \in X(\beta)}{\text{maximize}} \ \mu_D(x). \tag{6.10}$$

Following the conventional fuzzy approaches, to aggregate the multiple membership functions of the fuzzy goals for the expectations of the objective functions, Hulsurkar et al. (1997) adopted the minimum operator of Bellman and Zadeh (1970) defined by

$$\mu_D(x) = \min_{1 \leq l \leq k} \mu_l\left(z_l^E(x)\right)$$

and the product operator of Zimmermann (1978) defined as

$$\mu_D(x) = \prod_{l=1}^{k} \mu_l\left(z_l^E(x)\right).$$

As discussed in Chap. 4, however, it should be emphasized here that such approaches are preferable only when the DM feels that the minimum operator or the product operator is appropriate for the decision-making problem under consideration. In other words, in general decision situations, the DM does not always use the minimum operator or the product operator for aggregating the fuzzy goals. The most crucial problem in (6.10) is probably the identification of an appropriate aggregation function which well represents the DM's fuzzy preference. If $\mu_D(x)$ can be explicitly identified, then (6.10) can be reduced to a standard mathematical programming problem. However, this rarely happens, and then as an alternative way, it is recommended for the DM to examine multiple possible solutions in the set of M-Pareto optimal solutions through an interactive solution procedure to find a satisficing solution to (6.10).

In such an interactive fuzzy satisficing method, to generate a candidate for a satisficing solution which is also M-Pareto optimal, the DM is asked to specify the aspiration levels of achievement for all the membership function values, called the reference membership levels(Sakawa and Yano 1985c; Sakawa et al. 1987; Sakawa and Yano 1989; Sakawa and Yano 1990; Sakawa 1993).

For the DM's reference membership levels $\hat{\mu}_l, l = 1, \ldots, k$, the corresponding M-Pareto optimal solution, which is the nearest to the vector of the given reference membership levels in the minimax sense or better than it if the reference membership levels are attainable, is obtained by solving the minimax problem

$$\underset{x \in X(\beta)}{\text{minimize}} \max_{1 \leq l \leq k} \left\{\hat{\mu}_l - \mu_l\left(z_l^E(x)\right)\right\}. \tag{6.11}$$

By introducing an auxiliary variable v, (6.11) can be equivalently transformed to

$$\left.\begin{aligned}
&\text{minimize} \quad v \\
&\text{subject to} \quad \hat{\mu}_1 - \mu_1\left(z_1^E(x)\right) \leq v \\
&\qquad\qquad\quad \ldots\ldots\ldots \\
&\qquad\qquad\quad \hat{\mu}_k - \mu_k\left(z_k^E(x)\right) \leq v \\
&\qquad\qquad\quad x \in X(\beta).
\end{aligned}\right\} \tag{6.12}$$

For solving (6.12), the solution method based on the bisection method and the simplex method of linear programming presented in Chap. 4 is useful. To be more specific, if the value of v is fixed, the constraints of (6.12) can be reduced to a set of linear inequalities. Obtaining the optimal value v^* to (6.12) is equivalent to determining the minimum of v so that there exists a feasible solution x satisfying the constraints in (6.12). Since v satisfies

$$\max_{1 \leq l \leq k} \hat{\mu}_l - 1 \leq v \leq \min_{1 \leq l \leq k} \hat{\mu}_l,$$

we can find an optimal solution x^* corresponding to v^* by using the combined use of the bisection method and phase I of the simplex method.

After calculating the optimal value v^* to the problem

$$\left.\begin{aligned}
\text{minimize} \quad & v \\
\text{subject to} \quad & z_1^E(x) \leq \mu_1^{-1}(\hat{\mu}_1 - v) \\
& \cdots\cdots\cdots \\
& z_k^E(x) \leq \mu_k^{-1}(\hat{\mu}_k - v) \\
& x \in X(\boldsymbol{\beta})
\end{aligned}\right\} \qquad (6.13)$$

by the combined use of the bisection method and phase I of the simplex method, we solve the linear programming problem

$$\left.\begin{aligned}
\text{minimize} \quad & z_1^E(x) \\
\text{subject to} \quad & z_2^E(x) \leq \mu_2^{-1}(\hat{\mu}_2 - v^*) \\
& \cdots\cdots\cdots \\
& z_k^E(x) \leq \mu_k^{-1}(\hat{\mu}_k - v^*) \\
& x \in X(\boldsymbol{\beta}),
\end{aligned}\right\} \qquad (6.14)$$

where the first objective function $z_1(x)$ in (6.1) is supposed to be the most important to the DM.

For the obtained optimal solution x^* of (6.14), if there are inactive constraints in the first $(k-1)$ constraints, after replacing $\hat{\mu}_l$ for the inactive constraints with $\mu_l(z_l^E(x^*)) + v^*$, we resolve the revised problem. Furthermore, if the obtained x^* is not unique, the M-Pareto optimality test is performed by solving the linear programming problem

$$\left.\begin{aligned}
\text{maximize} \quad & w = \sum_{l=1}^{k} \varepsilon_l \\
\text{subject to} \quad & \mu_1\left(z_1^E(x)\right) - \varepsilon_1 = \mu_1\left(z_1^E(x^*)\right) \\
& \cdots\cdots\cdots \\
& \mu_k\left(z_k^E(x)\right) - \varepsilon_k = \mu_k\left(z_k^E(x^*)\right) \\
& x \in X(\boldsymbol{\beta}),\ \boldsymbol{\varepsilon} = (\varepsilon_1, \ldots, \varepsilon_k)^T \geq \mathbf{0}.
\end{aligned}\right\} \qquad (6.15)$$

6.1 Multiobjective Chance Constrained Programming

For the optimal solution x^o and ε^o to (6.15), (i) if $w = 0$, i.e., $\varepsilon_l^o = 0$ for $l = 1, \ldots, k$, x^* is Pareto optimal, and (ii) if $w > 0$, i.e., $\varepsilon_l^o > 0$ for at least one l, not x^* but x^o is M-Pareto optimal.

The DM should either be satisfied with the current M-Pareto optimal solution or continue to examine another solution by updating the reference membership levels. In order to help the DM update the reference membership levels, trade-off rates between a standing membership function, say $\mu_1\left(z_1^E(x)\right)$, and each of the other membership functions are quite useful. Such the trade-off rates are easily obtainable and are expressed as

$$-\frac{\partial \mu_1\left(z_1^E(x^*)\right)}{\partial \mu_l\left(z_l^E(x^*)\right)} = \pi_l \frac{\mu_1'\left(z_1^E(x^*)\right)}{\mu_l'\left(z_l^E(x^*)\right)}, \ l = 2, \ldots, k, \tag{6.16}$$

where μ_l' denotes the differential coefficient of μ_l and π_l, $l = 1, \ldots, k$ are the simplex multipliers in (6.14).

Following the preceding discussions, we summarize an interactive algorithm for deriving the satisficing solution for the DM from among the M-Pareto optimal solution set.

Procedure of Interactive Fuzzy Satisficing Method for the Expectation Model

Step 1   Ask the DM to specify the satisficing probability levels β_i, $i = 1, \ldots, m$.

Step 2   Calculate the individual minima and maxima of $E[\bar{c}_l]x$, $l = 1, \ldots, k$ by solving the linear programming problems (6.9).

Step 3   Ask the DM to specify the membership functions μ_l, $l = 1, \ldots, k$ for the expectations of the objective functions by using the individual minima and maxima obtained in step 2 as a reference.

Step 4   Set the initial reference membership levels at 1s, i.e., $\hat{\mu}_l = 1$, $l = 1, \ldots, k$, which can be viewed as the ideal values.

Step 5   For the current reference membership levels $\hat{\mu}_l$, solve the corresponding minimax problem (6.12). For the obtained optimal solution x^*, if there are inactive constraints in the first $(k-1)$ constraints of (6.12), replace $\hat{\mu}_l$ for the inactive constraints with $\mu_l(z_l^E(x^*)) + v^*$ and resolve the revised problem. Furthermore, if the solution x^* is not unique, perform the M-Pareto optimality test.

Step 6   The DM is supplied with the corresponding M-Pareto optimal solution and the trade-off rates between the membership functions. If the DM is satisfied with the current membership function values $\mu_l\left(z_l^E(x^*)\right)$, $l = 1, \ldots, k$, then stop. Otherwise, ask the DM to update the reference membership levels $\hat{\mu}_l$, $l = 1, \ldots, k$ by taking into account the current membership function values $\mu_l\left(z_l^E(x^*)\right)$ together with the trade-off rates $-\partial \mu_1 / \partial \mu_l$, $l = 2, \ldots, k$ (6.16), and return to step 5.

Observing that the trade-off rates $-\partial\mu_1/\partial\mu_l$, $l = 2,\ldots,k$ in step 6 indicate the decrement value of the membership function μ_1 with a unit increment of value of the membership function μ_l, the information of the trade-off rates can be used to estimate the local shape of $\left(\mu_1\left(z_1^E(\boldsymbol{x}^*)\right),\ldots,\mu_k\left(z_k^E(\boldsymbol{x}^*)\right)\right)$ around the current solution $\boldsymbol{x}^*$.

It should be stressed for the DM that any improvement of one membership function value can be achieved only at the expense of at least one of the other membership function values for the given satisficing probability levels β_i, $i = 1,\ldots,m$.

Example 6.1. Consider the following two-objective stochastic production planning problem which can be regarded as a stochastic version of Example 3.2:

$$\begin{aligned}
\text{minimize} \quad & z_1 = \bar{-3}x_1 - \bar{8}x_2 \\
\text{minimize} \quad & z_2 = \bar{5}x_1 + \bar{4}x_2 \\
\text{subject to} \quad & 2x_1 + 6x_2 \leq \overline{27} \\
& 3x_1 + 2x_2 \leq \overline{16} \\
& 4x_1 + x_2 \leq \overline{18} \\
& x_1 \geq 0, \quad x_2 \geq 0.
\end{aligned}$$

In this problem, suppose that the coefficients of the objective functions are represented as random variables and their means are $E[\bar{-3}] = -3$, $E[\bar{-8}] = -8$, $E[\bar{5}] = 5$, and $E[\bar{4}] = 4$ and the right-hand side constants of the constraints are also represented as normal random variables defined by $\overline{27} \sim N(27, 3^2)$, $\overline{16} \sim N(16, 2^2)$, and $\overline{18} \sim N(18, 2^2)$.

By solving the linear programming problems (6.9), we can obtain the individual minima and maxima of the expectations of the objective functions as $z_{1,\min}^E = -33.51328$, $z_{1,\max}^E = 0$, $z_{2,\min}^E = 0$, and $z_{2,\max}^E = 25.99421$, where $z_{l,\min}^E$ and $z_{l,\max}^E$ denote the minimum and the maximum of the expectation of the lth objective functions, respectively. By using Zimmermann's method (Zimmermann 1978), assume that the DM determines the linear membership functions as, for $l = 1, 2$,

$$\mu_l(z_l^E(\boldsymbol{x})) = \begin{cases} 1 & \text{if } z_l^E(\boldsymbol{x}) \leq z_{l,1}^E \\ \dfrac{z_l^E(\boldsymbol{x}) - z_{l,0}^E}{z_{l,1}^E - z_{l,0}^E} & \text{if } z_{l,1}^E < z_l^E(\boldsymbol{x}) \leq z_{l,0}^E \\ 0 & \text{if } z_{l,0}^E < z_l^E(\boldsymbol{x}), \end{cases}$$

where $z_{1,1}^E = -33.51328$, $z_{1,0}^E = 0$, $z_{2,1}^E = 0$, and $z_{2,0}^E = 25.99421$.

Suppose that the DM specifies the satisficing probability levels as $\beta_i = 0.8$, $i = 1, 2, 3$, and then the minimax problem (6.12) for the expectation model with the chance constraints is formulated as follows:

6.1 Multiobjective Chance Constrained Programming

Table 6.1 Interactive process in the expectation model

Iteration	First	Second	Third
$\hat{\mu}_1$	1.000	0.65	0.635
$\hat{\mu}_2$	1.000	0.6	0.595
$\mu_1\left(z_1^E(x)\right)$	0.608187	0.638596	0.632515
$\mu_2\left(z_2^E(x)\right)$	0.608187	0.588597	0.592515
$z_1^E(x)$	-20.37427	-21.39298	-21.18924
$z_2^E(x)$	10.18713	10.69649	10.59462

$$\begin{aligned}
\text{minimize} \quad & v \\
\text{subject to} \quad & \hat{\mu}_1 - \mu_1(z_1^E(x)) \leq v \\
& \hat{\mu}_2 - \mu_2(z_2^E(x)) \leq v \\
& 2x_1 + 6x_2 \leq 27 + 3\Phi^{-1}(0.2) = 24.475 \\
& 3x_1 + 2x_2 \leq 16 + 2\Phi^{-1}(0.2) = 14.317 \\
& 4x_1 + x_2 \leq 18 + 2\Phi^{-1}(0.2) = 16.317 \\
& x_1 \geq 0, \ x_2 \geq 0,
\end{aligned}$$

where Φ^{-1} is the inverse function of the distribution function Φ of the standard normal distribution $N(0, 1)$.

An example of an interactive process is given in Table 6.6. The interactive process starts from the initial reference membership levels $(\hat{\mu}_1, \hat{\mu}_2) = (1.0, 1.0)$ which yields a tentative solution with the same membership values for the two fuzzy goals of the expectations of the objective functions. In this example, after two updates of the reference membership levels, the DM's satisficing solution in which the first objective is given priority over the other one is obtained in the third iteration. At each iteration the minimax problem is solved by using a VBA program with the Excel solver. ◊

6.1.2 Variance Model

When the DM desires to simply minimize the expectations of the objective functions without concerning about the fluctuation of the realized values, the expectation model is appropriate. In contrast, if the DM prefers to decrease the fluctuation of the objective function values from the viewpoint of the stability of the obtained values, by minimizing the variances of the objective functions, the multiobjective stochastic programming problem can be reformulated as

$$\left.\begin{aligned}
\text{minimize} \quad & z_1^V(x) = Var\,[z_1(x)] = x^T V_1 x \\
& \cdots\cdots\cdots \\
\text{minimize} \quad & z_k^V(x) = Var\,[z_k(x)] = x^T V_k x \\
\text{subject to} \quad & x \in X(\boldsymbol{\beta}),
\end{aligned}\right\} \quad (6.17)$$

where $Var\,[z_l(x)]$ denotes the variance of $z_l(x)$, V_l is the variance–covariance matrix for the coefficient vector $\bar{c}_l$ of $z_l(x)$ and $X(\boldsymbol{\beta})$ is the feasible region of the chance constrained problem with the satisficing probability levels $\boldsymbol{\beta}$ (Sakawa et al. 2002).

Observing that (6.17) does not include any constraints with the expectations of the objective functions, the original variance model may yield undesirable values of the objective functions even if the fluctuation of the objective function values is minimized.

To cope with this difficulty, by incorporating the constraints with respect to the expectations into the original variance model, we formulate the variance model with the chance constraints and the expectation levels for the multiobjective stochastic programming problem represented as

$$\left.\begin{aligned}
\text{minimize} \quad & z_1^V(x) = Var\,[z_1(x)] = x^T V_1 x \\
& \cdots\cdots\cdots \\
\text{minimize} \quad & z_k^V(x) = Var\,[z_k(x)] = x^T V_k x \\
\text{subject to} \quad & E[\bar{C}]x \le \boldsymbol{\gamma} \\
& x \in X(\boldsymbol{\beta}),
\end{aligned}\right\} \quad (6.18)$$

where $E[\bar{C}] = (E[\bar{c}_1]^T, \ldots, E[\bar{c}_k]^T)^T$ and $\boldsymbol{\gamma} = (\gamma_1, \ldots, \gamma_k)^T$ is a vector of the expectation levels. Each γ_l is the maximum of the expectation level which is acceptable for the DM, and then the DM specifies the expectation levels $\boldsymbol{\gamma}$, taking into account the individual minima and maxima of $z_l^E(x)$ calculated by solving the linear programming problems

$$\left.\begin{aligned}
\underset{x\in X(\boldsymbol{\beta})}{\text{minimize}} \quad & z_l^E(x) = E[\bar{c}_l]x, \; l = 1,\ldots,k, \\
\underset{x\in X(\boldsymbol{\beta})}{\text{maximize}} \quad & z_l^E(x) = E[\bar{c}_l]x, \; l = 1,\ldots,k.
\end{aligned}\right\} \quad (6.19)$$

For notational convenience, let $X(\boldsymbol{\beta},\boldsymbol{\gamma})$ be the feasible region of (6.18), i.e.,

$$X(\boldsymbol{\beta},\boldsymbol{\gamma}) = \{x \mid E[\bar{C}]x \le \boldsymbol{\gamma},\, x \in X(\boldsymbol{\beta})\}.$$

Quite similar to the expectation model, in order to consider the imprecise nature of the DM's judgments for each objective function in (6.18), if the DM introduces the fuzzy goals such as "$z_l^V(x)$ should be substantially less than or equal to a certain value," (6.18) can be interpreted as

$$\underset{x\in X(\boldsymbol{\beta},\boldsymbol{\gamma})}{\text{maximize}} \; \{\mu_1\left(z_1^V(x)\right), \ldots, \mu_k\left(z_k^V(x)\right)\}, \quad (6.20)$$

where μ_l is a nonincreasing membership function to quantify the fuzzy goal for the lth objective function $z_k^V(x)$. To help the DM specify the membership function, it is desirable to calculate the individual minima of the variances of the objective functions by solving the quadratic programming problems

$$\underset{x\in X(\boldsymbol{\beta},\boldsymbol{\gamma})}{\text{minimize}} \; z_l^V(x) = x^T V_l x,\; l = 1,\ldots,k. \quad (6.21)$$

6.1 Multiobjective Chance Constrained Programming

For the DM's reference membership levels $\hat{\mu}_l$, $i = 1, \ldots, k$, the corresponding M-Pareto optimal solution is obtained by solving the minimax problem

$$\underset{x \in X(\beta,\gamma)}{\text{minimize}} \max_{1 \leq l \leq k} \left\{ \hat{\mu}_l - \mu_l \left(z_l^V(x) \right) \right\}. \tag{6.22}$$

If an optimal solution x^* to (6.22) is not unique, M-Pareto optimality of x^* is not guaranteed. As discussed in the expectation model, we can obtain the trade-off rates in the expectation model through the M-Pareto optimality test using linear programming techniques. Since (6.22) is not a linear programming problem, it is not easy to calculate the trade-off rates in the variance model. In order to obtain an M-Pareto optimal solution with less effort, solving the following augmented minimax problem is recommended:

$$\underset{x \in X(\beta,\gamma)}{\text{minimize}} \max_{1 \leq l \leq k} \left\{ \hat{\mu}_l - \mu_l \left(z_l^V(x) \right) + \rho \sum_{i=1}^{k} \left(\hat{\mu}_i - \mu_i \left(z_i^V(x) \right) \right) \right\}, \tag{6.23}$$

where ρ is a sufficiently small positive number. By introducing the auxiliary variable v, this problem can be equivalently transformed as

$$\left. \begin{aligned}
&\text{minimize} \quad v \\
&\text{subject to} \quad \hat{\mu}_1 - \mu_1 \left(z_1^V(x) \right) + \rho \sum_{i=1}^{k} \left\{ \hat{\mu}_i - \mu_i \left(z_i^V(x) \right) \right\} \leq v \\
&\qquad\qquad \cdots\cdots\cdots \\
&\qquad\qquad \hat{\mu}_k - \mu_k \left(z_k^V(x) \right) + \rho \sum_{i=1}^{k} \left\{ \hat{\mu}_i - \mu_i \left(z_i^V(x) \right) \right\} \leq v \\
&\qquad\qquad x \in X(\beta, \gamma).
\end{aligned} \right\} \tag{6.24}$$

For ease in computation, we assume that the membership function μ_l, $l = 1, \ldots, k$ are nonincreasing and concave. Let $g_l(x) = \hat{\mu}_l - \mu_l \left(z_l^V(x) \right)$. Since, for any x_1, $x_2 \in X(\beta, \gamma)$ and $\lambda \in [0, 1]$,

$$\begin{aligned}
g_l(\lambda x_1 + (1-\lambda)x_2) &= \hat{\mu}_l - \mu_l \left(z_l^V(\lambda x_1 + (1-\lambda)x_2) \right) \\
&\leq \hat{\mu}_l - \mu_l \left(\lambda z_l^V(x_1) + (1-\lambda) z_l^V(x_2) \right) \\
&\leq \hat{\mu}_l - \lambda \mu_l \left(z_l^V(x_1) \right) - (1-\lambda) \mu_l (z_l^V(x_2)) \\
&= \lambda \left(\hat{\mu}_l - \mu_l \left(z_l^V(x_1) \right) \right) + (1-\lambda) \left(\hat{\mu}_l - \mu_l \left(z_l^V(x_2) \right) \right) \\
&= \lambda g_l(x_1) + (1-\lambda) g_l(x_2), \tag{6.25}
\end{aligned}$$

$g_l(x)$ is convex. Since the sum of convex functions is also convex, it follows that each of the left-hand side functions of the constraints in (6.24), expressed as $g_l(x) + \rho \sum_{i=1}^{k} g_i(x)$, is also convex, which implies that (6.24) can be solved by a traditional convex programming technique such as the sequential quadratic programming method (Fletcher 1980; Gill et al. 1981; Powell 1983).

In order to derive a satisficing solution for the DM from among the M-Pareto optimal solution set, we present an interactive fuzzy satisficing method where the reference membership levels are repeatedly updated until the DM obtains a satisficing solution.

Procedure of Interactive Fuzzy Satisficing Method for the Variance Model

Step 1   Ask the DM to specify the satisficing probability levels β_i, $i = 1, \ldots, m$.
Step 2   Calculate the individual minima and maxima of $E[\bar{c}_l]x$, $l = 1, \ldots, k$ by solving linear programming problems (6.19).
Step 3   Ask the DM to specify the expectation levels γ_l, $l = 1, \ldots, k$.
Step 4   Calculate the individual minima of $z_l^V(x)$, $l = 1, \ldots, k$ by solving the quadratic programming problems (6.21).
Step 5   Ask the DM to specify the membership functions μ_l, $l = 1, \ldots, k$ with the individual minima $z_{l,\min}^V$ obtained in step 4 in mind.
Step 6   Set the initial reference membership levels at 1s, i.e., $\hat{\mu}_l = 1, l = 1, \ldots, k$, which can be viewed as the ideal values.
Step 7   For the current reference membership levels $\hat{\mu}_l$, $l = 1, \ldots, k$, solve the augmented minimax problem (6.23).
Step 8   The DM is supplied with the corresponding M-Pareto optimal solution. If the DM is satisfied with the current membership function values $\mu_l(z_l^V(x))$, $l = 1, \ldots, k$, then stop. Otherwise, ask the DM to update the reference membership levels $\hat{\mu}_l$, $l = 1, \ldots, k$, taking into account the current membership function values $\mu_l(z_l^V(x))$, and return to step 7.

Example 6.2. Consider the following two-objective stochastic production planning problem which is similar to that of Example 6.1:

$$\begin{aligned}
\text{minimize} \quad & z_1 = -\bar{3}x_1 - \bar{8}x_2 \\
\text{minimize} \quad & z_2 = \bar{5}x_1 + \bar{4}x_2 \\
\text{subject to} \quad & 2x_1 + 6x_2 \le \overline{27} \\
& 3x_1 + 2x_2 \le \overline{16} \\
& 4x_1 + x_2 \le \overline{18} \\
& x_1 \ge 0, \quad x_2 \ge 0,
\end{aligned}$$

where $E[-\bar{3}] = -3$, $E[-\bar{8}] = -8$, $E[\bar{5}] = 5$, and $E[\bar{4}] = 4$, $\overline{27} \sim N(27, 3^2)$, $\overline{16} \sim N(16, 2^2)$, and $\overline{18} \sim N(18, 2^2)$. Moreover, assume that the variance–covariance matrices of the coefficients of the objective functions z_1 and z_2 are given as $V_1 = \begin{bmatrix} 1 & 0.5 \\ 0.5 & 2 \end{bmatrix}$ and $V_2 = \begin{bmatrix} 2 & 0.5 \\ 0.5 & 1 \end{bmatrix}$, respectively.

6.1 Multiobjective Chance Constrained Programming

Assume that by using Zimmermann's method (Zimmermann 1978), the DM determines the linear membership functions as, for $l = 1, 2$,

$$\mu_l(z_l^V(x)) = \begin{cases} 1 & \text{if } z_l^V(x) \leq z_{l,1}^V \\ \dfrac{z_l^V(x) - z_{l,0}^V}{z_{l,1}^V - z_{l,0}^V} & \text{if } z_{l,1}^V < z_l^V(x) \leq z_{l,0}^V \\ 0 & \text{if } z_{l,0}^V < z_l^V(x), \end{cases}$$

where $z_1^V(x) = (x_1, x_2) \begin{bmatrix} 1 & 0.5 \\ 0.5 & 2 \end{bmatrix} \begin{pmatrix} x_1 \\ x_2 \end{pmatrix}$, $z_2^V(x) = (x_1, x_2) \begin{bmatrix} 2 & 0.5 \\ 0.5 & 1 \end{bmatrix} \begin{pmatrix} x_1 \\ x_2 \end{pmatrix}$, $z_{1,1}^V = 1.47845$, $z_{1,0}^V = 1.53125$, $z_{2,1}^V = 0.76563$, and $z_{2,0}^V = 0.92130$.

Moreover, assume that the DM specifies the satisficing probability levels as $\beta_i = 0.8$, $i = 1, 2, 3$ and provides the expectation levels as $\gamma_1 = -7$ and $\gamma_2 = 20$, taking into account the individual minima and maxima of $E[\bar{c}_l]x$, $l = 1, 2$. The augmented minimax problem (6.24) for the variance model with the chance constraints and the expectation levels is formulated as follows:

$$\begin{aligned}
\text{minimize} \quad & v \\
\text{subject to} \quad & \hat{\mu}_1 - \mu_1(z_1^V(x)) + \rho \sum_{l=1}^{2} \{\hat{\mu}_l - \mu_l(z_l^V(x))\} \leq v \\
& \hat{\mu}_2 - \mu_2(z_2^V(x)) + \rho \sum_{l=1}^{2} \{\hat{\mu}_l - \mu_l(z_l^V(x))\} \leq v \\
& 2x_1 + 6x_2 \leq 27 + 3\Phi^{-1}(0.2) = 24.475 \\
& 3x_1 + 2x_2 \leq 16 + 2\Phi^{-1}(0.2) = 14.317 \\
& 4x_1 + x_2 \leq 18 + 2\Phi^{-1}(0.2) = 16.317 \\
& E[\bar{c}_1]x = -3x_1 - 8x_2 \leq -7 \\
& E[\bar{c}_2]x = 5x_1 + 4x_2 \leq 20 \\
& x_1 \geq 0, \; x_2 \geq 0,
\end{aligned}$$

where Φ^{-1} is the inverse function of the distribution function Φ of the standard normal distribution $N(0, 1)$ and the parameter ρ is set at 0.000001.

An example of an interactive process is given in Table 6.2. Starting from the initial reference membership levels $(\hat{\mu}_1, \hat{\mu}_2) = (1.0, 1.0)$, we have the satisficing solution after two updates of the reference membership levels. At each iteration the augmented minimax problem is solved by using a VBA program with the Excel solver. ◊

Table 6.2 Interactive process in the variance model

Iteration	First	Second	Third
$\hat{\mu}_1$	1.00	0.75	0.75
$\hat{\mu}_2$	1.00	0.68	0.66
$\mu_1\left(z_1^V(x)\right)$	0.706131	0.742557	0.752597
$\mu_2\left(z_2^V(x)\right)$	0.706131	0.672557	0.662597
$z_1^V(x)$	1.493965	1.492042	1.491512
$z_2^V(x)$	0.811373	0.816599	0.818150

6.1.3 Probability Model

In contrast to the expectation and variance models discussed in the previous section, a certain target value is introduced for each of the objective function values in the probability model, and the probability that the objective function value is smaller than or equal to the target value is maximized, and then the objective functions of the probability model are represented as

$$P\left(\bar{c}_l x \le f_l\right), l = 1, \ldots, k, \qquad (6.26)$$

where f_l, $l = 1, \ldots, k$ are the target values given by the DM for the objective functions.

To help the DM specify the target values f_l, $l = 1, \ldots, k$, it is recommended to calculate the individual minima and maxima of $E[\bar{c}_l]x$, $l = 1, \ldots, k$ obtained by solving the linear programming problems

$$\left.\begin{array}{l} \displaystyle\minimize_{x \in X(\beta)} z_l^E(x) = E[\bar{c}_l]x, \ l = 1, \ldots, k, \\ \displaystyle\maximize_{x \in X(\beta)} z_l^E(x) = E[\bar{c}_l]x, \ l = 1, \ldots, k, \end{array}\right\} \qquad (6.27)$$

where $X(\beta)$ is the feasible region of the chance constraints with respect to the satisficing probability levels $\beta = (\beta_1, \ldots, \beta_m)$ defined by (6.5).

In the probability model for the multiobjective linear programming problems (Yano 2012), the problem maximizing the probability that the objective function value is smaller than a certain target value is formulated as

$$\left.\begin{array}{l} \text{maximize} \quad z_1^P(x) = P\left(\bar{c}_1 x \le f_1\right) \\ \qquad\qquad \ldots\ldots\ldots \\ \text{maximize} \quad z_k^P(x) = P\left(\bar{c}_k x \le f_k\right) \\ \text{subject to} \quad x \in X(\beta). \end{array}\right\} \qquad (6.28)$$

Assume that $\bar{c}_l = (\bar{c}_{l1}, \ldots, \bar{c}_{ln})$, $l = 1, \ldots, k$ is a multivariate normal random variable with the mean vector $E[\bar{c}_l]$ and the $n \times n$ variance–covariance matrix V_l and then the random variable

6.1 Multiobjective Chance Constrained Programming

$$\frac{\bar{c}_l x - E[\bar{c}_l] x}{\sqrt{x^T V_l x}} \tag{6.29}$$

is the standard normal random variable $N(0, 1)$. From this fact, it follows that

$$P(\bar{c}_l x \le f_l) = P\left(\frac{\bar{c}_l x - E[\bar{c}_l] x}{\sqrt{x^T V_l x}} \le \frac{f_l - E[\bar{c}_l] x}{\sqrt{x^T V_l x}}\right) = \Phi\left(\frac{f_l - E[\bar{c}_l] x}{\sqrt{x^T V_l x}}\right), \tag{6.30}$$

where Φ is the distribution function of the standard normal random variable.

Hence, the problem (6.28) can be equivalently transformed to

$$\left.\begin{array}{l} \text{maximize} \quad z_1^P(x) = \Phi\left(\dfrac{f_1 - E[\bar{c}_1] x}{\sqrt{x^T V_1 x}}\right) \\ \qquad\qquad \cdots\cdots\cdots \\ \text{maximize} \quad z_k^P(x) = \Phi\left(\dfrac{f_k - E[\bar{c}_k] x}{\sqrt{x^T V_k x}}\right) \\ \text{subject to} \quad x \in X(\beta). \end{array}\right\} \tag{6.31}$$

Taking into account the imprecise nature of the DM's judgments for each objective function in (6.31), by introducing the fuzzy goals such as "$z_l^P(x)$ should be substantially larger than or equal to a certain value," (6.31) can be interpreted as

$$\underset{x \in X(\beta)}{\text{maximize}} \left\{\mu_1\left(z_1^P(x)\right), \ldots, \mu_k\left(z_k^P(x)\right)\right\}, \tag{6.32}$$

where $\mu_l, l = 1, \ldots, k$ is a membership function to quantify the fuzzy goal for the lth objective function $z_l^P(x)$.

In order to find a candidate for the satisficing solution, the DM specifies the reference membership levels $\hat{\mu}_l, i = 1, \ldots, k$, and then by solving the minimax problem

$$\underset{x \in X(\beta)}{\text{minimize}} \max_{1 \le l \le k} \left\{\hat{\mu}_l - \mu_l\left(z_l^P(x)\right)\right\}, \tag{6.33}$$

an M-Pareto optimal solution corresponding to $\hat{\mu}_l, i = 1, \ldots, k$ is obtained. The problem (6.33) is equivalently expressed as

$$\left.\begin{array}{l} \text{minimize} \quad v \\ \text{subject to} \quad \hat{\mu}_1 - \mu_1\left(z_1^P(x)\right) \le v \\ \qquad\qquad \cdots\cdots\cdots \\ \qquad\qquad \hat{\mu}_k - \mu_k\left(z_k^P(x)\right) \le v \\ \qquad\qquad x \in X(\beta), \end{array}\right\} \tag{6.34}$$

and it is also rewritten as

$$
\begin{aligned}
&\text{minimize} \quad v \\
&\text{subject to} \quad \Phi^{-1}(\mu_1^{-1}(\hat{\mu}_1 - v))\sqrt{x^T V_1 x} \leq f_1 - E[\bar{c}_1]x \\
&\qquad\qquad\qquad \cdots\cdots\cdots\cdots \\
&\qquad\qquad \Phi^{-1}(\mu_k^{-1}(\hat{\mu}_k - v))\sqrt{x^T V_k x} \leq f_k - E[\bar{c}_k]x \\
&\qquad\qquad x \in X(\beta).
\end{aligned}
\quad (6.35)
$$

As in the expectation model, v in (6.35) satisfies

$$\hat{\mu}_{\min} \leq v \leq \hat{\mu}_{\max},$$

where

$$\hat{\mu}_{\min} = \max_{1 \leq l \leq k} \hat{\mu}_l - 1, \text{ and } \hat{\mu}_{\max} = \min_{1 \leq l \leq k} \hat{\mu}_l.$$

In order to solve (6.35) which is a nonlinear and nonconvex programming problem, we employ an algorithm based on the bisection method (Yano 2012). In this algorithm, the following functions with the parameter v for the constraints of (6.35) are defined:

$$g_l(x; v) = \Phi^{-1}(\mu_l^{-1}(\hat{\mu}_l - v))\sqrt{x^T V_l x} - (f_l - E[\bar{c}_l]x), \; l = 1, \ldots, k, \quad (6.36)$$

and it is assumed that the membership functions μ_l, $l = 1, \ldots, k$ are strictly increasing functions on $[p_{l\,\min}, p_{l\,\max}]$, where $\mu_l(p_{l\,\min}) = 0$, $\mu_l(p_{l\,\max}) = 1$, and $0.5 < p_{l\,\min} < p_{l\,\max} < 1$.

By this assumption, it follows that $\Phi^{-1}(\mu_l^{-1}(\hat{\mu}_l - v)) > 0, l = 1, \ldots, k$ holds. Thus, it is clear that, for any fixed $v \in [\hat{\mu}_{\min}, \hat{\mu}_{\max}]$, the constraint set

$$G(v) = \{x \in X(\beta) \mid g_l(x; v) \leq 0, \; l = 1, \ldots, k\} \quad (6.37)$$

is convex. Moreover, it is also obvious that if $v_1, v_2 \in [\hat{\mu}_{\min}, \hat{\mu}_{\max}]$ and $v_1 < v_2$, then it holds that $G(v_1) \subset G(v_2)$.

Assuming that $G(\hat{\mu}_{\max}) \neq \emptyset$ and $G(\hat{\mu}_{\min}) = \emptyset$, we present an algorithm based on the bisection method for v for obtaining an optimal solution (x^*, v^*) to (6.35) (Yano 2012).

Step 1 Set $v_0 = \hat{\mu}_{\min}$, $v_1 = \hat{\mu}_{\max}$ and $v \leftarrow (v_0 + v_1)/2$.
Step 2 Solve the following convex programming problem for the given v:

$$
\begin{aligned}
&\text{minimize} \quad g_1(x; v) \\
&\text{subject to} \quad g_l(x; v) \leq 0, \; l = 2, \ldots, k \\
&\qquad\qquad x \in X(\beta).
\end{aligned}
\quad (6.38)
$$

Let x^* denote an optimal solution to (6.38).

6.1 Multiobjective Chance Constrained Programming

Step 3 Let ϵ be a sufficiently small positive constant. If $|v_1 - v_0| < \epsilon$, then go to step 4. If $g_1(x^*; v) > 0$ or there exists some index $l \neq 1$ such that $g_l(x^*; v) > 0$, then set $v_0 \leftarrow v$, $v \leftarrow (v_0 + v_1)/2$, and go to step 2. Otherwise, set $v_1 \leftarrow v$, $v \leftarrow (v_0 + v_1)/2$, and go to step 2.

Step 4 Set $v^* \leftarrow v$, and the optimal solution (x^*, v^*) to (6.35) is obtained.

We present an interactive algorithm for deriving a satisficing solution of the DM from among the M-Pareto optimal solution set.

Procedure of Interactive Fuzzy Satisficing Method for the Probability Model

Step 1 Ask the DM to specify the satisficing probability levels β_i, $i = 1, \ldots, m$.

Step 2 Calculate the individual minima and maxima of $E[\bar{c}_l]x$, $l = 1, \ldots, k$ by solving linear programming problems (6.27).

Step 3 Ask the DM to specify the target values f_l, $l = 1, \ldots, k$.

Step 4 Ask the DM to specify the strictly increasing membership functions μ_l on $[p_{l\min}, p_{l\max}]$, where $\mu_l(p_{l\min}) = 0$, $\mu_l(p_{l\max}) = 1$, and $0.5 < p_{l\min} < p_{l\max} < 1$, $l = 1, \ldots, k$.

Step 5 Set the initial reference membership levels at 1s, which can be viewed as the ideal values, i.e., $\hat{\mu}_l = 1$, $l = 1, \ldots, k$.

Step 6 For the reference membership levels $\hat{\mu}_l$, $l = 1, \ldots, k$, solve the minimax problem (6.35) by using the above-mentioned algorithm.

Step 7 The DM is supplied with the corresponding M-Pareto optimal solution. If the DM is satisfied with the current membership function values $\mu_l(z_l^P(x))$, $l = 1, \ldots, k$, then stop. Otherwise, ask the DM to update the reference membership levels $\hat{\mu}_l$, $l = 1, \ldots, k$ by using the current membership function values $\mu_l(z_l^P(x))$ as a reference, and return to step 6.

Example 6.3. Consider the following two-objective stochastic production planning problem which is the same as that of Example 6.2:

$$\begin{aligned}
\text{minimize} \quad & z_1 = \bar{-3}x_1 - \bar{8}x_2 \\
\text{minimize} \quad & z_2 = \bar{5}x_1 + \bar{4}x_2 \\
\text{subject to} \quad & 2x_1 + 6x_2 \leq \overline{27} \\
& 3x_1 + 2x_2 \leq \overline{16} \\
& 4x_1 + x_2 \leq \overline{18} \\
& x_1 \geq 0, \quad x_2 \geq 0,
\end{aligned}$$

where $E[\bar{-3}] = -3$, $E[\bar{-8}] = -8$, $V_1 = \begin{bmatrix} 1 & 0.5 \\ 0.5 & 2 \end{bmatrix}$, $E[\bar{5}] = 5$, and $E[\bar{4}] = 4$, $V_2 = \begin{bmatrix} 2 & 0.5 \\ 0.5 & 1 \end{bmatrix}$, $\overline{27} \sim N(27, 3^2)$, $\overline{16} \sim N(16, 2^2)$, and $\overline{18} \sim N(18, 2^2)$.

Suppose that the DM specifies the satisficing probability levels as $\beta_i = 0.8$, $i = 1, 2, 3$ and provides the target values for the objective functions as $f_1 = -20$ and $f_2 = 15$.

The multiobjective problem based on the probability model with the chance constraints is formulated as follows:

$$\text{maximize} \quad z_1^P(x) = P(-\bar{3}x_1 - \bar{8}x_2 \leq -20) = \Phi \left(\frac{(-20) - (-3x_1 - 8x_2)}{\sqrt{(x_1, x_2) \begin{bmatrix} 1 & 0.5 \\ 0.5 & 2 \end{bmatrix} \begin{pmatrix} x_1 \\ x_2 \end{pmatrix}}} \right)$$

$$\text{maximize} \quad z_2^P(x) = P(\bar{5}x_1 + \bar{4}x_2 \leq 15) = \Phi \left(\frac{(15) - (5x_1 + 4x_2)}{\sqrt{(x_1, x_2) \begin{bmatrix} 2 & 0.5 \\ 0.5 & 1 \end{bmatrix} \begin{pmatrix} x_1 \\ x_2 \end{pmatrix}}} \right)$$

subject to
$2x_1 + 6x_2 \leq 27 + 3\Phi^{-1}(0.2) = 24.475$
$3x_1 + 2x_2 \leq 16 + 2\Phi^{-1}(0.2) = 14.317$
$4x_1 + x_2 \leq 18 + 2\Phi^{-1}(0.2) = 16.317$
$x_1 \geq 0, \ x_2 \geq 0,$

where Φ is the distribution function of the standard normal distribution $N(0, 1)$ and Φ^{-1} is its inverse function.

Moreover, assume that the DM employs the following linear membership function for each of $z_l^P(x)$, $l = 1, 2$:

$$\mu_l(z_l^P(x)) = \begin{cases} 1 & \text{if } z_l^P(x) \leq z_{l,1}^P \\ \dfrac{z_l^P(x) - z_{l,0}^P}{z_{l,1}^P - z_{l,0}^P} & \text{if } z_{l,1}^P < z_l^P(x) \leq z_{l,0}^P \\ 0 & \text{if } z_{l,0}^P < z_l^P(x), \end{cases}$$

where the parameters are specified as $z_{1,1}^P = 0.95$, $z_{1,0}^P = 0.65$, $z_{2,1}^P = 0.9$, and $z_{2,0}^P = 0.6$.

The minimax problem with the reference membership levels $\hat{\mu}_l$, $i = 1, 2$ for finding a candidate for the satisficing solution is formulated as

6.1 Multiobjective Chance Constrained Programming

Table 6.3 Interactive process in the probability model

Iteration	First	Second	Third
$\hat{\mu}_1$	1.00	0.72	0.7
$\hat{\mu}_2$	1.00	0.68	0.67
$z_1^P(x^*)$	0.857173	0.862352	0.861070
$z_2^P(x^*)$	0.807173	0.800352	0.802070
$\mu_1(z_1^P(x^*))$	0.690578	0.707839	0.703565
$\mu_2(z_2^P(x^*))$	0.690578	0.667839	0.673565

minimize $\quad v$

subject to $\quad \Phi^{-1}(\mu_1^{-1}(\hat{\mu}_1 - v))(x_1, x_2) \begin{bmatrix} 1 & 0.5 \\ 0.5 & 2 \end{bmatrix} \begin{pmatrix} x_1 \\ x_2 \end{pmatrix} \leq (-20) - (-3x_1 - 8x_2)$

$\Phi^{-1}(\mu_2^{-1}(\hat{\mu}_2 - v))(x_1, x_2) \begin{bmatrix} 2 & 0.5 \\ 0.5 & 1 \end{bmatrix} \begin{pmatrix} x_1 \\ x_2 \end{pmatrix} \leq (15) - (5x_1 + 4x_2)$

$2x_1 + 6x_2 \leq 27 + 3\Phi^{-1}(0.2) = 24.475$

$3x_1 + 2x_2 \leq 16 + 2\Phi^{-1}(0.2) = 14.317$

$4x_1 + x_2 \leq 18 + 2\Phi^{-1}(0.2) = 16.317$

$x_1 \geq 0, \; x_2 \geq 0$.

We can obtain an optimal solution to this problem by using a VBA program with the Excel solver based on the above-mentioned algorithm for solving (6.35).

An example of an interactive process is given in Table 6.3. Starting from the initial reference membership levels $(\hat{\mu}_1, \hat{\mu}_2) = (1.0, 1.0)$, we have the satisficing solution after two updates of the reference membership levels. $\Diamond$

So far, we have assumed that $\bar{c}_l = (\bar{c}_{l1}, \ldots, \bar{c}_{ln}), l = 1, \ldots, k$ is a multivariate normal random variable with the mean vector $E[\bar{c}_l]$ and the $n \times n$ variance–covariance matrix V_l. In contrast, Stancu-Minasian (1984) gives a different and somewhat simpler representation of stochastic objective functions. To be more specific, the lth objective function is expressed as

$$z_l(x) = \bar{c}_l x + \bar{v}_l, \tag{6.39}$$

where $\bar{c}_l$ is an n-dimensional coefficient row vector of random variables and $\bar{v}_l$ is also a random variable. Furthermore, it is assumed that $\bar{c}_l$ and $\bar{v}_l$ are expressed by

$$\bar{c}_l = c_l^1 + \bar{t}_l c_l^2 \quad \text{and} \quad \bar{v}_l = v_l^1 + \bar{t}_l v_l^2, \tag{6.40}$$

respectively. In particular, it is noted that $\bar{t}_l, l = 1, \ldots, k$ are mutually independent random variables with the continuous and strictly increasing probability distribution functions T_l, and c_l^1, c_l^2, v_l^1, and v_l^2 are not random variables but only real numbers.

In a similar way to (6.26), the objective function in the probability model with the special expression of random variables (6.40) is represented as

$$z_l^P(x) = P\left(\bar{c}_l x + \bar{v}_l \leq f_l\right), \tag{6.41}$$

where f_l is the given target value for the lth objective function.

If it holds that $c_l^2 x + v_l^2 > 0$ for any $x \in X(\beta)$, from the property of the distribution function T_l of the random variable $\bar{t}_l$, it is obvious that (6.41) can be transformed to

$$z_l^P(x) = P\left((c_l^1 + \bar{t}_l c_l^2)x + (v_l^1 + \bar{t}_l v_l^2) \leq f_l\right)$$
$$= P\left(\bar{t}_l \leq \frac{f_l - (c_l^1 x + v_l^1)}{(c_l^2 x + v_l^2)}\right)$$
$$= T_l\left(\frac{f_l - c_l^1 x - v_l^1}{c_l^2 x + v_l^2}\right). \tag{6.42}$$

Through a similar discussion in the probability model with the general expression of random variables, the minimax problem (6.34) is formulated, and it is rewritten as

$$\left.\begin{array}{l}\text{minimize} \quad v \\ \text{subject to} \quad T^{-1}(\mu_1^{-1}(\hat{\mu}_1 - v))(c_1^2 x + v_1^2) \leq f_1 - c_1^1 x - v_1^1 \\ \qquad\qquad \cdots\cdots\cdots\cdots \\ \qquad T^{-1}(\mu_k^{-1}(\hat{\mu}_k - v))(c_k^2 x + v_k^2) \leq f_k - c_k^1 x - v_k^1 \\ \quad x \in X(\beta). \end{array}\right\} \tag{6.43}$$

Since (6.43) can be solved by the combined use of phase I of the simplex method and the bisection method, we can give a similar interactive algorithm for the probability model with the special expression of random variables (6.39).

Example 6.4. Consider the following two-objective stochastic production planning problem which is similar to that of Example 6.2:

$$\begin{array}{rlrl}
\text{minimize} & z_1 & = & (-3 + \bar{t}_1)x_1 + (-8 + \sqrt{2}\bar{t}_2)x_2 \\
\text{minimize} & z_2 & = & (5 + \sqrt{2}\bar{t}_1)x_1 + (4 + \bar{t}_1)x_2 \\
\text{subject to} & & & 2x_1 + 6x_2 \leq \overline{27} \\
& & & 3x_1 + 2x_2 \leq \overline{16} \\
& & & 4x_1 + x_2 \leq \overline{18} \\
& & & x_1 \geq 0, \quad x_2 \geq 0,
\end{array}$$

where $\bar{t}_1$ and $\bar{t}_2$ are the standard normal random variables $N(0, 1)$ and the right-hand sides of the constraints are $\overline{27} \sim N(27, 3^2)$, $\overline{16} \sim N(16, 2^2)$, and $\overline{18} \sim N(18, 2^2)$.

6.1 Multiobjective Chance Constrained Programming

Table 6.4 Interactive process in the probability model with stochastic objective functions (6.39)

Iteration	First	Second	Third
$\hat{\mu}_1$	1.000	0.7	0.69
$\hat{\mu}_2$	1.000	0.66	0.66
$\mu_1(z_1^P(x))$	0.67382	0.69157	0.68717
$\mu_2(z_2^P(x))$	0.67382	0.65157	0.65717
$z_1^P(x)$	0.81846	0.82289	0.82179
$z_2^P(x)$	0.83476	0.83031	0.83143
$-\partial\mu_1/\partial\mu_2$	0.81085	0.78477	0.79120

Suppose that the DM specifies the satisficing probability levels as $\beta_i = 0.8$, $i = 1, 2, 3$ and provides the target values for the objective functions as $f_1 = -27$ and $f_2 = 20$. The multiobjective problem based on the probability model with the chance constraints is then formulated as

$$\text{maximize} \quad z_1^P(x) = P((-3 + \bar{t}_1)x_1 + (-8 + \sqrt{2}\bar{t}_2)x_2 \leq -27)$$
$$= \Phi\left(\frac{(-27) - (-3x_1 - 8x_2)}{x_1 + \sqrt{2}x_2}\right)$$
$$\text{maximize} \quad z_2^P(x) = P((5 + \sqrt{2}\bar{t}_1)x_1 + (4 + \bar{t}_1)x_2 \leq 20)$$
$$= \Phi\left(\frac{(20) - (5x_1 + 4x_2)}{\sqrt{2}x_1 + x_2}\right)$$
$$\text{subject to} \quad 2x_1 + 6x_2 \leq 27 + 3\Phi^{-1}(0.2) = 24.475$$
$$3x_1 + 2x_2 \leq 16 + 2\Phi^{-1}(0.2) = 14.317$$
$$4x_1 + x_2 \leq 18 + 2\Phi^{-1}(0.2) = 16.317$$
$$x_1 \geq 0, \ x_2 \geq 0,$$

where Φ is the distribution function of the standard normal distribution $N(0, 1)$ and Φ^{-1} is its inverse function.

Moreover, as in Example 6.3, suppose that the DM employs the linear membership functions for $z_l^P(x)$, $l = 1, 2$ with the parameters $z_{1,1}^P = 0.9$, $z_{1,0}^P = 0.65$, $z_{2,1}^P = 0.9$, and $z_{2,0}^P = 0.7$.

An example of an interactive process is given in Table 6.4. Since we solve this problem by combined use of the bisection method and the simplex method of linear programming, the trade-off rates between the membership functions can be calculated as shown in Table 6.4. ◇

6.1.4 Fractile Model

For a decision situation where the DM prefers to maximize the probability that each of the objective functions is smaller than or equal to a target value specified by the DM, the probability model is recommended. In contrast, if the DM rather wants to minimize the target variable under the condition that the probability that each of the objective functions is smaller than or equal to the target variable is guaranteed to be larger than or equal to a certain assured probability level specified by the DM, the fractile model is thought to be appropriate (Sakawa et al. 2001). It should be noted that the target variable is regarded as not a constant value of the goal as in the probability model but a variable to be minimized in the fractile model.

Replacing the minimization of the objective functions $z_l(x)$, $l = 1, \ldots, k$ in the multiobjective stochastic programming problem (6.4) with the minimization of the target variables f_l, $l = 1, \ldots, k$ under the probabilistic constraints with the assured probability levels $\theta_l \in (1/2, 1)$ specified by the DM, we consider the fractile model for the multiobjective stochastic programming problems formulated as

$$\left. \begin{aligned} &\text{minimize} && f_1 \\ &\quad\cdots\cdots \\ &\text{minimize} && f_k \\ &\text{subject to} && P(\bar{c}_1 x \leq f_1) \geq \theta_1 \\ &&& \cdots\cdots\cdots \\ &&& P(\bar{c}_k x \leq f_k) \geq \theta_k \\ &&& x \in X(\boldsymbol{\beta}), \end{aligned} \right\} \quad (6.44)$$

where each $\bar{c}_l$ is a normal random variable vector with the mean vector $E[\bar{c}_l]$ and the positive-definite variance–covariance matrix V_l.

From the fact that the random variable

$$\frac{\bar{c}_l x - E[\bar{c}_l] x}{\sqrt{x^T V_l x}}, \quad l = 1, \ldots, k \quad (6.45)$$

is the standardized normal random variable with mean 0 and variance 1^2, it follows that

$$P(\bar{c}_l x \leq f_l) = P\left(\frac{\bar{c}_l x - E[\bar{c}_l] x}{\sqrt{x^T V_l x}} \leq \frac{f_l - E[\bar{c}_l] x}{\sqrt{x^T V_l x}} \right) = \Phi\left(\frac{f_l - E[\bar{c}_l] x}{\sqrt{x^T V_l x}} \right), \quad (6.46)$$

where Φ is the distribution function of the standardized normal distribution. From (6.46), the probabilistic constraint

$$P(\bar{c}_l x \leq f_l) \geq \theta_l, \quad (6.47)$$

of (6.44) is equivalent to

$$\Phi\left(\frac{f_l - E[\bar{c}_l] x}{\sqrt{x^T V_l x}} \right) \geq \theta_l \Leftrightarrow \frac{f_l - E[\bar{c}_l] x}{\sqrt{x^T V_l x}} \geq \Phi^{-1}(\theta_l)$$

$$\Leftrightarrow f_l \geq E[\bar{c}_l] x + \Phi^{-1}(\theta_l) \sqrt{x^T V_l x}, \quad (6.48)$$

where Φ^{-1} is the inverse function of Φ.

6.1 Multiobjective Chance Constrained Programming

By substituting (6.48) for (6.47), the multiobjective stochastic programming problem for the fractile model (6.44) can be transformed into

$$\begin{aligned}
\text{minimize} \quad & f_1 \\
& \cdots \cdots \\
\text{minimize} \quad & f_k \\
\text{subject to} \quad & E[\bar{c}_1]x + \Phi^{-1}(\theta_1)\sqrt{x^T V_1 x} \leq f_1 \\
& \cdots \cdots \cdots \\
& E[\bar{c}_k]x + \Phi^{-1}(\theta_k)\sqrt{x^T V_k x} \leq f_k \\
& x \in X(\boldsymbol{\beta}).
\end{aligned} \quad (6.49)$$

Since in (6.49) the target variable f_l is minimized under the constraints that the target variables $f_l, l = 1, \ldots, k$ must be larger than or equal to

$$E[\bar{c}_l]x + \Phi^{-1}(\theta_l)\sqrt{x^T V_l x}, \quad (6.50)$$

it is obvious that minimizing f_l is equivalent to minimizing (6.50). Hence, (6.49) can be rewritten as

$$\begin{aligned}
\text{minimize} \quad & z_1^F(x) = E[\bar{c}_1]x + \Phi^{-1}(\theta_1)\sqrt{x^T V_1 x} \\
& \cdots \cdots \cdots \\
\text{minimize} \quad & z_k^F(x) = E[\bar{c}_k]x + \Phi^{-1}(\theta_k)\sqrt{x^T V_k x} \\
\text{subject to} \quad & x \in X(\boldsymbol{\beta}).
\end{aligned} \quad (6.51)$$

Since each of the objective functions in (6.51) is convex from the fact that $\Phi^{-1}(\theta_l) \geq 0$ for any $\theta_l \in [1/2, 1)$, it follows that (6.51) is a multiobjective convex programming problem.

In a way similar to the previous models, taking into account the imprecise nature of the DM's judgments for each objective function, we introduce the fuzzy goals such as "$z_l^F(x)$ should be substantially less than or equal to a certain value" and formulate the minimization problem of the fuzzy goals:

$$\underset{x \in X(\boldsymbol{\beta})}{\text{maximize}} \left\{ \mu_1\left(z_1^F(x)\right), \ldots, \mu_k\left(z_k^F(x)\right) \right\}, \quad (6.52)$$

where μ_l is assumed to be a concave membership function to quantify the fuzzy goal for the lth objective function in (6.51).

In order to find a candidate for the satisficing solution, the DM specifies the reference membership levels $\hat{\mu}_l, i = 1, \ldots, k$, and then by solving the augmented minimax problem

$$\underset{x \in X(\boldsymbol{\beta})}{\text{minimize}} \underset{1 \leq l \leq k}{\max} \left\{ \hat{\mu}_l - \mu_l(z_l^F(x)) + \rho \sum_{i=1}^{k} (\hat{\mu}_i - \mu_i(z_i^F(x))) \right\}, \quad (6.53)$$

an M-Pareto optimal solution corresponding to $\hat{\mu}_l$, $i = 1, \ldots, k$ is obtained. In (6.53), ρ is a sufficiently small positive number, and (6.53) is equivalently expressed as

$$
\left.\begin{aligned}
\text{minimize} \quad & v \\
\text{subject to} \quad & \hat{\mu}_1 - \mu_1(z_1^F(x)) + \rho \sum_{i=1}^{k} (\hat{\mu}_i - \mu_i(z_i^F(x))) \leq v \\
& \quad \cdots\cdots\cdots\cdots\cdots \\
& \hat{\mu}_k - \mu_k(z_k^F(x)) + \rho \sum_{i=1}^{k} (\hat{\mu}_i - \mu_i(z_i^F(x))) \leq v \\
& x \in X(\beta).
\end{aligned}\right\} \quad (6.54)
$$

As we illustrated that the augmented minimax problem (6.24) for the variance model is a convex programming problem, it can be shown that (6.54) is also convex under the assumption that each of membership functions μ_l, $l = 1, \ldots, k$ is nonincreasing and concave. Due to the convexity, an optimal solution to (6.54) can be found by using some convex programming technique such as the sequential quadratic programming method.

In the fractile model, we present an interactive algorithm for deriving a satisficing solution for the DM from among the M-Pareto optimal solution set.

Procedure of Interactive Fuzzy Satisficing Method for the Fractile Model

Step 1 Ask the DM to specify the satisficing probability levels β_i, $i = 1, \ldots, m$ and the assured probability levels $\theta_l \in [1/2, 1)$, $l = 1, \ldots, k$.
Step 2 Calculate the individual minima of $z_l^F(x)$, $l = 1, \ldots, k$.
Step 3 Ask the DM to specify the nonincreasing and concave membership functions μ_l by using the individual minima and maxima obtained in step 2 as a reference.
Step 4 Set the initial reference membership levels at 1s, which can be viewed as the ideal values, i.e., $\hat{\mu}_l = 1$, $l = 1, \ldots, k$.
Step 5 For the reference membership levels $\hat{\mu}_l$, $l = 1, \ldots, k$, solve the augmented minimax problem (6.54).
Step 6 The DM is supplied with the corresponding M-Pareto optimal solution. If the DM is satisfied with the current membership function values $\mu_l\left(z_l^F(x)\right)$, $l = 1, \ldots, k$, then stop. Otherwise, ask the DM to update the reference membership levels $\hat{\mu}_l$, $l = 1, \ldots, k$ by considering the current membership function values $\mu_l\left(z_l^F(x)\right)$, and return to step 5.

Example 6.5. Consider the following two-objective stochastic production planning problem which is the same as that of Example 6.2:

6.1 Multiobjective Chance Constrained Programming

$$\begin{aligned}
\text{minimize} \quad & z_1 = \bar{-3}x_1 - \bar{8}x_2 \\
\text{minimize} \quad & z_2 = \bar{5}x_1 + \bar{4}x_2 \\
\text{subject to} \quad & 2x_1 + 6x_2 \leq \overline{27} \\
& 3x_1 + 2x_2 \leq \overline{16} \\
& 4x_1 + x_2 \leq \overline{18} \\
& x_1 \geq 0, \quad x_2 \geq 0,
\end{aligned}$$

where $E[\bar{-3}] = -3$, $E[\bar{-8}] = -8$, $V_1 = \begin{bmatrix} 1 & 0.5 \\ 0.5 & 2 \end{bmatrix}$, $E[\bar{5}] = 5$, $E[\bar{4}] = 4$, $V_2 = \begin{bmatrix} 2 & 0.5 \\ 0.5 & 1 \end{bmatrix}$, $\overline{27} \sim N(27, 3^2)$, $\overline{16} \sim N(16, 2^2)$, and $\overline{18} \sim N(18, 2^2)$.

Suppose that the DM specifies the satisficing probability levels as $\beta_i = 0.8$, $i = 1, 2, 3$ and provides the assured probability levels as $\theta_i = 0.7$, $i = 1, 2$.

The multiobjective problem based on the fractile model with the chance constraints is formulated as

$$\text{minimize} \quad z_1^F(x) = (-3x_1 - 8x_2) + \Phi^{-1}(0.7) \sqrt{(x_1, x_2) \begin{bmatrix} 1 & 0.5 \\ 0.5 & 2 \end{bmatrix} \begin{pmatrix} x_1 \\ x_2 \end{pmatrix}}$$

$$\text{maximize} \quad z_2^F(x) = (5x_1 + 4x_2) + \Phi^{-1}(0.7) \sqrt{(x_1, x_2) \begin{bmatrix} 2 & 0.5 \\ 0.5 & 1 \end{bmatrix} \begin{pmatrix} x_1 \\ x_2 \end{pmatrix}}$$

$$\begin{aligned}
\text{subject to} \quad & 2x_1 + 6x_2 \leq 27 + 3\Phi^{-1}(0.2) = 24.475 \\
& 3x_1 + 2x_2 \leq 16 + 2\Phi^{-1}(0.2) = 14.317 \\
& 4x_1 + x_2 \leq 18 + 2\Phi^{-1}(0.2) = 16.317 \\
& x_1 \geq 0, \quad x_2 \geq 0,
\end{aligned}$$

where Φ^{-1} is the inverse function of the distribution function Φ of the standard normal distribution $N(0, 1)$.

The individual minima of the objective functions $z_l^F(x)$, $l = 1, 2$ are calculated as $z_{1,\min}^F = -30.37201$ and $z_{2,\min}^F = 0$ by using the Excel solver. Assume that by using Zimmermann's method (Zimmermann 1978), the DM identifies the following linear membership function:

$$\mu_l(z_l^F(x)) = \begin{cases} 1 & \text{if } z_l^F(x) \leq z_{l,1}^F \\ \dfrac{z_l^F(x) - z_{l,0}^F}{z_{l,1}^F - z_{l,0}^F} & \text{if } z_{l,1}^F < z_l^F(x) \leq z_{l,0}^F \\ 0 & \text{if } z_{l,0}^F < z_l^F(x), \end{cases}$$

where $z_{l,1}^F$ and $z_{l,0}^F$ are calculated as $z_{1,1}^F = -30.37201$, $z_{1,0}^F = 0$, $z_{2,1}^F = 0$, and $z_{2,0}^F = 28.98892$.

Table 6.5 Interactive process in the fractile model

Iteration	First	Second	Third
$\hat{\mu}_1$	1.00	0.65	0.65
$\hat{\mu}_2$	1.00	0.58	0.60
$z_1^F(x^*)$	-18.37305	-19.65916	-19.29170
$z_2^F(x^*)$	11.45255	12.25423	12.02518
$\mu_1(z_1^F(x^*))$	0.60493	0.64728	0.63518
$\mu_2(z_2^F(x^*))$	0.60493	0.57728	0.58518

The augmented minimax problem with the reference membership levels $\hat{\mu}_l$, $i = 1, 2$ for finding a candidate for the satisficing solution is formulated as

$$\begin{aligned}
\text{minimize} \quad & v \\
\text{subject to} \quad & \hat{\mu}_1 - \mu_1\left(z_1^F(x)\right) + \rho \sum_{l=1}^{2} \{\hat{\mu}_l - \mu_l\left(z_l^F(x)\right)\} \leq v \\
& \hat{\mu}_2 - \mu_2\left(z_2^F(x)\right) + \rho \sum_{l=1}^{2} \{\hat{\mu}_l - \mu_l\left(z_l^F(x)\right)\} \leq v \\
& 2x_1 + 6x_2 \leq 27 + 3\Phi^{-1}(0.2) = 24.475 \\
& 3x_1 + 2x_2 \leq 16 + 2\Phi^{-1}(0.2) = 14.317 \\
& 4x_1 + x_2 \leq 18 + 2\Phi^{-1}(0.2) = 16.317 \\
& x_1 \geq 0, \ x_2 \geq 0,
\end{aligned}$$

where the parameter ρ is set at 0.0001.

An example of an interactive process is given in Table 6.5. Starting from the initial reference membership levels $(\hat{\mu}_1, \hat{\mu}_2) = (1.0, 1.0)$, we have the satisficing solution after two updates of the reference membership levels. At each iteration the augmented minimax problem is solved by using the Excel solver. ◇

6.2 Multiobjective Simple Recourse Optimization

In the chance constrained problems which have been discussed in the previous sections, for random data variations, a mathematical model is formulated such that the violation of the constraints is permitted up to the specified probability levels. On the other hand, in a two-stage model including a simple recourse model as a special case, a shortage or an excess arising from the violation of the constraints is penalized, and then the expectation of the amount of the penalties for the constraint violation is minimized.

6.2 Multiobjective Simple Recourse Optimization

Consider a simple recourse model for the multiobjective stochastic programming problems

$$\left. \begin{array}{ll} \text{minimize} & z_1(x) = c_1 x \\ \quad\quad\quad \ldots\ldots \\ \text{minimize} & z_k(x) = c_k x \\ \text{subject to} & Ax = \bar{b} \\ & x \geq 0, \end{array} \right\} \quad (6.55)$$

where x is an n-dimensional decision variable column vector, c_l, $l = 1, 2, \ldots, k$ are n-dimensional coefficient row vectors, $A = [a_{ij}]$ is an $m \times n$ coefficient matrix, and $\bar{b}$ is an m-dimensional random variable column vector. It should be noted here that, in a simple recourse model, the random variables are involved only in the right-hand side of the constraints.

In a simple recourse model, it is assumed that the DM must make a decision before the realized values of the random variables involved in (6.55) are observed, and the penalty of the violation of the constraints is incorporated into the objective function in order to take into account the loss caused by random date variations.

To be more specific, by expressing the difference between Ax and $\bar{b}$ in (6.55) as two vectors $y^+ = (y_1^+, \ldots, y_m^+)^T$ and $y^- = (y_1^-, \ldots, y_m^-)^T$, the expectation of a recourse for the lth objective function is represented by

$$R_l(x) = E\left[\min_{y^+, y^-} (q_l^+ y^+ + q_l^- y^-) \,\middle|\, y^+ - y^- = \bar{b} - Ax\right], \quad (6.56)$$

where q_l^+ and q_l^- are m-dimensional constant row vectors. Thinking of each element of $y^+ = (y_1^+, \ldots, y_m^+)^T$ and $y^- = (y_1^-, \ldots, y_m^-)^T$ as a shortage and an excess of the constant in the left-hand side of the constraint such as the amount of the resource, respectively, we can regard each element of q_l^+ and q_l^- as the cost to compensate the shortage and the cost to dispose the excess, respectively.

For the multiobjective stochastic programming problem, the simple recourse problem is then formulated as

$$\left. \begin{array}{ll} \text{minimize} & c_1 x + R_1(x) \\ \quad\quad\quad \ldots\ldots \\ \text{minimize} & c_k x + R_k(x) \\ \text{subject to} & x \geq 0. \end{array} \right\} \quad (6.57)$$

Since q_l^+ and q_l^- are interpreted as penalty coefficients for the shortages and the excesses, it is quite natural to assume that $q_l^+ \geq 0$ and $q_l^- \geq 0$, and then it is evident that, for all $i = 1, \ldots, m$, the complementary relations

$$\hat{y}_i^+ > 0 \Rightarrow \hat{y}_i^- = 0,$$
$$\hat{y}_i^- > 0 \Rightarrow \hat{y}_i^+ = 0$$

should be satisfied for an optimal solution. With this observation in mind, we have

$$\hat{y}_i^+ = \bar{b}_i - \sum_{j=1}^n a_{ij} x_j, \ \hat{y}_i^- = 0 \ \text{if} \ \bar{b}_i \geq \sum_{j=1}^n a_{ij} x_j,$$

$$\hat{y}_i^+ = 0, \ \hat{y}_i^- = \sum_{j=1}^n a_{ij} x_j - \bar{b}_i \ \text{if} \ \bar{b}_i < \sum_{j=1}^n a_{ij} x_j.$$

Assume that $\bar{b}_i, \ i = 1, 2, \ldots, m$ are mutually independent. Let f_i and F_i denote the density functions and the distribution function of $\bar{b}_i$. The problem (6.56) can be then explicitly calculated as

$$R_l(x) = E\left[\min_{y^+, y^-}(q_l^+ y^+ + q_l^- y^-) \,|\, y^+ - y^- = \bar{b} - Ax\right]$$

$$= \sum_{i=1}^m q_{li}^+ \int_{\sum_{j=1}^n a_{ij} x_j}^{+\infty} \left(b_i - \sum_{j=1}^n a_{ij} x_j\right) f_i(b_i) db_i$$

$$+ \sum_{i=1}^m q_{li}^- \int_{-\infty}^{\sum_{j=1}^n a_{ij} x_j} \left(\sum_{j=1}^n a_{ij} x_j - b_i\right) f_i(b_i) db_i$$

$$= \sum_{i=1}^m q_{li}^+ E[\bar{b}_i] - \sum_{i=1}^m (q_{li}^+ + q_{li}^-) \int_{-\infty}^{\sum_{j=1}^n a_{ij} x_j} b_i f_i(b_i) db_i$$

$$- \sum_{i=1}^m q_{li}^+ \sum_{j=1}^n a_{ij} x_j + \sum_{i=1}^m (q_{li}^+ + q_{li}^-) \left(\sum_{j=1}^n a_{ij} x_j\right) F_i\left(\sum_{j=1}^n a_{ij} x_j\right).$$

(6.58)

Then, (6.57) can be rewritten as

$$\left.\begin{array}{ll}\text{minimize} & z_1^R(x) \\ \quad \cdots\cdots \\ \text{minimize} & z_k^R(x) \\ \text{subject to} & x \geq 0,\end{array}\right\} \quad (6.59)$$

6.2 Multiobjective Simple Recourse Optimization

where

$$z_l^R(x) = \sum_{i=1}^m q_{li}^+ E[\bar{b}_i] + \sum_{j=1}^n \left(c_{lj} - \sum_{i=1}^m a_{ij} q_{li}^+ \right) x_j$$

$$+ \sum_{i=1}^m (q_{li}^+ + q_{li}^-) \left\{ \left(\sum_{j=1}^n a_{ij} x_j \right) F_i \left(\sum_{j=1}^n a_{ij} x_j \right) - \int_{-\infty}^{\sum_{j=1}^n a_{ij} x_j} b_i f_i(b_i) db_i \right\}.$$

It should be noted here that (6.59) is a multiobjective convex programming problem due to the convexity of $z_l^R(x)$ (Wets 1966).

In a way similar to the chance constrained programming in the previous section, taking into account the imprecise nature of the DM's judgments for each objective function $z_l^R(x)$ in (6.59), by introducing the fuzzy goals such as "$z_l^R(x)$ should be substantially less than or equal to a certain value," (6.59) can be interpreted as

$$\left.\begin{array}{ll}\text{maximize} & \mu_1\left(z_1^R(x)\right) \\ \quad\quad \cdots\cdots \\ \text{maximize} & \mu_k\left(z_k^R(x)\right) \\ \text{subject to} & x \geq 0,\end{array}\right\} \quad (6.60)$$

where $\mu_l\left(z_l^R(x)\right)$ is a membership function to quantify the fuzzy goal for the lth objective function in (6.59). To help the DM specify the membership functions, it is recommended to calculate the individual minima of $z_l^R(x)$ by solving the convex programming problems

$$\underset{x \geq 0}{\text{minimize}}\ z_l^R(x),\ l = 1, \ldots, k. \quad (6.61)$$

In order to find a candidate for the satisficing solution, the DM specifies the reference membership levels $\hat{\mu}_l,\ l = 1, \ldots, k$, and then by solving the augmented minimax problem

$$\underset{x \geq 0}{\text{minimize}}\ \underset{1 \leq l \leq k}{\max} \left\{ \hat{\mu}_l - \mu_l\left(z_l^R(x)\right) + \rho \sum_{i=1}^k \left(\hat{\mu}_i - \mu_i\left(z_i^R(x)\right)\right) \right\}, \quad (6.62)$$

an M-Pareto optimal solution corresponding to $\hat{\mu}_l,\ l = 1, \ldots, k$ is obtained. In (6.62), ρ is a sufficiently small positive number, and (6.62) is equivalently expressed by

$$\left.\begin{array}{ll}\text{minimize} & v \\ \text{subject to} & \hat{\mu}_1 - \mu_1\left(z_1^R(x)\right) + \rho \sum_{i=1}^{k}\left(\hat{\mu}_i - \mu_i\left(z_i^R(x)\right)\right) \leq v \\ & \qquad \cdots\cdots\cdots \\ & \hat{\mu}_k - \mu_k\left(z_k^R(x)\right) + \rho \sum_{i=1}^{k}\left(\hat{\mu}_i - \mu_i\left(z_i^R(x)\right)\right) \leq v \\ & x \geq 0. \end{array}\right\} \quad (6.63)$$

It should be noted here that if each of the membership functions μ_l, $l = 1,\ldots,k$ is nonincreasing and concave, in a way similar to the variance and the fractile models, the convexity of the feasible region of (6.63) can be shown, and this means that (6.63) can be solved by using a conventional convex programming technique such as the sequential quadratic programming method.

We now present an interactive algorithm for deriving a satisficing solution for the DM, in which the reference membership levels μ_l, $l = 1,\ldots,k$ are repeatedly updated and (6.63) is solved until the DM is satisfied with an obtained solution.

Procedure of Interactive Fuzzy Satisficing Method for the Simple Recourse Model

Step 1 Calculate individual minima of $z_l^R(x)$, $l = 1, 2, \ldots, k$ in (6.59) by solving the convex programming problems (6.61).

Step 2 Ask the DM to subjectively specify the membership functions $\mu_l\left(z_l^R(x)\right)$ for the objective functions $z_l^R(x)$, which are nonincreasing and concave on the feasible region by using the individual minima $z_{l,\min}^R$ calculated in step 1 as a reference.

Step 3 Set the initial reference membership levels at 1s, which can be viewed as the ideal values, i.e., $\hat{\mu}_l = 1$, $l = 1,\ldots,k$.

Step 4 For the current reference membership levels $\hat{\mu}_l$, $l = 1,\ldots,k$, solve the augmented minimax problem (6.63).

Step 5 The DM is supplied with the corresponding M-Pareto optimal solution. If the DM is satisfied with the current membership function values $\mu_l\left(z_l^R(x)\right)$, $l = 1,\ldots,k$, then stop. Otherwise, ask the DM to update the reference membership levels $\hat{\mu}_l$, $l = 1,\ldots,k$ with the current membership function values $\mu_l\left(z_l^R(x)\right)$ in mind, and return to step 4.

Example 6.6. Consider the following two-objective production planning problem which is similar to the single-objective simple recourse problem given in Example 5.1:

6.2 Multiobjective Simple Recourse Optimization

$$\begin{aligned}
\text{minimize} \quad & z_1 = -3x_1 - 8x_2 \\
\text{minimize} \quad & z_2 = 5x_1 + 4x_2 \\
\text{subject to} \quad & 2x_1 + 6x_2 \leq 27 \\
& 3x_1 + 2x_2 \leq 16 \\
& 4x_1 + 1x_2 \leq 18 \\
& x_1 \geq 0,\ x_2 \geq 0.
\end{aligned}$$

In Example 5.1, we assume that the demands of products cannot be estimated exactly, and they might be only known as random variables estimated statistically. Taking into account the lost earnings for the shortage of the products or the holding costs for the excess of them, we formulate the production planning problem in a form of two-stage programming.

Let $\bar{h}_1$ and $\bar{h}_2$ be random variables for the demands of products P_1 and P_2, respectively. Assume that the random variables $\bar{h}_1$ and $\bar{h}_2$ are uniform distributions on the interval $[0, 5]$. Thus, their density functions are $f_1(h_1) = \frac{1}{5}$ and $f_2(h_2) = \frac{1}{5}$, and their distribution functions are $F_1(h_1) = \frac{1}{5}h_1$ and $F_2(h_2) = \frac{1}{5}h_2$. Introducing variables y_i^+ and y_i^- representing the shortage and the excess of product P_i, the following equations hold:

$$x_i + y_i^+ - y_i^- = \bar{h}_i,\ i = 1, 2.$$

Suppose that the lost earnings for one unit of the shortage of products P_1 and P_2 are 5 and 7, respectively, and that the holding costs for one unit of the excess of products P_1 and P_2 are 5 and 4, respectively. Taking the volumes y_i^- of unsold products into account, the total profit becomes $3(x_1 - y_1^-) + 8(x_2 - y_2^-)$. Setting $q_1^+ = 5,\ q_2^+ = 7,\ q_1^- = 5 + 3 = 8$, and $q_2^- = 4 + 8 = 12$, we can formulate the following two-objective simple recourse problem:

$$\begin{aligned}
\text{minimize} \quad & z_1^R(x) = -3x_1 - 8x_2 + E[5y_1^+ + 7y_2^+ + 8y_1^- + 12y_2^-] \\
\text{minimize} \quad & z_2^R(x) = 5x_1 + 4x_2 \\
\text{subject to} \quad & x_1 + y_1^+ - y_1^- = \bar{h}_1 \\
& x_2 + y_2^+ - y_2^- = \bar{h}_2 \\
& 2x_1 + 6x_2 \leq 27 \\
& 3x_1 + 2x_2 \leq 16 \\
& 4x_1 + 1x_2 \leq 18 \\
& x_1 \geq 0,\ x_2 \geq 0.
\end{aligned}$$

In this formulation, we do not introduce the penalty for the shortage and the excess of the products in the second objective function z_2 representing the amount of pollution.

Table 6.6 Interactive process in the simple recourse optimization

Iteration	First	Second	Third
$\hat{\mu}_1$	1.000	0.66	0.66
$\hat{\mu}_2$	1.000	0.62	0.61
$\mu_1\left(z_1^R(x^*)\right)$	0.63600	0.65541	0.66022
$\mu_2\left(z_2^R(x^*)\right)$	0.63600	0.61541	0.61022
$z_1^R(x^*)$	−7.34405	−7.56816	−7.62373
$z_2^R(x^*)$	19.93405	20.21319	20.28351

As in Example 5.1, we can transform the above two-objective simple recourse problem into the following two-objective convex programming problem:

$$\text{minimize } z_1^R(x) = \frac{13}{10}x_1^2 + \frac{19}{10}x_2^2 - 8x_1 - 15x_2 + 30$$
$$\text{minimize } z_2^R(x) = 5x_1 + 4x_2$$
$$\text{subject to } 2x_1 + 6x_2 \leq 27$$
$$3x_1 + 2x_2 \leq 16$$
$$4x_1 + 1x_2 \leq 18$$
$$x_1 \geq 0, \ x_2 \geq 0.$$

It should be noted that since the feasible region of the two-objective simple recourse problem is included in the quadrangle $\{(x_1, x_2) \in \mathbb{R} \mid 0 \leq x_1 \leq 5, \ 0 \leq x_2 \leq 5\}$, we need not consider any points in the outside of this quadrangle in the aspect of computation. Referring to the minimum and maximum of each objective function, assume that the DM identifies the following linear membership function:

$$\mu_l(z_l^R(x)) = \begin{cases} 1 & \text{if } z_l^R(x) \leq z_{l,1}^R \\ \dfrac{z_l^R(x) - z_{l,0}^R}{z_{l,1}^R - z_{l,0}^R} & \text{if } z_{l,1}^R < z_l^R(x) \leq z_{l,0}^R \\ 0 & \text{if } z_{l,0}^R < z_l^R(x), \end{cases}$$

where $z_{1,1}^R = -11.54724$, $z_{1,0}^R = 0$, $z_{2,1}^R = 15$, and $z_{2,0}^R = 28.55515$.

An example of an interactive process is given in Table 6.6. Starting from the initial reference membership levels $(\hat{\mu}_1, \hat{\mu}_2) = (1.0, 1.0)$, the augmented minimax problem for finding a candidate for the satisficing solution is solved by using the Excel solver. We have the satisficing solution after two updates of the reference membership levels. ◊

By taking into account the imprecision of the DM's judgment for stochastic objective functions and constraints in multiobjective problems, fuzzy multiobjective stochastic programming is formulated in this chapter. For extended studies, Sakawa and his colleagues deal with the experts' ambiguous understanding of the realized

6.2 Multiobjective Simple Recourse Optimization

values of the random parameters as well as the randomness of parameters involved in objective functions and constraints (Sakawa et al. 2011). Moreover, for resolving conflict of decision-making problems in hierarchical managerial or public organizations where there exist two DMs who have different priorities in making decisions, they also discuss fuzzy stochastic two-level programming problems (Sakawa and Matsui 2013b,c,d,e; Sakawa et al. 2011).

Problems

6.1 Consider the stochastic multiobjective linear programming problem

$$
\begin{aligned}
\text{minimize} \quad & z_1 = -\bar{5}x_1 - \bar{5}x_2 \\
\text{minimize} \quad & z_2 = \bar{5}x_1 + \bar{1}x_2 \\
\text{subject to} \quad & 5x_1 + 7x_2 \leq \overline{12} \\
& 9x_1 + x_2 \leq \overline{10} \\
& -5x_1 + 3x_2 \leq \bar{3} \\
& x_1 \geq 0, \quad x_2 \geq 0,
\end{aligned}
$$

and assume that the means of the stochastic coefficients of z_1 are $E[-\bar{5}] = -5$ and $E[-\bar{5}] = -5$ and those of z_2 are $E[\bar{5}] = 5$ and $E[\bar{1}] = 1$. Moreover, it is also assume that the right-hand side constants of the constraints are normal random variables defined by $\overline{12} \sim N(12, 2^2)$, $\overline{10} \sim N(10, 1^2)$ and $\bar{3} \sim N(3, 0.5^2)$.

Suppose that the DM employs the chance constrained programming with the satisficing probability levels $\beta_1 = \beta_2 = 0.8$ and adopts the interactive fuzzy satisficing method for the expectation model.

Show an example of the interactive process in the interactive fuzzy satisficing method for the expectation model, assuming that the DM specifies the linear membership functions for the fuzzy goals of the two-objective functions as follows:

fuzzy goal of $z_1^E(\boldsymbol{x}) = -5x_1 - 5x_2$: $\mu_1(0) = 0$, $\mu_1(-8.69) = 1$,
fuzzy goal of $z_2^E(\boldsymbol{x}) = 5x_1 + x_2$: $\mu_2(5.45) = 0$, $\mu_2(0) = 1$.

6.2 Consider the stochastic multiobjective linear programming problem given in the Problem 6.1, and further assume that the variance–covariance matrices of the stochastic coefficients of z_1 and z_2 are given as $V_1 = \begin{bmatrix} 1 & 0.5 \\ 0.5 & 1 \end{bmatrix}$ and $V_2 = \begin{bmatrix} 1 & 0.1 \\ 0.1 & 0.2 \end{bmatrix}$.

Suppose that the DM employs the chance constrained programming with the satisficing probability levels $\beta_1 = \beta_2 = 0.8$ and adopts the interactive fuzzy satisficing method for the variance model with the expectation levels $z_1^E(\boldsymbol{x}) = -5x_1 - 5x_2 \leq -5$ and $z_2^E(\boldsymbol{x}) = 5x_1 + x_2 \leq 3$.

Show an example of the interactive process in the interactive fuzzy satisficing method for the variance model, assuming that the DM specifies the linear membership functions for the fuzzy goals of the two-objective functions as follows:

fuzzy goal of $z_1^V(x) = (x_1, x_2) \begin{bmatrix} 1 & 0.5 \\ 0.5 & 1 \end{bmatrix} \begin{pmatrix} x_1 \\ x_2 \end{pmatrix}$: $\mu_1(0.91) = 0$, $\mu_1(0.75) = 1$,

fuzzy goal of $z_2^V(x) = (x_1, x_2) \begin{bmatrix} 1 & 0.1 \\ 0.1 & 0.2 \end{bmatrix} \begin{pmatrix} x_1 \\ x_2 \end{pmatrix}$: $\mu_2(0.35) = 0$, $\mu_2(0.19) = 1$.

6.3 Consider the stochastic multiobjective linear programming problem given in the problem 6.2; however, assume that the right-hand side constants of the constraints are slightly different normal random variables defined by $\bar{12} \sim N(12, 3^2)$, $\bar{10} \sim N(10, 2^2)$, and $\bar{3} \sim N(3, 2^2)$.

Suppose that the DM employs the chance constrained programming with the satisficing probability levels $\beta_1 = \beta_2 = 0.8$ and adopts the interactive fuzzy satisficing method for the probability model with the target values $f_1 = -5$ and $f_2 = 3$ for the stochastic objective functions $z_1 = -\bar{5}x_1 - \bar{5}x_2$ and $z_2 = \bar{5}x_1 + \bar{1}x_2$.

Show an example of the interactive process in the interactive fuzzy satisficing method for the probability model, assuming that the DM specifies the linear membership functions for the fuzzy goals of the two-objective functions as follows:

fuzzy goal of $z_1^P(x) = P(-\bar{5}x_1 - \bar{5}x_2 \leq -5)$: $\mu_1(0.65) = 0$, $\mu_1(0.95) = 1$,
fuzzy goal of $z_2^P(x) = P(\bar{5}x_1 + \bar{1}x_2 \leq 3)$: $\mu_2(0.60) = 0$, $\mu_2(0.90) = 1$.

6.4 Consider the following stochastic multiobjective linear programming problem which is similar to that of the Problem 6.2:

$$\begin{aligned}
\text{minimize } z_1 &= (-5 + \bar{t}_1)x_1 + (-5 + \bar{t}_2)x_2 \\
\text{minimize } z_2 &= (5 + \bar{t}_1)x_1 + (1 + \sqrt{0.2}\bar{t}_1)x_2 \\
\text{subject to } & 5x_1 + 7x_2 \leq \bar{12} \\
& 9x_1 + x_2 \leq \bar{10} \\
& -5x_1 + 3x_2 \leq \bar{3} \\
& x_1 \geq 0, \quad x_2 \geq 0,
\end{aligned}$$

where $\bar{t}_1$ and $\bar{t}_2$ are the standard normal random variables $N(0, 1)$ and the right-hand sides of the constraints are $\bar{12} \sim N(12, 2^2)$, $\bar{10} \sim N(10, 1^2)$, and $\bar{3} \sim N(3, 0.5^2)$.

Suppose that the DM employs the chance constrained programming with the satisficing probability levels $\beta_1 = \beta_2 = 0.8$ and adopts the interactive fuzzy satisficing method for the probability model with the target values $f_1 = -5$ and $f_2 = 3$ for the stochastic objective functions $z_1 = (-5 + \bar{t}_1)x_1 + (-5 + \bar{t}_2)x_2$ and $z_2 = (5 + \bar{t}_1)x_1 + (1 + \sqrt{0.2}\bar{t}_1)x_2$.

6.2 Multiobjective Simple Recourse Optimization

Show an example of the interactive process in the interactive fuzzy satisficing method for the probability model, assuming that the DM specifies the linear membership functions for the fuzzy goals of the two-objective functions as follows:

fuzzy goal of $z_1^P(x) = P((-5+\bar{t}_1)x_1+(-5+\bar{t}_2)x_2 \leq -5)$: $\mu_1(0.65) = 0$, $\mu_1(0.95) = 1$,
fuzzy goal of $z_2^P(x) = P((5+\bar{t}_1)x_1+(1+\sqrt{0.2\bar{t}_1})x_2 \leq 3)$: $\mu_2(0.60) = 0$, $\mu_2(0.90) = 1$.

6.5 Consider the stochastic multiobjective linear programming problem given in the Problem 6.2.

Suppose that the DM employs the chance constrained programming with the satisficing probability levels $\beta_1 = \beta_2 = 0.8$ and adopts the interactive fuzzy satisficing method for the fractile model with the assured probability levels $\theta_1 = \theta_2 = 0.6$ for the probabilities $P(-\bar{5}x_1 - \bar{5}x_2 \leq -5)$ and $P(\bar{5}x_1 + \bar{1}x_2 \leq 3)$.

Show an example of the interactive process in the interactive fuzzy satisficing method for the fractile model, assuming that the DM specifies the linear membership functions for the fuzzy goals of the two-objective functions as follows:

fuzzy goal of $z_1^F(x) = (-5x_1 - 5x_2) + \Phi^{-1}(0.6)\sqrt{(x_1, x_2)\begin{bmatrix} 1 & 0.5 \\ 0.5 & 1 \end{bmatrix}\begin{pmatrix} x_1 \\ x_2 \end{pmatrix}}$:

$\mu_1(0) = 0$, $\mu_1(-8.312) = 1$,

fuzzy goal of $z_1^F(x) = (5x_1 + x_2) + \Phi^{-1}(0.6)\sqrt{(x_1, x_2)\begin{bmatrix} 1 & 0.1 \\ 0.1 & 0.2 \end{bmatrix}\begin{pmatrix} x_1 \\ x_2 \end{pmatrix}}$:

$\mu_2(5.719) = 0$, $\mu_2(0) = 1$.

6.6 Consider the following two-objective production planning problem:

$$\begin{aligned}
\text{minimize} \quad & z_1 = -5x_1 - 5x_2 \\
\text{minimize} \quad & z_2 = 5x_1 + x_2 \\
\text{subject to} \quad & 5x_1 + 7x_2 \leq 12 \\
& 9x_1 + 1x_2 \leq 10 \\
& -5x_1 + 3x_2 \leq 3 \\
& x_1 \geq 0, x_2 \geq 0.
\end{aligned}$$

In the above problem, x_1 and x_2 denote the amounts of production for products P_1 and P_2, respectively, and z_1 and z_2 represent the opposite of the total profit and the amount of pollution to be minimized, respectively.

Suppose that the demands of products P_1 and P_2 are represented by random variables which are the same uniform distribution on the interval [0, 2]. Moreover, suppose that the lost earnings for one unit of the shortage of products P_1 and P_2 are 5 and 7, respectively, and that the holding costs for one unit of the excess of products P_1 and P_2 are 5 and 4, respectively.

Suppose that the DM introduce the penalty of the violation of the constraints only into the first objective z_1 and adopts the interactive fuzzy satisficing method for the simple recourse model.

Show an example of the interactive process in the interactive fuzzy satisficing method for the simple recourse model, assuming that the DM specifies the linear membership functions for the fuzzy goals of the two-objective functions as follows:

fuzzy goal of $z_1^R(x) = -5x_1 - 5x_2 + R_1(x)$: $\mu_1(12) = 0$, $\mu_1(-2.256) = 1$,
fuzzy goal of $z_2^R(x) = 5x_1 + x_2$: $\mu_2(5.868) = 0$, $\mu_2(0) = 1$.

Chapter 7
Purchase and Transportation Planning for Food Retailing

In this chapter, in order to demonstrate linear programming techniques given in the previous chapters by applying to a real-world decision-making problem, we consider purchase and transportation planning for food retailing in Japan and formulate a linear programming problem where the profit of a food retailer is maximized. The food retailer deals with vegetables and fruits which are purchased at the central wholesale markets in several cities and transports them by truck from each of the central wholesaler markets to the food retailer's storehouse. Having examined the optimal solution to the formulated linear programming problem, in view of the recent global warming issue, a single-objective problem is reformulated as a multiobjective one. Furthermore, introducing the fuzzy goals of the decision maker, interactive fuzzy programming approaches are employed. Extensions to fuzzy stochastic multiobjective programming are also discussed.

7.1 Linear Programming Formulation

It has been recognized that many people in Japan buy vegetables and fruits in food supermarkets, and the food supermarkets usually purchase such fresh produce at central wholesale markets. In Japan, 80 % of vegetables and 60 % of fruits are distributed by way of wholesale markets (Ministry of Agriculture, Forestry and Fisheries of Japan 2008), and this fact means that the wholesale markets have been fulfilling an efficient intermediary role connecting consumers and farm producers.

Since Japanese consumers tend to buy small amounts of vegetables and fruits frequently, food retailers such as supermarkets must provide a wide range of fresh products every day. To cope with Japanese consumers' behavior and to increase the profits, food retailers purchase vegetables and fruits in wholesale markets in several cities across Japan.

In this chapter, we consider purchase and transportation planning of a food retailer in Japan and formulate a decision problem of the food retailer as a linear

Fig. 7.1 Purchase and transportation

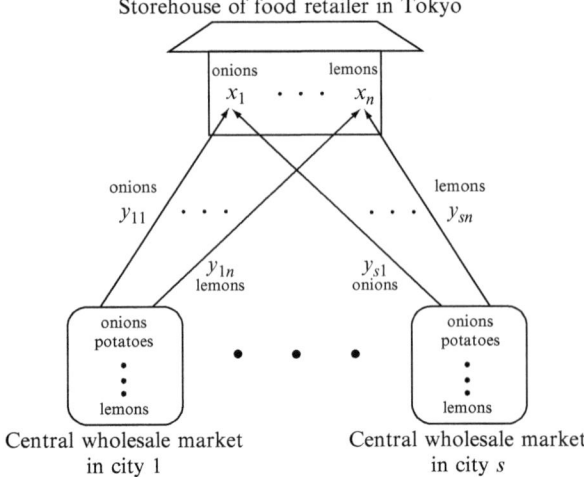

programming problem such that its profit is maximized (Sakawa et al. 2013a). In this problem, the food retailer first specifies the upper and the lower limits of the purchase volume of each vegetable or fruit and determines its purchase quantity at each of the central wholesale markets in several major cities in Japan. Although the food retailer in this application is a hypothetical decision maker, data used in the mathematical modeling are realistic.

The food retailer deals with n kinds of vegetables and fruits which are purchased at the central wholesale markets in s cities and transports them by truck from each of the central wholesaler markets to the food retailer's storehouse in Tokyo as shown in Fig. 7.1.

Let y_{ji}, $j = 1,\ldots,s$, $i = 1,\ldots,n$ denote the purchase volume of food i at the central wholesale market in city j, and let $x_i = \sum_{j=1}^{s} y_{ji}$ denote the total purchase volume of food i. We assume that all the purchased foods do not remain unsold. Thus, the total purchase volume x_i of food i is equal to the sales amount of food i at the store in Tokyo. For concise representation, on occasion the decision variables are expressed by $x = (x_1,\ldots,x_n)^T$ and $y = (y_1^T,\ldots,y_s^T)^T$, $y_j = (y_{j1},\ldots,y_{jn})^T$, $j = 1,\ldots,s$.

The food retailer maximizes the net profit represented by

$$z(x,y) = \sum_{i=1}^{n} c_i x_i - \sum_{j=1}^{s}\sum_{i=1}^{n} d_{ji} y_{ji} - \sum_{j=1}^{s}\sum_{i=1}^{n} b_{ji} y_{ji}, \qquad (7.1)$$

where c_i is the retail price of food i, d_{ji} is the purchase price of food i at the central wholesale market in city j, and b_{ji} is the transportation cost per unit of food i from city j. The first term of the objective function (7.1) is the total sales, the second term is the total purchasing expense, and the third term is the total cost of transportation.

7.1 Linear Programming Formulation

Subject to the following constraints, the objective function (7.1) of the food retailer is maximized. For any food i, the total purchase volume of food i must be larger than or equal to the lower limit D_i^L and be smaller than or equal to the upper limit D_i^U from the business judgment of the food retailer taking into account the demands and the inventories. The constraints for the upper and lower limits of the total purchase volumes are then represented by

$$D_i^L \le x_i \le D_i^U, \; i = 1, \ldots, n. \tag{7.2}$$

Since the total purchase volume of food i is equal to the sum of purchase volumes of food i at the central wholesale markets in all the cities, the following equation holds:

$$\sum_{j=1}^{s} y_{ji} = x_i, \; i = 1, \ldots, n. \tag{7.3}$$

Moreover, there are limitations on financial resources of the food retailer for purchasing foods at the central wholesaler markets, and they are expressed by

$$\sum_{i=1}^{n} d_{ji} y_{ji} \le o_j, \; j = 1, \ldots, s, \tag{7.4}$$

where o_j is the budget cap in city j.

Let W be the capacity of the storehouse of the food retailer, and let v_i be the cubic volume per unit of food i. The constraint for the storehouse is represented by

$$\sum_{i=1}^{n} v_i x_i \le W. \tag{7.5}$$

A linear programming problem for the purchase and transportation planning, in which the objective function (7.1) is maximized under the constraints described above (7.2)–(7.5), is formulated as

$$\left.\begin{array}{l} \text{maximize } z(\boldsymbol{x},\boldsymbol{y}) = \displaystyle\sum_{i=1}^{n} c_i x_i - \sum_{j=1}^{s}\sum_{i=1}^{n} d_{ji} y_{ji} - \sum_{j=1}^{s}\sum_{i=1}^{n} b_{ji} y_{ji} \\ \text{subject to } D_i^L \le x_i \le D_i^U, \; i = 1, \ldots, n \\ \qquad\displaystyle\sum_{j=1}^{s} y_{ji} = x_i, \; i = 1, \ldots, n \\ \qquad\displaystyle\sum_{i=1}^{n} d_{ji} y_{ji} \le o_j, \; j = 1, \ldots, s \\ \qquad\displaystyle\sum_{i=1}^{n} v_i x_i \le W \\ \qquad \boldsymbol{x} \ge \boldsymbol{0}, \; \boldsymbol{y} \ge \boldsymbol{0}. \end{array}\right\} \tag{7.6}$$

Table 7.1 Retail and purchase prices (yen/kg)

		Food 1	Food 2	Food 3	Food 4	Food 5	Food 6
Retail price		90	111	99	82	105	180
	City 1 d_{1i}	55	57	100	102	104	156
	City 2 d_{2i}	78	87	113	95	115	187
	City 3 d_{3i}	73	90	98	85	114	169
Purchase	City 4 d_{4i}	83	105	103	83	113	178
price	City 5 d_{5i}	95	117	104	86	111	189
	City 6 d_{6i}	111	110	88	71	97	189
	City 7 d_{7i}	92	81	87	72	104	179
	City 8 d_{8i}	85	106	72	60	88	151
		Food 7	Food 8	Food 9	Food 10	Food 11	Food 12
Retail price		259	162	347	275	358	245
	City 1 d_{1i}	288	229	349	339	421	221
	City 2 d_{2i}	270	168	394	284	336	250
	City 3 d_{3i}	274	186	312	335	342	231
Purchase	City 4 d_{4i}	276	188	429	296	373	226
price	City 5 d_{5i}	273	170	365	289	377	258
	City 6 d_{6i}	260	173	317	287	368	274
	City 7 d_{7i}	248	138	300	257	315	265
	City 8 d_{8i}	217	93	249	242	260	249
		Food 13	Food 14	Food 15	Food 16		
Retail price		140	887	183	276		
	City 1 d_{1i}	157	926	195	294		
	City 2 d_{2i}	165	867	198	353		
	City 3 d_{3i}	149	743	168	283		
Purchase	City 4 d_{4i}	115	872	159	290		
price	City 5 d_{5i}	147	934	193	290		
	City 6 d_{6i}	147	939	156	310		
	City 7 d_{7i}	176	693	168	301		
	City 8 d_{8i}	186	782	150	231		

We assume that the food retailer sells 16 vegetables and fruits, i.e., $n = 16$, and it purchases them at the central wholesale markets in eight cities in Japan, i.e., $s = 8$. The retail and the purchase prices of the 16 items are shown in Table 7.1, where the purchase prices are the averages of actual prices in the central wholesale markets in March 2008. Foods i, $i = 1, \ldots, 16$ represent onions, potatoes, cabbage, Japanese radish, Chinese cabbage, carrots, cucumbers, lettuce, tomatoes, spinach, eggplant, apples, bananas, strawberries, mandarin oranges, and lemons, respectively; and cities j, $j = 1, \ldots, 8$ stand for Sapporo, Sendai, Niigata, Kanazawa, Tokyo, Osaka, Hiroshima, and Miyazaki, respectively. The eight cities in Japan is shown in Fig. 7.2. The retail prices are specified such that the cost to sales ratios range from 50% to 75% and the average cost to sales ratio of the 16 items is about 60%.

7.1 Linear Programming Formulation

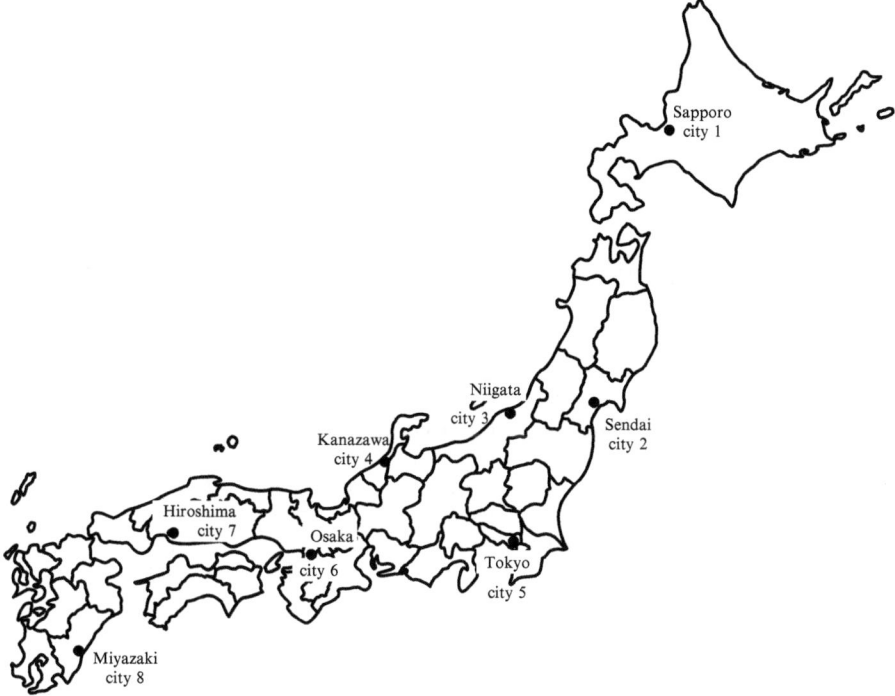

Fig. 7.2 Cities in Japan

The fresh foods are transported from each of the eight cities to the storehouse of the food retailer in Tokyo by truck. The transportation cost per unit b_{ji} of food i from city j to the storehouse is given in Table 7.2, and it is calculated under the assumption that the capacity of a truck is 8 tons, express toll highways are utilized, and the cost of fuel is ¥116/L. The capacity of the storehouse is $150\,\text{m}^2 \times 2\,\text{m}$, and the cubic volumes of food i per kilogram are shown in Table 7.3.

The lower limit D_i^L of the total purchase volume of food i is determined by reference to the demand of 10,000 households, and the upper limit D_i^U is set from 1.1 to 1.4 times the quantities of the lower limit D_i^L; these figures are shown in Table 7.4. The budget caps o_j on purchases in the eight cities are given in Table 7.5.

An optimal solution to (7.6) with the parameters shown in Tables 7.1–7.5 is given in Table 7.6. As seen in Table 7.6, the profits of the food retailer is $z(x, y) = $ ¥2,195,972, and the corresponding total sales, the purchasing expense, and the transportation cost are $\sum_{i=1}^{n} c_i x_i = $ 15,569,354, $\sum_{j=1}^{s} \sum_{i=1}^{n} d_{ji} y_{ji} = $ 13,000,000, and $\sum_{j=1}^{s} \sum_{i=1}^{n} b_{ji} y_{ji} = $ 373,327, respectively. The total purchase volumes of foods 1 and 2 reach the upper limits D_i^U, that of food 9 is between the upper limit D_i^U and the lower limit D_i^L, and those of the rest of the foods are at the lower limit D_i^L. The purchasing expenses in all the cities reach the budget caps. Although the wholesale prices d_{ji} of foods in city 5, Tokyo, are greater than the

Table 7.2 Transportation costs (yen/kg)

	Food 1	Food 2	Food 3	Food 4	Food 5	Food 6
City 1 b_{1i}	12.476020	7.984653	12.476020	2.694820	9.980816	5.988490
City 2 b_{2i}	2.834936	1.814359	2.834936	0.612346	2.267949	1.360769
City 3 b_{3i}	2.837123	1.815758	2.837123	0.612818	2.269698	1.361819
City 4 b_{4i}	3.882100	2.484544	3.882100	0.838534	3.105680	1.863408
City 5 b_{5i}	0.202730	0.129747	0.202730	0.043790	0.162184	0.097310
City 6 b_{6i}	4.553846	2.914462	4.553846	0.983631	3.643077	2.185846
City 7 b_{7i}	6.225852	3.984545	6.225852	1.344784	4.980682	2.988409
City 8 b_{8i}	10.273461	6.575015	10.273461	2.219068	8.218769	4.931261
	Food 7	Food 8	Food 9	Food 10	Food 11	Food 12
City 1 b_{1i}	2.495204	59.884896	4.990408	39.923264	24.952040	9.980816
City 2 b_{2i}	0.566987	13.607693	1.133974	9.071796	5.669872	2.267949
City 3 b_{3i}	0.567425	13.618188	1.134849	9.078792	5.674245	2.269698
City 4 b_{4i}	0.776420	18.634078	1.552840	12.422719	7.764199	3.105680
City 5 b_{5i}	0.040546	0.973104	0.081092	0.648736	0.405460	0.162184
City 6 b_{6i}	0.910769	21.858463	1.821539	14.572308	9.107693	3.643077
City 7 b_{7i}	1.245170	29.884090	2.490341	19.922727	12.451704	4.980682
City 8 b_{8i}	2.054692	49.312615	4.109385	32.875076	20.546923	8.218769
	Food 13	Food 14	Food 15	Food 16		
City 1 b_{1i}	2.495204	4.990408	3.742806	3.742806		
City 2 b_{2i}	0.566987	1.133974	0.850481	0.850481		
City 3 b_{3i}	0.567425	1.134849	0.851137	0.851137		
City 4 b_{4i}	0.776420	1.552840	1.164630	1.164630		
City 5 b_{5i}	0.040546	0.081092	0.060819	0.060819		
City 6 b_{6i}	0.910769	1.821539	1.366154	1.366154		
City 7 b_{7i}	1.245170	2.490341	1.867756	1.867756		
City 8 b_{8i}	2.054692	4.109385	3.082038	3.082038		

Table 7.3 Cubic volumes of foods (cm^3/kg)

	Food 1	Food 2	Food 3	Food 4	Food 5	Food 6
Cubic volume v_i	5,000	3,200	5,000	1,080	4,000	2,400
	Food 7	Food 8	Food 9	Food 10	Food 11	Food 12
Cubic volume v_i	1,000	24,000	2,000	16,000	10,000	4,000
	Food 13	Food 14	Food 15	Food 16		
Cubic volume v_i	1,000	2,000	1,500	1,500		

retail prices c_i, the food retailer buys foods 5, 10, 12, 14, and 16 in order to fill the lower limits of the total purchase volumes. Basically, the food retailer buys highly profitable foods in the corresponding cities. That is, food i is likely to be purchased in city j if $(c_i - d_{ji} - b_{ji})/c_i$ is relatively larger. For example, one finds that the food retailer purchases food 3, cabbage, in city 8, Miyazaki, to be most profitable, as expected. From the same reason, food 13, bananas, is purchased in city 4, Kanazawa.

7.1 Linear Programming Formulation

Table 7.4 Lower and upper limits of the foods (kg)

	Food 1	Food 2	Food 3	Food 4	Food 5	Food 6
Lower limit D_i^L	4,000	4,000	2,000	5,000	10,000	2,000
Upper limit D_i^U	5,000	5,000	2,400	6,000	14,000	2,500
	Food 7	Food 8	Food 9	Food 10	Food 11	Food 12
Lower limit D_i^L	800	1,500	3,000	3,000	1,200	6,000
Upper limit D_i^U	1,000	2,000	4,000	3,600	1,500	6,600
	Food 13	Food 14	Food 15	Food 16		
Lower limit D_i^L	12,500	6,000	4,000	1,000		
Upper limit D_i^U	14,500	7,500	4,800	1,300		

Table 7.5 Caps on purchase for eight cities (yen)

	City 1	City 2	City 3	City 4	City 5	City 6
Cap o_j	2,000,000	1,500,000	1,500,000	1,500,000	1,500,000	1,500,000
	City 7	City 8				
Cap o_j	1,500,000	2,000,000				

It follows that by purchasing foods and transporting them as shown in Table 7.6, the food retailer can maximize its profit (7.1) satisfying the constraints (7.2)–(7.5).

It is also important to examine variations of the solutions when some parameters are changed from the managerial viewpoints of the food retailer. Changes in the cost of fuel for truck transportation are an issue of considerable concern for the management of the food retailer. Although we assume that the cost of fuel is ¥116/L, taking into account the fact that the highest fuel price in 2008 in Japan was ¥148/L, it is significant to compute an optimal solution to the purchase and transportation planning problem (7.6) in the case of the high fuel cost by changing the cost of fuel from ¥116 to ¥150. In this case, it is found that the optimal solution is the same as before. Therefore, the total sales and the purchasing expense are $\sum_{i=1}^{n} c_i x_i = 15{,}569{,}354$ and $\sum_{j=1}^{s} \sum_{i=1}^{n} d_{ji} y_{ji} = 13{,}000{,}000$, respectively, which are the same as the figures given in Table 7.6, but the transportation cost changes to $\sum_{j=1}^{s} \sum_{i=1}^{n} b_{ji} y_{ji} = 373{,}327$ which is increased by ¥26,007. As a result, due to the increase of the transportation cost, the net profit of the food retailer decreases by ¥26,008 from $z(x, y) = $ ¥2,195,972 to ¥2,170,019.

From another viewpoint, we study the sensitivity of the optimal solution. By increasing the upper limits D_i^U of foods such that their total purchase volumes reach the upper limits, larger purchase volumes will be expected, and it is interesting for the food retailer to examine the variation of the optimal solution and the net profit by the effect of this increase. Since the total purchase volume of food 2, potatoes, reaches its upper limit, after changing its upper limit from $D_2^U = 5{,}000$ to $D_2^U = 5{,}100$, we solve the linear programming problem (7.6) again, and the result is given in Table 7.7, where the revised upper limit of D_2^U of food 2 is shown in boldface and the values changed from the original solution are marked with asterisks. Since, from

Table 7.6 Result of the purchase problem for food retailing

	Food 1	Food 2	Food 3	Food 4	Food 5	Food 6
Total purchase volume (kg): x_i	5,000	5,000	2,000	5,000	10,000	2,000
Purchase volume at city 1 (kg): y_{1i}	5,000	5,000	0	0	0	2,000
Purchase volume at city 2 (kg): y_{2i}	0	0	0	0	0	0
Purchase volume at city 3 (kg): y_{3i}	0	0	0	0	0	0
Purchase volume at city 4 (kg): y_{4i}	0	0	0	0	0	0
Purchase volume at city 5 (kg): y_{5i}	0	0	0	0	969	0
Purchase volume at city 6 (kg): y_{6i}	0	0	0	0	9,031	0
Purchase volume at city 7 (kg): y_{7i}	0	0	0	0	0	0
Purchase volume at city 8 (kg): y_{8i}	0	0	2,000	5,000	0	0
Lower limit (kg): D_i^L	4,000	4,000	2,000	5,000	10,000	2,000
Sum of purchase volumes (kg): $\sum_{j=1}^{8} y_{ji}$	5,000	5,000	2,000	5,000	10,000	2,000
Upper limit (kg): D_i^U	5,000	5,000	2,400	6,000	14,000	2,500

	Food 7	Food 8	Food 9	Food 10	Food 11	Food 12
Total purchase volume: x_i	800	1,500	3,722	3,000	1,200	6,000
Purchase volume at city 1: y_{1i}	0	0	0	0	0	5,104
Purchase volume at city 2: y_{2i}	0	0	0	0	0	0
Purchase volume at city 3: y_{3i}	0	0	0	0	0	0
Purchase volume at city 4: y_{4i}	0	0	0	0	0	277
Purchase volume at city 5: y_{5i}	0	0	0	3,000	0	619
Purchase volume at city 6: y_{6i}	0	0	0	0	0	0
Purchase volume at city 7: y_{7i}	0	0	0	0	0	0
Purchase volume at city 8: y_{8i}	800	1,500	3,722	0	1,200	0
Lower limit: D_i^L	800	1,500	3,000	3,000	1,200	6,000
Sum of purchase volumes: $\sum_{j=1}^{8} y_{ji}$	800	1,500	3,722	3,000	1,200	6,000
Upper limit: D_i^U	1,000	2,000	4,000	3,600	1,500	6,600

7.1 Linear Programming Formulation

	Food 13	Food 14	Food 15	Food 16	Amount	Cap
Total purchase volume: x_i	12,500	6,000	4,000	1,000	—	—
Purchase volume at city 1: y_{1i}	0	0	0	0	2,000,000	2,000,000
Purchase volume at city 2: y_{2i}	0	1,730	0	0	1,500,000	1,500,000
Purchase volume at city 3: y_{3i}	0	2,019	0	0	1,500,000	1,500,000
Purchase volume at city 4: y_{4i}	12,500	0	0	0	1,500,000	1,500,000
Purchase volume at city 5: y_{5i}	0	87	0	982	1,500,000	1,500,000
Purchase volume at city 6: y_{6i}	0	0	4,000	0	1,500,000	1,500,000
Purchase volume at city 7: y_{7i}	0	2,164	0	0	1,500,000	1,500,000
Purchase volume at city 8: y_{8i}	0	0	0	18	2,000,000	2,000,000
Lower limit: D_i^L	12,500	6,000	4,000	1,000	—	—
Sum of purchase volumes: $\sum_{j=1}^{8} y_{ji}$	12,500	6,000	4,000	1,000	—	—
Upper limit: D_i^U	14,500	7,500	4,800	1,300	—	—

Usage of storehouse (cm^3): $\sum_{i=1}^{16} v_i x_i = 261{,}443{,}801$; capacity (cm^3): $W = 300{,}000{,}000$

Total sales (yen): $\sum_{i=1}^{n} c_i x_i = 15{,}569{,}354$; purchasing expense (yen): $\sum_{j=1}^{s} \sum_{i=1}^{n} d_{ji} y_{ji} = 13{,}000{,}000$

Transportation cost (yen): $\sum_{j=1}^{s} \sum_{i=1}^{n} b_{ji} y_{ji} = 373{,}327$; net profit (yen): $z(\boldsymbol{x}, \boldsymbol{y}) = 2{,}195{,}972$

Table 7.7 Sensitivity analysis for the upper limit of the total purchase volume

	Food 1	Food 2	Food 3	Food 4	Food 5	Food 6
Total purchase volume (kg): x_i	5,000	5,100*	2,000	5,000	10,000	2,000
Purchase volume at city 1 (kg): y_{1i}	5,000	5,100*	0	0	0	2,000
Purchase volume at city 2 (kg): y_{2i}	0	0	0	0	0	0
Purchase volume at city 3 (kg): y_{3i}	0	0	0	0	0	0
Purchase volume at city 4 (kg): y_{4i}	0	0	0	0	0	0
Purchase volume at city 5 (kg): y_{5i}	0	0	0	0	969	0
Purchase volume at city 6 (kg): y_{6i}	0	0	0	0	9,031	0
Purchase volume at city 7 (kg): y_{7i}	0	0	0	0	0	0
Purchase volume at city 8 (kg): y_{8i}	0	0	2,000	5,000	0	0
Lower limit (kg): D_i^L	4,000	4,000	2,000	5,000	10,000	2,000
Sum of purchase volumes (kg): $\sum_{j=1}^{8} y_{ji}$	5,000	5,100*	2,000	5,000	10,000	2,000
Upper limit (kg): D_i^U	5,000	5,100	2,400	6,000	14,000	2,500

	Food 7	Food 8	Food 9	Food 10	Food 11	Food 12
Total purchase volume: x_i	800	1,500	3,722	3,000	1,200	6,000
Purchase volume at city 1: y_{1i}	0	0	0	0	0	5,078*
Purchase volume at city 2: y_{2i}	0	0	0	0	0	0
Purchase volume at city 3: y_{3i}	0	0	0	0	0	0
Purchase volume at city 4: y_{4i}	0	0	0	0	0	277
Purchase volume at city 5: y_{5i}	0	0	0	3,000	0	645*
Purchase volume at city 6: y_{6i}	0	0	0	0	0	0
Purchase volume at city 7: y_{7i}	0	0	0	0	0	0
Purchase volume at city 8: y_{8i}	800	1,500	3,701*	0	1,200	0
Lower limit: D_i^L	800	1,500	3,000	3,000	1,200	6,000
Sum of purchase volumes: $\sum_{j=1}^{8} y_{ji}$	800	1,500	3,701*	3,000	1,200	6,000
Upper limit: D_i^U	1,000	2,000	4,000	3,600	1,500	6,600

7.1 Linear Programming Formulation

	Food 13	Food 14	Food 15	Food 16	Amount	Cap
Total purchase volume: x_i	12,500	6,000	4,000	1,000	–	–
Purchase volume at city 1: y_{1i}	0	0	0	0	2,000,000	2,000,000
Purchase volume at city 2: y_{2i}	0	1,730	0	0	1,500,000	1,500,000
Purchase volume at city 3: y_{3i}	0	2,019	0	0	1,500,000	1,500,000
Purchase volume at city 4: y_{4i}	12,500	0	0	0	1,500,000	1,500,000
Purchase volume at city 5: y_{5i}	0	87	0	959*	1,500,000	1,500,000
Purchase volume at city 6: y_{6i}	0	0	4,000	0	1,500,000	1,500,000
Purchase volume at city 7: y_{7i}	0	2,164	0	0	1,500,000	1,500,000
Purchase volume at city 8: y_{8i}	0	0	0	41*	2,000,000	2,000,000
Lower limit: D_i^L	12,500	6,000	4,000	1,000	–	–
Sum of purchase volumes: $\sum_{j=1}^{8} y_{ji}$	12,500	6,000	4,000	1,000	–	–
Upper limit: D_i^U	14,500	7,500	4,800	1,300	–	–

Usage of storehouse (cm^3): $\sum_{i=1}^{16} v_i x_i = 261{,}721{,}227*$; capacity (cm^3): $W = 300{,}000{,}000$

Total sales (yen): $\sum_{i=1}^{n} c_i x_i = 15{,}573{,}088*$, purchasing expense (yen): $\sum_{j=1}^{s}\sum_{i=1}^{n} d_{ji} y_{ji} = 13{,}000{,}000$

Transportation cost (yen): $\sum_{j=1}^{s}\sum_{i=1}^{n} b_{ji} y_{ji} = 373{,}855*$; net profit (yen): $z(\mathbf{x},\mathbf{y}) = 2{,}199{,}233*$

the increase of the upper limit of the total purchase volume, the total sales increases by ¥3,734, the purchasing expense is the same as before, and the transportation cost increases only by ¥527, the net profit of the food retailer increases by ¥3,207 to $z_1(x,y) =$ ¥2,199,233. Specifically, the expansion of the upper limit of food 2 increases the purchase volume of food 2 in city 1, decreases that of food 12 in city 1 due to its budget cap, increases that of food 12 and decreases that of food 16 in city 5, and finally increases that of food 16 and decreases that of food 16 in city 8.

The budget cap of each city is also a managerial parameter that the food retailer can change at its discretion. In particular, it can purchase almost all foods in city 8, Miyazaki, at lower prices compared to the other cities. Therefore, it should be interesting for the food retailer to examine the variation of the optimal solution and the net profit by the increase of the budget cap in city 8. After increasing the budget cap in city 8 from ¥2,000,000 to ¥2,100,000, we solve the linear programming problem (7.6) again, and the result is given in Table 7.8, where the budget cap of city 8 is shown in boldface and the values changed from the original solution are marked with asterisks. Since, from the increase of the budget cap in city 8, the total sales increases by ¥137,499, the purchasing expense increases by ¥100,000 equal to the increase of the budget cap, and the transportation cost increases only by ¥1,333, the net profit of the food retailer increases by ¥36,167 to $z_1(x,y) =$ ¥2,232,193. Specifically, the expansion of the budget cap of city 8 increases the purchase volume of food 9 in city 8 up to the upper limit, decreases that of food 16 in city 5 and increases it in city 8, increases that of food 12 in city 5 and decreases it in city 4, and finally increases that of food 13 in city 4.

7.2 Multiobjective Linear Programming Formulation

Recall that the objective function of the purchase and transportation planning problem (7.6) considered thus far is represented by

$$z(x,y) = \sum_{i=1}^{n} c_i x_i - \sum_{j=1}^{s}\sum_{i=1}^{n} d_{ji} y_{ji} - \sum_{j=1}^{s}\sum_{i=1}^{n} b_{ji} y_{ji},$$

where the first term is the total sales, the second term is the purchasing expense, and the third term is the transportation cost.

However, considering the recent global warming issue, suppose that the manager of the food retailer who is the DM of the purchase and transportation planning problem desires to cut back on the emission of gases that causes global warming. In such circumstances, the purchase and transportation planning problem should be reformulated as a multiobjective linear programming problem (Sakawa et al. 2013).

To be more explicit, in the purchase and transportation planning problem, considering the reduction of the greenhouse gas emission, we divide the original objective function into two parts; one is the sales profit excluding the transportation

7.2 Multiobjective Linear Programming Formulation

Table 7.8 Sensitivity analysis for the budget cap

	Food 1	Food 2	Food 3	Food 4	Food 5	Food 6
Total purchase volume (kg): x_i	5,000	5,000	2,000	5,000	10,000	2,000
Purchase volume at city 1 (kg): y_{1i}	5,000	5,000	0	0	0	2,000
Purchase volume at city 2 (kg): y_{2i}	0	0	0	0	0	0
Purchase volume at city 3 (kg): y_{3i}	0	0	0	0	0	0
Purchase volume at city 4 (kg): y_{4i}	0	0	0	0	0	0
Purchase volume at city 5 (kg): y_{5i}	0	0	0	0	969	0
Purchase volume at city 6 (kg): y_{6i}	0	0	0	0	9,031	0
Purchase volume at city 7 (kg): y_{7i}	0	0	0	0	0	0
Purchase volume at city 8 (kg): y_{8i}	0	0	2,000	5,000	0	0
Lower limit (kg): D_i^L	4,000	4,000	2,000	5,000	10,000	2,000
Sum of purchase volumes (kg): $\sum_{j=1}^{8} y_{ji}$	5,000	5,000	2,000	5,000	10,000	2,000
Upper limit (kg): D_i^U	5,000	5,000	2,400	6,000	14,000	2,500

	Food 7	Food 8	Food 9	Food 10	Food 11	Food 12
Total purchase volume: x_i	800	1,500	3,722	3,000	1,200	6,000
Purchase volume at city 1: y_{1i}	0	0	0	0	0	5,104
Purchase volume at city 2: y_{2i}	0	0	0	0	0	0
Purchase volume at city 3: y_{3i}	0	0	0	0	0	0
Purchase volume at city 4: y_{4i}	0	0	0	0	0	127*
Purchase volume at city 5: y_{5i}	0	0	0	3,000	0	769*
Purchase volume at city 6: y_{6i}	0	0	0	0	0	0
Purchase volume at city 7: y_{7i}	0	0	0	0	0	0
Purchase volume at city 8: y_{8i}	800	1,500	4,000*	0	1,200	0
Lower limit: D_i^L	800	1,500	3,000	3,000	1,200	6,000
Sum of purchase volumes: $\sum_{j=1}^{8} y_{ji}$	800	1,500	4,000*	3,000	1,200	6,000
Upper limit: D_i^U	1,000	2,000	4,000	3,600	1,500	6,600

(continued)

Table 7.8 (continued)

	Food 13	Food 14	Food 15	Food 16	Amount	Cap
Total purchase volume: x_i	12,500	6,000	4,000	1,000	–	–
Purchase volume at city 1: y_{1i}	0	0	0	0	2,000,000	2,000,000
Purchase volume at city 2: y_{2i}	0	1,730	0	0	1,500,000	1,500,000
Purchase volume at city 3: y_{3i}	0	2,019	0	0	1,500,000	1,500,000
Purchase volume at city 4: y_{4i}	12,794*	0	0	0	1,500,000	1,500,000
Purchase volume at city 5: y_{5i}	0	87	0	849*	1,500,000	1,500,000
Purchase volume at city 6: y_{6i}	0	0	4,000	0	1,500,000	1,500,000
Purchase volume at city 7: y_{7i}	0	2,164	0	0	1,500,000	1,500,000
Purchase volume at city 8: y_{8i}	0	0	0	151*	2,100,000*	2,100,000
Lower limit: D_i^L	12,500	6,000	4,000	1,000	–	–
Sum of purchase volumes: $\sum_{j=1}^{8} y_{ji}$	12,794*	6,000	4,000	1,000	–	–
Upper limit: D_i^U	14,500	7,500	4,800	1,300	–	–

Usage of storehouse (cm^3): $\sum_{i=1}^{16} v_i x_i = 261{,}721{,}227^*$; capacity (cm^3): $W = 300{,}000{,}000$

Total sales (yen): $\sum_{i=1}^{n} c_i x_i = 15{,}706{,}853^*$; purchasing expense (yen): $\sum_{j=1}^{s}\sum_{i=1}^{n} d_{ji} y_{ji} = 13{,}100{,}000^*$

Transportation cost (yen): $\sum_{j=1}^{s}\sum_{i=1}^{n} b_{ji} y_{ji} = 374{,}660^*$; net profit (yen): $z(\boldsymbol{x}, \boldsymbol{y}) = 2{,}232{,}193^*$

7.2 Multiobjective Linear Programming Formulation

cost, namely, the difference between the total sales and the purchasing expense, while the other is the transportation cost. For convenience sake, we refer to the sales profit excluding the transportation cost simply as the sales profit.

Thus, the DM should not only maximize the sales profit

$$z_1(x,y) = \sum_{i=1}^{n} c_i x_i - \sum_{j=1}^{s}\sum_{i=1}^{n} d_{ji} y_{ji} \qquad (7.7)$$

but also minimize the transportation cost

$$z_2(x,y) = \sum_{j=1}^{s}\sum_{i=1}^{n} b_{ji} y_{ji}. \qquad (7.8)$$

The purchase and transportation planning problem considering the reduction of the greenhouse gas emission can be then formulated as the following two-objective linear programming problem:

$$\left.\begin{array}{l} \text{maximize } z_1(x,y) = \displaystyle\sum_{i=1}^{n} c_i x_i - \sum_{j=1}^{s}\sum_{i=1}^{n} d_{ji} y_{ji} \\[2pt] \text{minimize } z_2(x,y) = \displaystyle\sum_{j=1}^{s}\sum_{i=1}^{n} b_{ji} y_{ji} \\[2pt] \text{subject to } D_i^L \le x_i \le D_i^U,\ i=1,\ldots,n \\[2pt] \displaystyle\sum_{j=1}^{s} y_{ji} = x_i,\ i=1,\ldots,n \\[2pt] \displaystyle\sum_{i=1}^{n} d_{ji} y_{ji} \le o_j,\ j=1,\ldots,s \\[2pt] \displaystyle\sum_{i=1}^{n} v_i x_i \le W \\[2pt] x \ge 0,\ y \ge 0. \end{array}\right\} \qquad (7.9)$$

First, from the standpoint of the weighting problem

$$\operatorname*{minimize}_{x \in X}\ \sum_{i=1}^{k} w_i z_i(x)$$

examined in Chap. 3, we deal with the two-objective linear programming problem (7.9) for the purchase and transportation planning. By assigning the weighting coefficients w_1 and w_2 to the opposite of the sales profit $-z_1$ and the transportation cost z_2, respectively, (7.9) can be reformulated as a weighting problem

Table 7.9 Objective function values of weighting problem

	Original problem	Weighting problem	
		$(w_1, w_2) = (1, 5)$	$(w_1, w_2) = (3, 2)$
Total sales	15,569,299	15,225,062	15,585,498
Purchasing expense	13,000,000	13,000,000	13,000,000
Transportation cost	373,327	217,626	397,216
Net profit $z = z_1 - z_2$	2,195,972	2,007,436	2,188,282
Sales profit z_1	2,569,299	2,225,062	2,585,498
Transportation cost z_2	373,327	217,626	397,216

$$\left.\begin{aligned}
\text{minimize} \quad & -w_1 z_1(x, y) + w_2 z_2(x, y) \\
& = w_1 \left(-\sum_{i=1}^{n} c_i x_i + \sum_{j=1}^{s}\sum_{i=1}^{n} d_{ji} y_{ji} \right) + w_2 \left(\sum_{j=1}^{s}\sum_{i=1}^{n} b_{ji} y_{ji} \right) \\
\text{subject to} \quad & D_i^L \leq x_i \leq D_i^U, \quad i = 1, \ldots, n \\
& \sum_{j=1}^{s} y_{ji} = x_i, \quad i = 1, \ldots, n \\
& \sum_{i=1}^{n} d_{ji} y_{ji} \leq o_j, \quad j = 1, \ldots, s \\
& \sum_{i=1}^{n} v_i x_i \leq W \\
& x \geq 0, \, y \geq 0.
\end{aligned}\right\}$$

(7.10)

Assume that emphasizing the reduction of the greenhouse gas emission, the DM assesses the weighting coefficients as $(w_1, w_2) = (1, 5)$. This means that the weighting coefficient of the transportation cost is five times as large as that of the sales profit. The corresponding linear programming problem yields an optimal solution as is shown in the third column in Table 7.9. Conversely, if the DM chooses $(w_1, w_2) = (3, 2)$ by giving a slightly high weight to the sales profit, the corresponding linear programming problem yields an optimal solution as is shown in the fourth column in Table 7.9.

Compare the optimal solution of the original linear programming problem (7.6) with that of the weighting problem (7.10) with the weighting vector $(w_1, w_2) = (1, 5)$. Since this weighting vector gives heavy weight to the transportation cost z_2, as can be seen from Table 7.9, the resulting transportation cost of ¥217,626 is a relatively low value compared with that of ¥373,327 of the original problem. This means that the solution is improved from the viewpoint of the transportation cost z_2 responsible for the greenhouse gas emission, whereas the sales profit z_1 is reduced (worsened) from ¥2,569,299 to ¥2,225,062. Hence, these two-objective functions conflict with each other. On the other hand, the weighting vector $(w_1, w_2) = (3, 2)$ gives the contrasting result. That is, the sales profit z_1 is increased (improved)

7.2 Multiobjective Linear Programming Formulation

from ¥2,569,299 to ¥2,585,498, whereas the transportation cost z_2 is increased (worsened) from ¥373,327 to ¥397,216. Also, for two weighting problems with $(w_1, w_2) = (1, 5)$ and $(w_1, w_2) = (3, 2)$, the value of the original objective function z representing the net profit is, respectively, ¥2,007,436 and ¥2,188,282, both of which are smaller than ¥2,195,972 of the original linear programming problem. This fact obviously follows from the optimality of the solution to the original problem.

Next, proceed to interactive multiobjective linear programming. From a practical point of view, it is appropriate to derive a satisficing solution in the Pareto optimal solution set of the two-objective linear programming problem (7.9) from the DM by using the reference point method (Wierzbicki 1980). In the reference point method, when the DM specifies the reference point $(\hat{z}_1, \ldots, \hat{z}_k)$, the corresponding Pareto optimal solution, which is, in the minimax sense, nearest to the reference point or better than that if the reference point is attainable, is obtained by solving the following minimax problem:

$$\underset{x \in X}{\text{minimize}} \ \max_{i=1,\ldots,k} \{z_i(x) - \hat{z}_i\}.$$

For the two-objective linear programming problem (7.9), when the DM specifies the reference point $(\hat{z}_1, \hat{z}_2)$, the corresponding minimax problem is formulated as

$$\left.\begin{aligned}
\text{minimize } & \max\{-z_1(x,y) - \hat{z}_1, z_2(x,y) - \hat{z}_2\} \\
&= \max\left\{-\sum_{i=1}^{n} c_i x_i + \sum_{j=1}^{s}\sum_{i=1}^{n} d_{ji} y_{ji} - \hat{z}_1, \sum_{j=1}^{s}\sum_{i=1}^{n} b_{ji} y_{ji} - \hat{z}_2\right\} \\
\text{subject to } & D_i^L \le x_i \le D_i^U, \ i = 1, \ldots, n \\
& \sum_{j=1}^{s} y_{ji} = x_i, \ i = 1, \ldots, n \\
& \sum_{i=1}^{n} d_{ji} y_{ji} \le o_j, \ j = 1, \ldots, s \\
& \sum_{i=1}^{n} v_i x_i \le W \\
& x \ge 0, \ y \ge 0.
\end{aligned}\right\} \quad (7.11)$$

By introducing the auxiliary variable v, the minimax problem (7.11) is transformed to

Table 7.10 Interactive two-objective linear programming

	Original problem	$\hat{z}_1 = -2{,}600{,}000$ $\hat{z}_2 = 300{,}000$	$\hat{z}_1 = -2{,}600{,}000$ $\hat{z}_2 = 260{,}000$	$\hat{z}_1 = -2{,}630{,}000$ $\hat{z}_2 = 260{,}000$
Total sales	15,569,299	15,545,614	15,523,388	15,540,058
Purchasing expense	13,000,000	13,000,000	13,000,000	13,000,000
Transportation cost	373,327	354,386	336,612	349,942
Net profit $z = z_1 - z_2$	2,195,972	2,191,228	2,186,776	2,190,116
Sales profit z_1	2,569,299	2,545,614	2,523,388	2,540,058
Transportation cost z_2	373,327	354,386	336,612	349,942

$$\left. \begin{aligned} & \text{minimize } v \\ & \text{subject to } -\sum_{i=1}^{n} c_i x_i + \sum_{j=1}^{s}\sum_{i=1}^{n} d_{ji} y_{ji} - \hat{z}_1 \leq v \\ & \phantom{\text{subject to }} \sum_{j=1}^{s}\sum_{i=1}^{n} b_{ji} y_{ji} - \hat{z}_2 \leq v \\ & \phantom{\text{subject to }} D_i^L \leq x_i \leq D_i^U, \ i = 1, \ldots, n \\ & \phantom{\text{subject to }} \sum_{j=1}^{s} y_{ji} = x_i, \ i = 1, \ldots, n \\ & \phantom{\text{subject to }} \sum_{i=1}^{n} d_{ji} y_{ji} \leq o_j, \ j-1, \ldots, s \\ & \phantom{\text{subject to }} \sum_{i=1}^{n} v_i x_i \leq W \\ & \phantom{\text{subject to }} x \geq 0, \ y \geq 0. \end{aligned} \right\} \quad (7.12)$$

We give an example of the interactive process in the reference point method in Table 7.10 including the optimal value of the original problem (7.6), and a possible scenario is as follows. Suppose that as related information, observing that the individual maximum of the sales profit z_1 is ¥2,585,822 and the individual minimum of the transportation cost z_2 is ¥201,042 and for the optimal solution of the original linear programming problem (7.6) the sales profit and the transportation cost are ¥2,569,299 and ¥373,327, respectively, the DM specifies the reference point as $(\hat{z}_1, \hat{z}_2) = (-2{,}600{,}000,\ 300{,}000)$.

As shown in Table 7.10, to the initial reference point, the sales profit z_1 and the transportation cost z_2 are calculated as $z_1 = $ ¥2,545,614 and $z_2 = $ ¥354,386. On the basis of this information, assume that the DM updates the reference point to $(\hat{z}_1, \hat{z}_2) = (-2{,}600{,}000,\ 260{,}000)$ to reduce the transportation cost at the expense of the sales profit. For the updated reference point, the corresponding Pareto optimal solution is obtained, and then the sales profit z_1 and the transportation cost z_2 change to $z_1 = $ ¥2,523,388 and $z_2 = $ ¥336,612 as the DM expected. Assume further that the DM updates the reference point to $(\hat{z}_1, \hat{z}_2) = (-2{,}630{,}000,\ 260{,}000)$ in order to slightly improve the sales profit. As the DM desired, then the sales profit increases to $z_1 = $ ¥2,540,058 with the increased (worsened) transportation cost of $z_2 = $ ¥349,942.

7.3 Fuzzy Multiobjective Linear Programming Formulation

In this section, handling the imprecision or vagueness in human judgments, we employ interactive fuzzy programming approaches (Sakawa et al. 2013). To start, consider fuzzy linear programming incorporating the DM's fuzzy goal and fuzzy constraints for the purchase and transportation planning problem.

As calculated previously, the optimal value of the purchase and transportation planning problem is $z(x, y) = 2{,}195{,}972$. Using this value as a reference, assume that the DM is totally satisfied if the sales profit is larger than or equal to ¥2,300,000 and is totally unsatisfied if the sales profit is less than or equal to ¥2,100,000. Furthermore, assume that the rate of increased membership satisfaction is constant. Such a fuzzy goal of the DM can be determined by the linear membership function

$$\mu_G(z(x,y)) = \begin{cases} 1 & \text{if } z(x,y) \geq 2.3 \times 10^6 \\ \dfrac{z(x,y) - 2.1 \times 10^6}{0.2 \times 10^6} & \text{if } 2.1 \times 10^6 \leq z(x,y) \leq 2.3 \times 10^6 \\ 0 & \text{if } z(x,y) \leq 2.1 \times 10^6. \end{cases} \quad (7.13)$$

As for the constraints, since the constraints on the financial resources in each city

$$\sum_{i=1}^{n} d_{ji} y_{ji} \leq o_j, \ j = 1, \ldots, s$$

can be slightly softened, assume that the DM is totally satisfied if the expense is less than or equal to o_j and is totally unsatisfied if the expense is larger than or equal to $o_j + \bar{o}_j$ and the rate of increased membership satisfaction is constant. Such fuzzy constraints can be determined by the linear membership function

$$\mu_{C_j}(z(x,y)) = \begin{cases} 1 & \text{if } z(x,y) \leq o_j \\ 1 - \dfrac{z(x,y) - o_j}{\bar{o}_j} & \text{if } o_j \leq z(x,y) \leq o_j + \bar{o}_j \\ 0 & \text{if } z(x,y) \geq o_j + \bar{o}_j. \end{cases} \quad (7.14)$$

Parameter values for the above-mentioned fuzzy goal and the fuzzy constraints are summarized in Table 7.11.

Following the fuzzy decision by Bellman and Zadeh (1970), finding the maximum decision is to choose (x^*, y^*) such that

$$\mu_D(x^*, y^*) = \max_{(x,y) \in \tilde{X}} \min\{\mu_G(z(x,y)), \mu_{C_1}(z(x,y)), \ldots, \mu_{C_s}(z(x,y))\}, \quad (7.15)$$

Table 7.11 Fuzzy goal and fuzzy constraints

	$\mu = 0$	$\mu = 1$
μ_G	2,100,000	2,300,000
$\mu_{C_1}: (o_1 + \bar{o}_1, o_1)$	2,100,000	2,000,000
$\mu_{C_2}: (o_2 + \bar{o}_2, o_2)$	1,600,000	1,500,000
$\mu_{C_3}: (o_3 + \bar{o}_3, o_3)$	1,600,000	1,500,000
$\mu_{C_4}: (o_4 + \bar{o}_4, o_4)$	1,600,000	1,500,000
$\mu_{C_5}: (o_5 + \bar{o}_5, o_5)$	1,600,000	1,500,000
$\mu_{C_6}: (o_6 + \bar{o}_6, o_6)$	1,600,000	1,500,000
$\mu_{C_7}: (o_7 + \bar{o}_7, o_7)$	1,600,000	1,500,000
$\mu_{C_8}: (o_8 + \bar{o}_8, o_8)$	2,100,000	2,000,000

where

$$\bar{X} = \left\{ (x, y) \,\Big|\, D_i^L \leq x_i \leq D_i^U, \sum_{j=1}^{s} y_{ji} = x_i, i = 1, \ldots, n, \sum_{i=1}^{n} v_i x_i \leq W, \right.$$

$$\left. x \geq 0, y \geq 0 \right\} \quad (7.16)$$

denotes the feasible region of the problem which consists of the upper and lower limits of the total purchase volumes (7.2), the total purchase volume constraints (7.3), the storehouse constraint (7.5), and the nonnegativity conditions. Since the financial resources constraints (7.4) are relaxed as fuzzy constraints, they are not included in the conditions of the feasible region (7.16).

By introducing the auxiliary variable λ, the problem (7.15) can be transformed into the equivalent conventional linear programming problem

$$\left. \begin{aligned} & \text{maximize } \lambda \\ & \text{subject to } \frac{z(x, y)}{2.0 \times 10^5} - \lambda \geq 10.5 \\ & \frac{\sum_{i=1}^{n} d_{ji} y_{ji}}{\bar{o}_j} + \lambda \leq 1 + \frac{o_j}{\bar{o}_j}, \, j = 1, \ldots, s \\ & D_i^L \leq x_i \leq D_i^U, \, i = 1, \ldots, n \\ & \sum_{j=1}^{s} y_{ji} = x_i, \, i = 1, \ldots, n \\ & \sum_{i=1}^{n} v_i x_i \leq W \\ & x \geq 0, y \geq 0. \end{aligned} \right\} \quad (7.17)$$

Solving (7.17) by the simplex method of linear programming yields the optimal value $\lambda = 0.741$. In Table 7.12, the optimal solutions to the fuzzy linear programming problem (7.17) and the original nonfuzzy linear programming problem (7.6) are given.

7.3 Fuzzy Multiobjective Linear Programming Formulation

Table 7.12 Solutions of original and fuzzy linear programming problems

	Original problem	Fuzzy problem
Objective values z	2,195,972	2,248,211
Total sales	15,569,299	15,832,739
Purchasing expense	13,000,000	13,207,156
Transportation cost	373,327	377,372
Purchase volume		
City 1	2,000,000	2,025,894
City 2	1,500,000	1,525,894
City 3	1,500,000	1,525,894
City 4	1,500,000	1,525,894
City 5	1,500,000	1,525,894
City 6	1,500,000	1,525,894
City 7	1,500,000	1,525,894
City 8	2,000,000	2,025,894

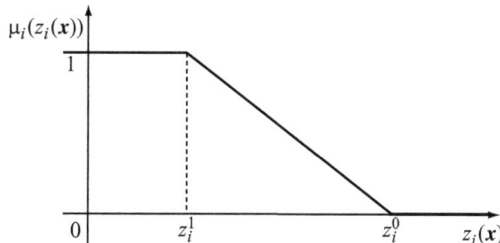

Fig. 7.3 Linear membership function

As can been seen in Table 7.12, in the fuzzy linear programming problem, since all of the original rigid constraints are softened by allowing a small amount of violation, each city spends additional ¥25,894 compared with the original rigid constraints, and the purchase cost and the transportation costs are slightly increased. As a result, it follows that the net profit is increased by ¥52,239 from ¥2,195,972 to ¥2,248,211 at the expense of additional ¥25,894 for each city.

Next, moving on to the purchase and transportation planning problem considering the reduction of the greenhouse gas emission, assume that the DM has a fuzzy goal for each of the objective functions in the two-objective linear programming problem (7.9). In a minimization problem, the DM may have a fuzzy goal expressed as "the ith objective function $z_i(x)$ should be substantially less than or equal to some value." As considered in Chap. 4, this type of statement can be quantified by eliciting the corresponding membership function from the DM. Fig. 7.3 illustrates a possible shape of the linear membership function, where z_i^0 and z_i^1 denote the values of the ith objective function $z_i(x)$ such that the membership function values are 0 and 1, respectively.

Although it is possible to incorporate the fuzzy constraints into the multiobjective linear programming problem, for the sake of simplicity, we only consider a fuzzy two-objective linear programming problem incorporating only fuzzy goals for the sales profit and the transportation cost based on the multiobjective linear programming problem formulated in the previous section.

For notational convenience, let X denote the feasible region of the problem satisfying the upper and lower limits of the total purchase volumes (7.2), the total purchase volume constraints (7.3), the financial resources constraints (7.4), the storehouse constraint (7.5), and the nonnegativity conditions. The fuzzy two-objective linear programming problem for the purchase and transportation planning is then formally formulated as

$$\left.\begin{array}{l} \text{fuzzy max } z_1(\boldsymbol{x},\boldsymbol{y}) = \sum_{i=1}^{n} c_i x_i - \sum_{j=1}^{s}\sum_{i=1}^{n} d_{ji} y_{ji} \\ \text{fuzzy min } z_2(\boldsymbol{x},\boldsymbol{y}) = \sum_{j=1}^{s}\sum_{i=1}^{n} b_{ji} y_{ji} \\ \text{subject to } (\boldsymbol{x},\boldsymbol{y}) \in X. \end{array}\right\} \quad (7.18)$$

For the objective functions $z_1(\boldsymbol{x},\boldsymbol{y})$ and $z_2(\boldsymbol{x},\boldsymbol{y})$, suppose that the DM specifies the fuzzy goals $\mu_1(z_1(\boldsymbol{x},\boldsymbol{y}))$ and $\mu_2(z_2(\boldsymbol{x},\boldsymbol{y}))$, and then (7.18) is rewritten as

$$\left.\begin{array}{l} \text{maximize } \mu_1(z_1(\boldsymbol{x},\boldsymbol{y})) \\ \text{maximize } \mu_2(z_2(\boldsymbol{x},\boldsymbol{y})) \\ \text{subject to } (\boldsymbol{x},\boldsymbol{y}) \in X. \end{array}\right\} \quad (7.19)$$

We employ the method for determining a linear membership function suggested by Zimmermann (1978) that we already explained in Chap. 4 and briefly review it. For an objective function $z_i(\boldsymbol{x})$ to be minimized, he introduced the linear membership function

$$\mu_i(z_i(\boldsymbol{x})) = \begin{cases} 0 & \text{if } z_i(\boldsymbol{x}) \geq z_i^0 \\ \dfrac{z_i(\boldsymbol{x}) - z_i^0}{z_i^1 - z_i^0} & \text{if } z_i^0 \geq z_i(\boldsymbol{x}) \geq z_i^1 \\ 1 & \text{if } z_i(\boldsymbol{x}) \leq z_i^1, \end{cases} \quad (7.20)$$

where z_i^0 and z_i^1 denote the values of the objective function $z_i(\boldsymbol{x})$ such that the membership function values are 0 and 1, respectively.

Using the individual minimum

$$z_i^{\min} = z_i(\boldsymbol{x}^{io}) = \min_{x \in X} z_i(\boldsymbol{x}), \ i = 1,\ldots,k, \quad (7.21)$$

7.3 Fuzzy Multiobjective Linear Programming Formulation

together with

$$z_i^m = \max\{z_i(x^{1o}), \ldots, z_i(x^{i-1,o}), z_i(x^{i+1,o}), \ldots, z_i(x^{ko})\}, \ i = 1, \ldots, k, \quad (7.22)$$

he determined the linear membership function (7.20) by choosing $z_i^1 = z_i^{\min}$ and $z_i^0 = z_i^m$.

For the purchase and transportation planning problem, since the individual maximum of the sales profit $z_1(x, y)$ is ¥2,585,822 and the individual minimum of the transportation cost $z_2(x, y)$ is ¥201,042, using the method by Zimmermann, we have $z_1^1 = 2{,}585{,}822$ and $z_2^1 = 201{,}042$, and, from (7.22), z_1^0 and z_2^0 are calculated as $z_1^0 = 2{,}117{,}800$ and $z_2^0 = 404{,}218$.

Assuming that the DM has a fuzzy goal for each of the objective functions in multiobjective linear programming problems, Sakawa et al. (1987) proposed an interactive fuzzy satisficing method (see also Sect. 4.4) for deriving a satisficing solution for the DM from among an M-Pareto optimal solution set, in which the DM is supplied with the trade-off information obtained from an optimal solution to the minimax problem. In their method, after identifying the membership functions $\mu_i(z_i(x))$, $i = 1, \ldots, k$ of the fuzzy goals for the objective functions $z_i(x)$, $i = 1, \ldots, k$, the DM is asked to specify the reference membership levels which are interpreted as the aspiration levels for the membership functions. As discussed in Sect. 4.4, the reference membership value can be considered as a natural extension of the reference point in the reference point method by Wierzbicki (1980).

For the fuzzy two-objective linear programming problem (7.19), find a satisficing solution for the DM using the interactive fuzzy satisficing method. In the interactive fuzzy satisficing method, after the DM specifies the reference membership levels $(\hat{\mu}_1, \hat{\mu}_2, \ldots, \hat{\mu}_k)$, the corresponding M-Pareto optimal solution, which is nearest to the point of the reference membership levels in the minimax sense or better than that if the reference membership levels are attainable, is obtained by solving the following minimax problem:

$$\underset{x \in X}{\text{minimize}} \ \max_{i=1,\ldots,k} \{\hat{\mu}_i - \mu_i(z_i(x))\}.$$

By introducing the auxiliary variable v, it can be transformed to

$$\text{minimize } v$$
$$\text{subject to } \hat{\mu}_i - \mu_i(z_i(x)) \leq v, \ i = 1, 2, \ldots, k$$
$$x \in X.$$

For the fuzzy two-objective linear programming problem (7.19), given the reference membership levels $(\hat{\mu}_1, \hat{\mu}_2)$, the corresponding minimax problem is formulated as

Table 7.13 Interactive process

	Reference membership levels ($\hat{\mu}_1, \hat{\mu}_2$)		
	(1,1)	(0.8, 1)	(0.8, 0.9)
μ_1	0.649	0.533	0.592
μ_2	0.649	0.733	0.692
Total sales	15,421,714	15,367,391	15,394,873
Purchasing expense	13,000,000	13,000,000	13,000,000
Transportation cost	272,284	255,231	263,618
Net profit $z = z_1 - z_2$	2,149,430	2,112,160	2,131,255
Sales profit z_1	2,421,714	2,367,391	2,394,873
Transportation cost z_2	272,284	255,231	263,618

$$\left.\begin{aligned}
&\text{minimize } v \\
&\text{subject to } \hat{\mu}_1 - \frac{z_1(\boldsymbol{x},\boldsymbol{y}) - z_1^0}{z_1^1 - z_1^0} \leq v \\
&\qquad\qquad \hat{\mu}_2 - \frac{z_2(\boldsymbol{x},\boldsymbol{y}) - z_2^0}{z_2^1 - z_2^0} \leq v \\
&\qquad\qquad D_i^L \leq x_i \leq D_i^U, \ i = 1,\ldots,n \\
&\qquad\qquad \sum_{j=1}^{s} y_{ji} = x_i, \ i = 1,\ldots,n \\
&\qquad\qquad \sum_{i=1}^{n} d_{ji} y_{ji} \leq o_j, \ j = 1,\ldots,s \\
&\qquad\qquad \sum_{i=1}^{n} v_i x_i \leq W \\
&\qquad\qquad \boldsymbol{x} \geq \boldsymbol{0}, \ \boldsymbol{y} \geq \boldsymbol{0}.
\end{aligned}\right\} \quad (7.23)$$

Table 7.13 shows an example of the interactive process for the fuzzy two-objective linear programming problem (7.19) using the interactive fuzzy satisficing method. For the initial reference membership levels $(\hat{\mu}_1, \hat{\mu}_2) = (1, 1)$, the corresponding minimax problem (7.23) is solved and an M-Pareto optimal solution is obtained. The DM is then supplied with the membership function values $(\mu_1, \mu_2) = (0.649, 0.649)$ and the related information including the objective function values $(z_1, z_2) = (2,421,714, 272,284)$. Suppose that the DM is not satisfied with the membership function values and updates the reference membership levels to $(\hat{\mu}_1, \hat{\mu}_2) = (0.8, 1)$, expecting the improvement of the satisfaction level for μ_2 at the expense of μ_1. It is significant to note here that, for the initial reference membership levels $(1, 1)$, since two fuzzy goals are treated equally as in the fuzzy decision by Bellman and Zadeh (1970), the same membership value of 0.649 is obtained for each membership function. Solving the minimax problem (7.23) to the updated reference membership levels $(\hat{\mu}_1, \hat{\mu}_2) = (0.8, 1)$, we obtain an M-Pareto optimal solution, and the membership function values $(\mu_1, \mu_2) = (0.533, 0.733)$ and the objective function values $(z_1, z_2) = (2{,}367{,}391,\ 255{,}231)$ are also calculated. Suppose that the DM is still not satisfied with these membership function values and updates the reference membership levels to $(\hat{\mu}_1, \hat{\mu}_2) = (0.8, 0.9)$, expecting slight increase of the satisfaction level for μ_1. For the updated reference membership levels, the corresponding minimax problem yields an M-Pareto optimal solution, and the corresponding membership function values $(\mu_1, \mu_2) = (0.592, 0.692)$ and the objective function values $(z_1, z_2) = (2{,}394{,}873,\ 263{,}618)$ are obtained. At the third interaction, assuming that the DM is satisfied with the values of the membership functions, the procedure stops.

7.4 Fuzzy Multiobjective Linear Stochastic Programming Formulation

In this section we deal with fuzzy multiobjective stochastic programming for the purchase and transportation planning problem. To be clear about parameters depending on stochastic events, review them in the original linear programming problem (7.6) or in the two-objective linear programming problem (7.9) for the purchase and transportation planning for food retailing. The two-objective problem (7.9) is given again in the following:

$$\text{maximize } z_1(\boldsymbol{x}, \boldsymbol{y}) = \sum_{i=1}^{n} c_i x_i - \sum_{j=1}^{s}\sum_{i=1}^{n} d_{ji} y_{ji}$$

$$\text{minimize } z_2(\boldsymbol{x}, \boldsymbol{y}) = \sum_{j=1}^{s}\sum_{i=1}^{n} b_{ji} y_{ji}$$

$$\text{subject to } D_i^L \leq x_i \leq D_i^U,\ i = 1, \ldots, n$$

$$\sum_{j=1}^{s} y_{ji} = x_i,\ i = 1, \ldots, n$$

$$\sum_{i=1}^{n} d_{ji} y_{ji} \leq o_j,\ j = 1, \ldots, s$$

$$\sum_{i=1}^{n} v_i x_i \leq W$$

$$\boldsymbol{x} \geq \boldsymbol{0},\ \boldsymbol{y} \geq \boldsymbol{0}.$$

The parameters in the problem and their meanings are summarized as follows:

c_i: the retail price of food i
d_{ji}: the purchase price of food i at the central wholesale market in city j
b_{ji}: the transportation cost per unit of food i from city j
D_i^L: the lower limit of the total purchase volume of food i

D_i^U: the upper limit of the total purchase volume of food i
o_j: the budget cap in city j
v_i: the cubic volume per unit of food i
W: the capacity of the storehouse of the food retailer

Since the retail price c_i, the lower and upper limits of the total purchase volume D_i^L and D_i^U, and the budget cap o_j are determined by the food retailer, they do not depend on any stochastic event. All foods in this problem are packed in package boxes, and therefore the cubic volume of food v_i is not generally affected by stochastic events. In contrast, the purchase price d_{ji} fluctuates depending on market circumstances, and the transportation cost b_{ji} changes by fluctuations in the fuel price. From these reasons, in this section we treat the two parameters d_{ji} and b_{ji} as random variables $\bar{d}_{ji}$ and $\bar{b}_{ji}$. To be more specific, we assume that $\bar{d}_{ji}$ and $\bar{b}_{ji}$ are normal random variables, i.e., $\bar{d}_{ji} \sim N(d_{ji}, \sigma_{d_{ji}}^2)$ and $\bar{b}_{ji} \sim N(b_{ji}, \sigma_{b_{ji}}^2)$, and the standard deviations are determined by $\sigma_{d_{ji}} = 0.01 d_{ji}$ and $\sigma_{b_{ji}} = 0.01 b_{ji}$. Moreover, assume that $\bar{d}_{ji}, \bar{b}_{ji}, j = 1, \ldots, s, i = 1, \ldots, n$ are independent of each other.

Incorporating these stochastic parameters $\bar{d}_{ji}$ and $\bar{b}_{ji}$, the two-objective problem (7.9) is rewritten as

$$\left.\begin{aligned}
\text{maximize } & z_1(\boldsymbol{x},\boldsymbol{y}) = \sum_{i=1}^n c_i x_i - \sum_{j=1}^s \sum_{i=1}^n \bar{d}_{ji} y_{ji} \\
\text{minimize } & z_2(\boldsymbol{x},\boldsymbol{y}) = \sum_{j=1}^s \sum_{i=1}^n \bar{b}_{ji} y_{ji} \\
\text{subject to } & D_i^L \leq x_i \leq D_i^U, \ i = 1, \ldots, n \\
& \sum_{j=1}^s y_{ji} = x_i, \ i = 1, \ldots, n \\
& \sum_{i=1}^n \bar{d}_{ji} y_{ji} \leq o_j, \ j = 1, \ldots, s \\
& \sum_{i=1}^n v_i x_i \leq W \\
& \boldsymbol{x} \geq \boldsymbol{0}, \ \boldsymbol{y} \geq \boldsymbol{0}.
\end{aligned}\right\} \quad (7.24)$$

As an initial attempt, we apply the expectation model with the chance constraints discussed in Sects. 5.3.1 and 6.1.1 to the fuzzy two-objective stochastic programming problem for the purchase and transportation planning (7.24).

Since the constraints of (7.24) include the stochastic parameters $\bar{d}_{ji}$, the constraints

$$\sum_{i=1}^n \bar{d}_{ji} y_{ji} \leq o_j, \ j = 1, \ldots, s$$

7.4 Fuzzy Multiobjective Linear Stochastic Programming Formulation

are reformulated as the chance constraints

$$P\left(\sum_{i=1}^{n} \bar{d}_{ji} y_{ji} \leq o_j\right) \geq \beta_j, \; j = 1, \ldots, s. \tag{7.25}$$

From the assumption that $\bar{d}_{ji} \sim N(d_{ji}, \sigma_{d_{ji}}^2)$, the random variable

$$\frac{o_j - \sum_{i=1}^{n} \bar{d}_{ji} y_{ji} - \left(o_j - \sum_{i=1}^{n} d_{ji} y_{ji}\right)}{\sqrt{\sum_{i=1}^{n} \sigma_{d_{ji}}^2 y_{ji}^2}} \tag{7.26}$$

is the standard normal random variable $N(0, 1)$, and therefore it follows that

$$P\left(\sum_{i=1}^{n} \bar{d}_{ji} y_{ji} \leq o_j\right)$$

$$= P\left(\frac{o_j - \sum_{i=1}^{n} \bar{d}_{ji} y_{ji} - \left(o_j - \sum_{i=1}^{n} d_{ji} y_{ji}\right)}{\sqrt{\sum_{i=1}^{n} \sigma_{d_{ji}}^2 y_{ji}^2}} \geq \frac{-\left(o_j - \sum_{i=1}^{n} d_{ji} y_{ji}\right)}{\sqrt{\sum_{i=1}^{n} \sigma_{d_{ji}}^2 y_{ji}^2}}\right)$$

$$= 1 - \Phi\left(\frac{\sum_{i=1}^{n} d_{ji} y_{ji} - o_j}{\sqrt{\sum_{i=1}^{n} \sigma_{d_{ji}}^2 y_{ji}^2}}\right), \tag{7.27}$$

where Φ is the distribution function of the standard normal distribution $N(0, 1)$. The chance constraints (7.25) can be then transformed into

$$\sum_{i=1}^{n} d_{ji} y_{ji} - o_j - \Phi^{-1}(1 - \beta_j) \sqrt{\sum_{i=1}^{n} \sigma_{d_{ji}}^2 y_{ji}^2} \leq 0, \; j = 1, \ldots, s. \tag{7.28}$$

In the expectation model, the objective functions are represented by their expectations, and then (7.24) is reformulated as

$$\left.\begin{aligned}
\text{minimize } & z_1^E(\pmb{x},\pmb{y}) = E(-z_1(\pmb{x},\pmb{y})) = -\sum_{i=1}^{n} c_i x_i + \sum_{j=1}^{s}\sum_{i=1}^{n} d_{ji} y_{ji} \\
\text{minimize } & z_2^E(\pmb{x},\pmb{y}) = E(z_2(\pmb{x},\pmb{y})) = \sum_{j=1}^{s}\sum_{i=1}^{n} b_{ji} y_{ji} \\
\text{subject to } & D_i^L \le x_i \le D_i^U, \ i = 1,\ldots,n \\
& \sum_{j=1}^{s} y_{ji} = x_i, \ i = 1,\ldots,n \\
& \sum_{i=1}^{n} d_{ji} y_{ji} - o_j - \Phi^{-1}(1-\beta_j)\sqrt{\sum_{i=1}^{n} \sigma_{d_{ji}}^2 y_{ji}^2} \le 0, \ j = 1,\ldots,s \\
& \sum_{i=1}^{n} v_i x_i \le W \\
& \pmb{x} \ge \pmb{0}, \ \pmb{y} \ge \pmb{0}.
\end{aligned}\right\} \quad (7.29)$$

In order to consider the imprecise nature of the DM's judgments for each expectation of the objective functions $z_1^E(\pmb{x},\pmb{y})$ and $z_2^E(\pmb{x},\pmb{y})$ in (7.29), when we introduce the fuzzy goals such as "the expectation should be substantially less than or equal to a certain value," (7.29) can be interpreted as

$$\underset{(\pmb{x},\pmb{y})\in X(\pmb{\beta})}{\text{maximize}} \{\mu_1\left(z_1^E(\pmb{x},\pmb{y})\right), \mu_2\left(z_2^E(\pmb{x},\pmb{y})\right)\}, \quad (7.30)$$

where μ_1 and μ_2 are membership functions to quantify the fuzzy goals for the objective functions $z_1^E(\pmb{x},\pmb{y})$ and $z_2^E(\pmb{x},\pmb{y})$ in (7.29) and $X(\pmb{\beta})$ denotes the feasible region of (7.29).

In the interactive fuzzy satisficing method, to generate a candidate for a satisficing solution which is also M-Pareto optimal, the DM is asked to specify the aspiration levels of achievement for all the membership function values, called the reference membership levels (Sakawa 1993; Sakawa and Yano 1985c, 1989, 1990; Sakawa et al. 1987).

For a pair of reference membership levels $\hat{\mu}_1$ and $\hat{\mu}_2$ for the membership functions $\mu_1\left(z_1^E(\pmb{x},\pmb{y})\right)$ and $\mu_2\left(z_2^E(\pmb{x},\pmb{y})\right)$ in (7.30), the corresponding M-Pareto optimal solution, which is the nearest to the vector of the given reference membership levels in the minimax sense or better than it if the reference membership levels are attainable, is obtained by solving the minimax problem

$$\underset{(\pmb{x},\pmb{y})\in X(\pmb{\beta})}{\text{minimize}} \max\{\hat{\mu}_1 - \mu_1\left(z_1^E(\pmb{x},\pmb{y})\right), \hat{\mu}_2 - \mu_2(z_2^E(\pmb{x},\pmb{y}))\}. \quad (7.31)$$

Assume that the membership functions are linear, i.e.,

7.4 Fuzzy Multiobjective Linear Stochastic Programming Formulation

$$\mu_1\left(z_1^E(x,y)\right) = \frac{-\sum_{i=1}^{n} c_i x_i + \sum_{j=1}^{s}\sum_{i=1}^{n} d_{ji} y_{ji} - z_1^0}{z_1^1 - z_1^0}$$

$$\mu_2\left(z_2^E(x,y)\right) = \frac{\sum_{j=1}^{s}\sum_{i=1}^{n} b_{ji} y_{ji} - z_2^0}{z_2^1 - z_2^0},$$

and by introducing an auxiliary variable v, (7.31) can be finally expressed as

minimize v

subject to
$$\hat{\mu}_1 - \frac{-\sum_{i=1}^{n} c_i x_i + \sum_{j=1}^{s}\sum_{i=1}^{n} d_{ji} y_{ji} - z_1^0}{z_1^1 - z_1^0} \leq v$$

$$\hat{\mu}_2 - \frac{\sum_{j=1}^{s}\sum_{i=1}^{n} b_{ji} y_{ji} - z_2^0}{z_2^1 - z_2^0} \leq v$$

$$D_i^L \leq x_i \leq D_i^U, \; i = 1,\ldots,n$$

$$\sum_{j=1}^{s} y_{ji} = x_i, \; i = 1,\ldots,n$$

$$\sum_{i=1}^{n} d_{ji} y_{ji} - o_j - \Phi^{-1}(1-\beta_j)\sqrt{\sum_{i=1}^{n} \sigma_{d_{ji}}^2 y_{ji}^2} \leq 0, \; j = 1,\ldots,s$$

$$\sum_{i=1}^{n} v_i x_i \leq W$$

$$x \geq 0, \; y \geq 0.$$

(7.32)

In Table 7.14, we give an example of the interactive process in the interactive fuzzy satisficing method for the expectation model, in which the minimax problem (7.32) is repeatedly solved. Assume that the parameters of the fuzzy goals are specified as $z_1^0 = -2,300,000$, $z_1^1 = -2,600,000$, $z_2^0 = 320,000$, and $z_2^1 = 300,000$ and the satisficing probability levels as $\beta_j = 0.8$, $j = 1,\ldots,s$. In this interactive process, by setting the initial reference membership levels $(1,1)$, a solution with the same membership values $\mu_1 = \mu_2 = 0.543$ is obtained. The constraints of (7.32) include the chance constraints $\sum_{i=1}^{n} d_{ji} y_{ji} - o_j - \Phi^{-1}(1-\beta_j)\sqrt{\sum_{i=1}^{n} \sigma_{d_{ji}}^2 y_{ji}^2} \leq 0$, $j = 1,\ldots,s$, and therefore the feasible region of the problem diminishes compared with those of the original problem (7.6) and the minimax problem (7.23) for the interactive fuzzy satisficing method given in the previous section. From this fact, the purchasing expense becomes smaller than 13,000,000 which is that of the

Table 7.14 Interactive process in expectation model

	Reference membership levels $(\hat{\mu}_1, \hat{\mu}_2)$		
	(1,1)	(0.52, 0.58)	(0.53, 0.58)
μ_1	0.543	0.538	0.539
μ_2	0.543	0.598	0.589
Total sales	15,378,118	15,376,566	15,376,825
Purchasing expense	12,915,168	12,915,170	12,915,170
Transportation cost	309,137	308,040	308,223
Net profit $z = z_1 - z_2$	2,153,813	2,153,355	2,153,431
Sales profit z_1	2,462,950	2,461,395	2,461,654
Transportation cost z_2	309,137	308,040	308,223

original problem (7.6) and the minimax problem (7.23). By revising the reference membership levels to $(0.52, 0.58)$ which give weight to the second fuzzy goal slightly, we obtain a solution decreasing the membership function value μ_1 of the fuzzy goal for the sales profit z_1 and increasing the membership function value μ_2 of the fuzzy goal for the transportation cost z_2. In the third iteration, one finds that the membership values are changed inversely by increasing $\hat{\mu}_1$ by 0.01 and leaving $\hat{\mu}_2$ unchanged.

Next, under the same assumption as that of the expectation model, consider the fractile model with the chance constraints dealt with in Sects. 5.3.4 and 6.1.4 for applying to the fuzzy two-objective stochastic programming problem for the purchase and transportation planning, and it is formulated as follows:

$$\left.\begin{aligned}
& \text{minimize } f_1 \\
& \text{minimize } f_2 \\
& \text{subject to } P\left(-\sum_{i=1}^{n} c_i x_i + \sum_{j=1}^{s}\sum_{i=1}^{n} \bar{d}_{ji} y_{ji} \leq f_1\right) \geq \theta_1 \\
& \quad P\left(\sum_{j=1}^{s}\sum_{i=1}^{n} \bar{b}_{ji} y_{ji} \leq f_2\right) \geq \theta_2 \\
& \quad D_i^L \leq x_i \leq D_i^U, \ i = 1,\ldots,n \\
& \quad \sum_{j=1}^{s} y_{ji} = x_i, \ i = 1,\ldots,n \\
& \quad \sum_{i=1}^{n} d_{ji} y_{ji} - o_j - \Phi^{-1}(1-\beta_j)\sqrt{\sum_{i=1}^{n} \sigma_{d_{ji}}^2 y_{ji}^2} \leq 0, \ j = 1,\ldots,s \\
& \quad \sum_{i=1}^{n} v_i x_i \leq W \\
& \quad x \geq 0, \ y \geq 0.
\end{aligned}\right\} \quad (7.33)$$

7.4 Fuzzy Multiobjective Linear Stochastic Programming Formulation

In this model, f_1 and f_2 are variables to be minimized, and therefore we call them the target variables. The target values f_1 and f_2 are minimized under the condition that the probability that each of the objective functions is smaller than or equal to the target variable is guaranteed to be larger than or equal to a certain assured probability level θ_i specified by the DM.

As discussed in Sect. 6.1.4, (7.33) can be transformed to

$$\left.\begin{aligned}
\text{minimize}\quad & z_1^F(\boldsymbol{x},\boldsymbol{y}) = -\sum_{i=1}^{n} c_i x_i + \sum_{j=1}^{s}\sum_{i=1}^{n} d_{ji} y_{ji} \\
& \qquad\qquad\qquad + \Phi^{-1}(\theta_1)\sqrt{\sum_{j=1}^{s}\sum_{i=1}^{n}\sigma_{d_{ji}}^2 y_{ji}^2} \\
\text{minimize}\quad & z_2^F(\boldsymbol{x},\boldsymbol{y}) = \sum_{j=1}^{s}\sum_{i=1}^{n} b_{ji} y_{ji} + \Phi^{-1}(\theta_2)\sqrt{\sum_{j=1}^{s}\sum_{i=1}^{n}\sigma_{b_{ji}}^2 y_{ji}^2} \\
\text{subject to}\quad & D_i^L \le x_i \le D_i^U,\ i=1,\ldots,n \\
& \sum_{j=1}^{s} y_{ji} = x_i,\ i=1,\ldots,n \\
& \sum_{i=1}^{n} d_{ji} y_{ji} - o_j - \Phi^{-1}(1-\beta_j)\sqrt{\sum_{i=1}^{n}\sigma_{d_{ji}}^2 y_{ji}^2} \le 0,\ j=1,\ldots,s \\
& \sum_{i=1}^{n} v_i x_i \le W \\
& \boldsymbol{x} \ge \boldsymbol{0},\ \boldsymbol{y} \ge \boldsymbol{0}.
\end{aligned}\right\} \tag{7.34}$$

As assumed in the expectation model, we also suppose that the DM incorporates the fuzzy goals represented by the following linear membership functions for the objective functions in (7.34):

$$\mu_1(z_1^F(\boldsymbol{x},\boldsymbol{y})) = \frac{1}{z_1^1 - z_1^0}\left\{-\sum_{i=1}^{n} c_i x_i + \sum_{j=1}^{s}\sum_{i=1}^{n} d_{ji} y_{ji} + \Phi^{-1}(\theta_1)\sqrt{\sum_{j=1}^{s}\sum_{i=1}^{n}\sigma_{d_{ji}}^2 y_{ji}^2} - z_1^0\right\} \tag{7.35}$$

$$\mu_2(z_2^F(\boldsymbol{x},\boldsymbol{y})) = \frac{1}{z_2^1 - z_2^0}\left\{\sum_{j=1}^{s}\sum_{i=1}^{n} b_{ji} y_{ji} + \Phi^{-1}(\theta_2)\sqrt{\sum_{j=1}^{s}\sum_{i=1}^{n}\sigma_{b_{ji}}^2 y_{ji}^2} - z_2^0\right\}. \tag{7.36}$$

In the interactive fuzzy satisficing method, for the DM's reference membership levels $\hat{\mu}_1$ and $\hat{\mu}_2$ to the membership functions (7.35) and (7.36), the corresponding M-Pareto optimal solution is obtained by solving the minimax problem

$$\operatorname*{minimize}_{(\boldsymbol{x},\boldsymbol{y})\in X(\beta)}\ \max\left\{\hat{\mu}_1 - \mu_1\left(z_1^F(\boldsymbol{x},\boldsymbol{y})\right),\ \hat{\mu}_2 - \mu_2\left(z_2^F(\boldsymbol{x},\boldsymbol{y})\right)\right\}. \tag{7.37}$$

Table 7.15 Interactive process in fractile model

	Reference membership levels ($\hat{\mu}_1, \hat{\mu}_2$)		
	(1,1)	(0.65, 0.55)	(0.68, 0.55)
μ_1	0.578	0.606	0.615
μ_2	0.578	0.506	0.485
Total sales	15,332,363	15,336,499	15,337,734
Purchasing expense	12,915,147	12,915,041	12,915,004
Transportation cost	281,857	284,011	284,657
Net profit $z = z_1 - z_2$	2,135,359	2,137,447	2,422,730
Sales profit z_1	2,417,216	2,421,458	2,138,073
Transportation cost z_2	281,857	284,011	284,657

By introducing an auxiliary variable v, (7.37) can be expressed as

$$\begin{aligned}
\text{minimize } & v \\
\text{subject to } & \hat{\mu}_1 - \frac{1}{z_1^1 - z_1^0}\left\{-\sum_{i=1}^n c_i x_i + \sum_{j=1}^s \sum_{i=1}^n d_{ji} y_{ji} + \Phi^{-1}(\theta_1)\sqrt{\sum_{j=1}^s \sum_{i=1}^n \sigma_{d_{ji}}^2 y_{ji}^2} - z_1^0\right\} \leq v \\
& \hat{\mu}_2 - \frac{1}{z_2^1 - z_2^0}\left\{\sum_{j=1}^s \sum_{i=1}^n b_{ji} y_{ji} + \Phi^{-1}(\theta_2)\sqrt{\sum_{j=1}^s \sum_{i=1}^n \sigma_{b_{ji}}^2 y_{ji}^2} - z_2^0\right\} \leq v \\
& D_i^L \leq x_i \leq D_i^U, \ i = 1,\ldots,n \\
& \sum_{j=1}^s y_{ji} = x_i, \ i = 1,\ldots,n \\
& \sum_{i=1}^n d_{ji} y_{ji} - o_j - \Phi^{-1}(1-\beta_j)\sqrt{\sum_{i=1}^n \sigma_{d_{ji}}^2 y_{ji}^2} \leq 0, \ j = 1,\ldots,s \\
& \sum_{i=1}^n v_i x_i \leq W \\
& x \geq 0, \ y \geq 0.
\end{aligned}$$
(7.38)

In Table 7.15, we give an example of the interactive process in the interactive fuzzy satisficing method for the fractile model, in which the minimax problem (7.38) is repeatedly solved. Assume that the minimum probability levels are specified as $\theta_1 = 0.8$ and $\theta_2 = 0.8$, the parameters of the fuzzy goals as $z_1^0 = -2,300,000$, $z_1^1 = -2,450,000$, $z_2^0 = 300,000$, and $z_2^1 = 270,000$ and the satisficing probability levels as $\beta_j = 0.8$, $j = 1,\ldots,s$. As seen in the table, since the minimax problem (7.38) also includes the same chance constraints as that of the minimax problem (7.32) in

7.4 Fuzzy Multiobjective Linear Stochastic Programming Formulation

the expectation model, the purchasing expense becomes smaller than 13,000,000 which is that of the original problem (7.6) and the minimax problem (7.23). One finds that, according to the updates of the reference membership levels serving the DM's intention, the membership values are manipulated adequately. It is noted that we cannot calculate the minimax problem (7.38) by using the Excel solver because it has 130 variables, and in order to handle this difficulty, only for this problem, we use "What's BEST Version 11 Superclass" which is a commercially available add-in to Excel for solving mathematical programming problems.

Finally, we try to apply the probability model with the chance constraints dealt with in Sects. 5.3.3 and 6.1.3 to the fuzzy two-objective stochastic programming problem for the purchase and transportation planning. To mitigate the computational load, we employ a simpler representation of random variables. To be more specific, we assume that $\bar{d}_{ji} = d_{ji} + \sigma_{d_{ji}} \bar{t}_1$ and $\bar{b}_{ji} = b_{ji} + \sigma_{b_{ji}} \bar{t}_2$, where $\bar{t}_1 \sim N(0,1)$ and $\bar{t}_2 \sim N(0,1)$, and the standard deviations are determined by $\sigma_{d_{ji}} = 0.01 d_{ji}$ and $\sigma_{b_{ji}} = 0.01 b_{ji}$. Moreover, assume that $\bar{t}_1$ and $\bar{t}_2$ are independent of each other.

Since we employ the simpler representation of random variables $\bar{d}_{ji} = d_{ji} + \sigma_{d_{ji}} \bar{t}_1$, the chance constraints also become simple as follows.

The chance constraints

$$P\left(\sum_{i=1}^{n} \bar{d}_{ji} y_{ji} \le o_j\right) = P\left(\sum_{i=1}^{n} (d_{ji} + \sigma_{d_{ji}} \bar{t}_1) y_{ji} \le o_j\right) \ge \beta_j, \quad j = 1, \ldots, s$$

are transformed to

$$\Phi\left(\frac{o_j - \sum_{i=1}^{n} d_{ji} y_{ji}}{\sum_{i=1}^{n} \sigma_{d_{ji}} y_{ji}}\right) \ge \beta_j, \quad j = 1, \ldots, s, \quad (7.39)$$

where Φ is the distribution function of the standard normal distribution $N(0,1)$ and (7.39) can be also represented by linear constraints

$$\sum_{i=1}^{n} d_{ji} y_{ji} - o_j + \Phi^{-1}(\beta_j) \sum_{i=1}^{n} \sigma_{d_{ji}} y_{ji} \le 0, \quad j = 1, \ldots, s. \quad (7.40)$$

In the probability model, the objective function is the probability that the objective function value is smaller than or equal to the target value, and then the two-objective stochastic programming problem (7.24) is reformulated as

$$\left.\begin{array}{l}\text{maximize } z_1^P(\pmb{x},\pmb{y}) = P\left(-\sum_{i=1}^{n}c_i x_i + \sum_{j=1}^{s}\sum_{i=1}^{n}\bar{d}_{ji}y_{ji} \le f_1\right)\\ \text{maximize } z_2^P(\pmb{x},\pmb{y}) = P\left(\sum_{j=1}^{s}\sum_{i=1}^{n}\bar{b}_{ji}y_{ji} \le f_2\right)\\ \text{subject to } D_i^L \le x_i \le D_i^U, \ i=1,\dots,n\\ \sum_{j=1}^{s}y_{ji} = x_i, \ i=1,\dots,n\\ \sum_{i=1}^{n}d_{ji}y_{ji} - o_j + \Phi^{-1}(\beta_j)\sum_{i=1}^{n}\sigma_{d_{ji}}y_{ji} \le 0, \ j=1,\dots,s\\ \sum_{i=1}^{n}v_i x_i \le W\\ \pmb{x} \ge \pmb{0}, \ \pmb{y} \ge \pmb{0},\end{array}\right\} \quad (7.41)$$

where f_1 and f_2 are the target values specified by the DM for the objective function values, and it is noted that the objective functions in (7.41) can be transformed to

$$z_1^P(\pmb{x},\pmb{y}) = \Phi\left(\frac{f_1 - \left(-\sum_{i=1}^{n}c_i x_i + \sum_{j=1}^{s}\sum_{i=1}^{n}d_{ji}y_{ji}\right)}{\sum_{j=1}^{s}\sum_{i=1}^{n}\sigma_{d_{ji}}y_{ji}}\right), \quad (7.42)$$

$$z_2^P(\pmb{x},\pmb{y}) = \Phi\left(\frac{f_2 - \sum_{j=1}^{s}\sum_{i=1}^{n}b_{ji}y_{ji}}{\sum_{j=1}^{s}\sum_{i=1}^{n}\sigma_{b_{ji}}y_{ji}}\right). \quad (7.43)$$

Assume that the DM incorporates the fuzzy goals for the objective functions in (7.41) and the membership functions of the fuzzy goals are specified as the following linear ones:

$$\mu_1(z_1^P(\pmb{x},\pmb{y})) = \frac{1}{z_1^1 - z_1^0}\left\{\Phi\left(\frac{f_1 - \left(-\sum_{i=1}^{n}c_i x_i + \sum_{j=1}^{s}\sum_{i=1}^{n}d_{ji}y_{ji}\right)}{\sum_{j=1}^{s}\sum_{i=1}^{n}\sigma_{d_{ji}}y_{ji}}\right) - z_1^0\right\} \quad (7.44)$$

7.4 Fuzzy Multiobjective Linear Stochastic Programming Formulation

$$\mu_2(z_2^P(\boldsymbol{x},\boldsymbol{y})) = \frac{1}{z_2^1 - z_2^0} \left\{ \Phi \left(\frac{f_2 - \sum_{j=1}^{s}\sum_{i=1}^{n} b_{ji} y_{ji}}{\sum_{j=1}^{s}\sum_{i=1}^{n} \sigma_{b_{ji}} y_{ji}} \right) - z_2^0 \right\} \tag{7.45}$$

In the interactive fuzzy satisficing method, for the DM's reference membership levels $\hat{\mu}_1$ and $\hat{\mu}_2$ to the membership functions (7.44) and (7.45), the corresponding M-Pareto optimal solution is obtained by solving the minimax problem:

$$\underset{(\boldsymbol{x},\boldsymbol{y})\in X(\boldsymbol{\beta})}{\text{minimize}} \ \max \left\{ \hat{\mu}_1 - \mu_1\left(z_1^P(\boldsymbol{x},\boldsymbol{y})\right), \ \hat{\mu}_2 - \mu_2\left(z_2^P(\boldsymbol{x},\boldsymbol{y})\right) \right\}. \tag{7.46}$$

By introducing an auxiliary variable v, (7.46) can be expressed as

$$\begin{aligned}
& \text{minimize } v \\
& \text{subject to } \hat{\mu}_1 - \frac{1}{z_1^1 - z_1^0}\left\{ \Phi\left(\frac{f_1 - (-\sum_{i=1}^{n} c_i x_i + \sum_{j=1}^{s}\sum_{i=1}^{n} d_{ji} y_{ji})}{\sum_{j=1}^{s}\sum_{i=1}^{n} \sigma_{d_{ji}} y_{ji}} \right) - z_1^0 \right\} \leq v \\
& \hat{\mu}_2 - \frac{1}{z_2^1 - z_2^0}\left\{ \Phi\left(\frac{f_2 - \sum_{j=1}^{s}\sum_{i=1}^{n} b_{ji} y_{ji}}{\sum_{j=1}^{s}\sum_{i=1}^{n} \sigma_{b_{ji}} y_{ji}} \right) - z_2^0 \right\} \leq v \\
& D_i^L \leq x_i \leq D_i^U, \ i = 1,\ldots,n \\
& \sum_{j=1}^{s} y_{ji} = x_i, \ i = 1,\ldots,n \\
& \sum_{i=1}^{n} d_{ji} y_{ji} - o_j + \Phi^{-1}(\beta_j) \sum_{i=1}^{n} \sigma_{d_{ji}} y_{ji} \leq 0, \ j = 1,\ldots,s \\
& \sum_{i=1}^{n} v_i x_i \leq W \\
& \boldsymbol{x} \geq \boldsymbol{0}, \ \boldsymbol{y} \geq \boldsymbol{0},
\end{aligned} \tag{7.47}$$

and it can be further transformed to

Table 7.16 Interactive process in probability model

	Reference membership levels $(\hat{\mu}_1, \hat{\mu}_2)$		
	(1,1)	(0.65, 0.59)	(0.65, 0.63)
μ_1	0.602	0.605	0.603
μ_2	0.602	0.545	0.583
z_1^P	0.781	0.782	0.781
z_2^P	0.831	0.814	0.825
$-\partial\mu_1/\partial\mu_2$	0.0654	0.0615	0.0640
Total sales	15,291,276	15,291,745	15,291,435
Purchasing expense	12,891,502	12,891,502	12,891,502
Transportation cost	277,348	277,527	277,409
Net profit $z = z_1 - z_2$	2,122,426	2,122,716	2,122,525
Sales profit z_1	2,399,774	2,400,243	2,399,933
Transportation cost z_2	277,348	277,527	277,409

$$\begin{aligned}
&\text{minimize } v \\
&\text{subject to } \Phi^{-1}\left((\hat{\mu}_1 - v)(z_1^1 - z_1^0) + z_1^0\right) \sum_{j=1}^{s}\sum_{i=1}^{n} \sigma_{d_{ji}} y_{ji} \\
&\qquad\qquad \leq f_1 - \left(-\sum_{i=1}^{n} c_i x_i + \sum_{j=1}^{s}\sum_{i=1}^{n} d_{ji} y_{ji}\right) \\
&\Phi^{-1}\left((\hat{\mu}_2 - v)(z_2^1 - z_2^0) + z_2^0\right) \sum_{j=1}^{s}\sum_{i=1}^{n} \sigma_{b_{ji}} y_{ji} \leq f_2 - \sum_{j=1}^{s}\sum_{i=1}^{n} b_{ji} y_{ji} \\
&D_i^L \leq x_i \leq D_i^U, \ i = 1, \ldots, n \\
&\sum_{j=1}^{s} y_{ji} = x_i, \ i = 1, \ldots, n \\
&\sum_{i=1}^{n} d_{ji} y_{ji} - o_j + \Phi^{-1}(\beta_j) \sum_{i=1}^{n} \sigma_{d_{ji}} y_{ji} \leq 0, \ j = 1, \ldots, s \\
&\sum_{i=1}^{n} v_i x_i \leq W \\
&\mathbf{x} \geq \mathbf{0}, \ \mathbf{y} \geq \mathbf{0}.
\end{aligned}$$

(7.48)

As described in Sect. 6.1.3, (7.48) can be solved by the combined use of phase I of the simplex method and the bisection method.

We give an example of the interactive process in the probability model in Table 7.16 when the satisficing probability levels β_j, the target values f_i, and the parameters z_i^0 and z_i^1 of the fuzzy goals are specified as follows:

7.4 Fuzzy Multiobjective Linear Stochastic Programming Formulation

$$\beta_j = 0.8, \quad j = 1, \ldots, 8,$$
$$f_1 = -2,300,000, \quad f_2 = 280,000,$$
$$z_1^0 = 0.60, \quad z_1^1 = 0.90,$$
$$z_2^0 = 0.65, \quad z_2^1 = 0.95.$$

After finding the minimum value v^* of (7.48) through the bisection method, in order to uniquely determine an optimal solution x^* corresponding to v^*, the linear fractional programming problem is solved. At this moment the trade-off rate between the membership functions is obtained, and it is also given in the table. As seen in the table, according to the updates of the reference membership levels serving the DM's intention, the membership values are manipulated adequately. It is found that the purchasing expense becomes smaller than 13,000,000 which is that of the original problem (7.6) and the minimax problem (7.23) because the chance constraints $\sum_{i=1}^{n} d_{ji} y_{ji} - o_j + \Phi^{-1}(\beta_j) \sum_{i=1}^{n} \sigma_{d_{ji}} y_{ji} \leq 0, j = 1, \ldots, s$ are included in the constraints of (7.48).

Sakawa and his colleagues study some extensions of the linear programming formulation for food retailing discussed in this chapter so as to cope with outsourcing of the purchase operations and multi-store operation in multiple regions, and these models are formulated two-level and three-level linear programming problems in a noncooperative way (Sakawa and Nishizaki 2009; Sakawa et al. 2012, 2013b).

Problems

7.1 A grocery store purchases two kinds of fresh foods at two central wholesale markets, and the fresh foods are transported by truck from each of the central wholesaler markets to the grocery store in Tokyo. Let x_1 and x_2 denote the total purchase volume of foods 1 and 2, respectively. Moreover, let y_{11} and y_{12} denote the purchase volumes of foods 1 and 2 at the central wholesale market in city 1, and let y_{21} and y_{22} denote the purchase volumes of foods 1 and 2 at the central wholesale market in city 2. Assume that the retail prices of foods 1 and 2 are, respectively, 90 and 111, and also assume that the purchase prices of foods 1 and 2 are, respectively, 50 and 57 at the central wholesale market in city 1, and those of foods 1 and 2 are, respectively, 60 and 50 at the central wholesale market in city 2. Moreover, assume that the transportation costs per unit of foods 1 and 2 from city 1 are, respectively, 20 and 8, and those of foods 1 and 2 from city 2 are, respectively, 8 and 10. The grocery store desires to maximize the net profit which is defined by the total sales minus the total purchasing expense and the total transportation cost. Show the objective function z of the grocery store.

7.2 Consider the purchase and transportation planning given in problem 7.1. From the business judgment of the grocery store taking into account the demands and the inventories, suppose that the total purchase volumes of food 1 must be larger than or equal to the lower limit 4,000 and be smaller than or equal to the upper limit 5,000 and that of food 2 must be also larger than or equal to the lower limit 4,000 and be smaller than or equal to the upper limit 5,000. Show the constraints for the upper and lower limits of the total purchase volumes for foods 1 and 2.

7.3 Consider the purchase and transportation planning given in problem 7.1. Food 1 is purchased in the central wholesaler markets in cities 1 and 2, and the sum of these purchase volumes is the total purchase volume of foods 1. As for food 2, the same is true. Formulate these relations.

7.4 Consider the purchase and transportation planning given in problem 7.1. There are limitations on financial resources of the grocery store for purchasing foods at the central wholesaler markets. Suppose that the budget caps in cities 1 and 2 are 220,000 and 200,000, respectively. Show the constraints on the financial resources in the two cities.

7.5 By using the objective function formulated in problem 7.1 and the constraints defined in problems 7.2–7.4, formulate the purchase and transportation planning as a linear programming problem, and solve it using the Excel solver.

7.6 Consider the purchase and transportation planning given in problem 7.1. Let the difference between the total sales and the purchasing expense be the first objective function z_1 and let the transportation cost be the second objective function z_2. By using the two-objective functions z_1 and z_2 and the constraints defined in problems 7.2–7.4, formulate the purchase and transportation planning as a two-objective linear programming problem. After setting the weighting coefficients as $(w_1, w_2) = (1, 4)$ and formulating the weighting problem for the two-objective linear programming problem for the purchase and transportation planning, solve it using the Excel solver.

7.7 For the two-objective linear programming problem for the purchase and transportation planning, after determining the membership functions of the fuzzy goals for the objective functions z_1 and z_2 by using the method by Zimmermann, formulate the problem for finding the maximum decision following the fuzzy decision by Bellman and Zadeh, and solve it using the Excel solver.

7.8 Consider the purchase and transportation planning given in problem 7.1. Assume that the purchase prices fluctuate depending on market circumstances, and therefore the purchase prices of foods 1 and 2 from city 1 are, respectively, $\overline{50} = 50 + 5\bar{t}_1$ and $\overline{57} = 57 + 6\bar{t}_1$, and those of foods 1 and 2 from city 2 are, respectively, $\overline{60} = 60 + 5\bar{t}_1$ and $\overline{50} = 50 + 5\bar{t}_1$, where $\bar{t}_1 \sim N(0, 1)$. Also assume that the transportation costs change by fluctuations in the fuel price, and therefore the transportation costs per unit of foods 1 and 2 from city 1 are, respectively, $\overline{20} = 20 + 3\bar{t}_2$ and $\bar{8} = 8 + 2\bar{t}_2$, and those of foods 1 and 2 from city 2 are, respectively, $\bar{8} = 8 + 2\bar{t}_2$ and $\overline{10} = 10 + 2\bar{t}_2$, where $\bar{t}_2 \sim N(0, 1)$. Moreover, assume that the random variables $\bar{t}_1$ and $\bar{t}_2$ are independent of each other.

Suppose that the grocery store adopts the interactive fuzzy satisficing method for the probability model with the satisficing probability levels $\beta_1 = \beta_2 = 0.8$ and the target values $f_1 = -390,000$ and $f_2 = 110,000$.

Formulate the minimax problem for the interactive fuzzy satisficing method for the probability model, assuming that the grocery store employs the membership functions for the fuzzy goals formulated in problem 7.7.

Appendix A
Linear Algebra

A.1 Vector

A vector is an ordered array of numbers. It can be either a row or a column of elements. Vectors are usually denoted by lowercase boldface letters. An n-dimensional column vector x is an array of n real numbers $x_1, \ldots, x_n$, i.e.,

$$x = \begin{pmatrix} x_1 \\ \vdots \\ x_n \end{pmatrix}.$$

Here $x_i, i = 1, \ldots, n$ is called the ith component, or element, of the vector x. The zero vector, or null vector, denoted by $\mathbf{0}$, is a vector with all components equal to zeros. The unit vector has a 1 in the ith position and 0s elsewhere. A unit vector will generally be denoted by e_i, where 1 appears in the ith position. To be more precise,

$$\mathbf{0} = \begin{pmatrix} 0 \\ 0 \\ \vdots \\ 0 \end{pmatrix}, \ e_1 = \begin{pmatrix} 1 \\ 0 \\ \vdots \\ 0 \end{pmatrix}, \ e_2 = \begin{pmatrix} 0 \\ 1 \\ \vdots \\ 0 \end{pmatrix}, \ldots, \ e_n = \begin{pmatrix} 0 \\ 0 \\ \vdots \\ 1 \end{pmatrix}.$$

The collection of all n-dimensional vectors is called n-dimensional vector space. Compared with vectors, real numbers are called scalars.

The two vectors

$$x = \begin{pmatrix} x_1 \\ \vdots \\ x_n \end{pmatrix}, \ y = \begin{pmatrix} y_1 \\ \vdots \\ y_n \end{pmatrix}.$$

are called equal, denoted by $x = y$, if and only if all components are equal, i.e., $x_i = y_i, i = 1, \ldots, n$.

Two vectors of the same size can be added, where addition is performed componentwise.

(i) Addition

For two n-dimensional vectors x and y, the addition denoted by $x + y$ is defined by

$$x + y = \begin{pmatrix} x_1 + y_1 \\ \vdots \\ x_n + y_n \end{pmatrix}.$$

(ii) Scalar Multiplication

The multiplication of a vector by a scalar is performed by simply multiplying each element in the vector by the scalar. For a scalar α and an n-dimensional vector x, the scalar multiplication is defined by

$$\alpha x = \begin{pmatrix} \alpha x_1 \\ \vdots \\ \alpha x_n \end{pmatrix}.$$

Associated with column vectors, a row array of n numbers $z_1, \ldots, z_n$ is called an n-dimensional row vector and denoted by

$$z = (z_1, \ldots, z_n)$$

The addition and scalar multiplication for row vectors are the same as for column vectors.

To present a column vector

$$x = \begin{pmatrix} x_1 \\ \vdots \\ x_n \end{pmatrix}$$

as a row vector, the transpose of x, denoted by

$$x^T = (x_1, \ldots, x_n),$$

is taken.

The notion of linear dependence or independence plays an important role in the linear algebra.

(iii) Linear Dependence and Independence

A collection of m vectors $x^1, \ldots, x^m$ of dimension n is called linearly dependent if there exist some scalars $\lambda_1, \ldots, \lambda_m$ that are not all zero such that

$$\lambda_1 x^1 + \ldots + \lambda_m x^m = \mathbf{0}.$$

A.1 Vector

Otherwise, called linearly independent. That is, a collection of vectors $x^1, \ldots, x^m$ is called linearly independent if

$$\sum_{i=1}^{n} \lambda_i x_i = 0 \text{ implies that } \lambda_i = 0 \text{ for } i = 1, \ldots, m.$$

Introducing the inner product of two n-dimensional column vectors x and y, the concepts of norms and orthogonality can be defined.

(iv) Inner Product

The inner product of two n-dimensional column vectors x and y, denoted by (x,y) or $x^T y$, is defined by

$$(x,y) = x^T y = (x_1, \ldots, x_n) \begin{pmatrix} y_1 \\ \vdots \\ y_n \end{pmatrix} = \sum_{i=1}^{n} x_i y_i.$$

(v) Norm

The norm of an n-dimensional vector x, denoted by $\|x\|$, is a measure of the size of x and is defined by

$$\|x\| = \sqrt{(x,x)} = \sqrt{x_1^2 + \cdots + x_n^2}.$$

(vi) Orthogonality

If $(x,y) = 0$, two vectors x and y are called orthogonal to each other. The n-dimensional vector space where the concepts of the inner product and the norm are introduced is called an n-dimensional (real) Euclidian space and denoted by $\mathbb{R}^n$. As a special case, $\mathbb{R}^1$ represents the set of all real numbers, and it is also simply written as $\mathbb{R}$.

(vii) Outer Product

Compared with the inner product, the outer product is defined by

$$xy^T = \begin{pmatrix} x_1 \\ \vdots \\ x_n \end{pmatrix} (y_1, \ldots, y_n) = \begin{bmatrix} x_1 y_1 & \cdots & x_1 y_n \\ \vdots & \ddots & \vdots \\ x_n y_1 & \cdots & x_n y_n \end{bmatrix}.$$

For two vectors x and y of the same size, the following inequality, referred to as the Cauchy–Schwarz inequality, holds

$$(x, y) \leq \|x\| \|y\|.$$

Noting that

$$(\alpha x + y, \alpha x + y) = (x, x)\alpha^2 + 2(x, y)\alpha + (y, y) \geq 0$$

for any real number α, the discriminant of the quadratic equation for α is nonpositive, i.e.,

$$((x, y))^2 - (x, x)(y, y) \leq 0,$$

which implies the Cauchy–Schwarz inequality.

A.2 Matrix

An $m \times n$ matrix A is a rectangular array of mn numbers arranged in m horizontal rows and n vertical columns:

$$\begin{bmatrix} a_{11} & a_{12} & \cdots & a_{1n} \\ \vdots & \vdots & \ddots & \vdots \\ a_{m1} & a_{m2} & \cdots & a_{mn} \end{bmatrix}.$$

The number in the ith row and jth column of A is denoted by a_{ij} and is called the (i, j) element of A or (i, j) entry of A. Using a general element a_{ij}, in an abbreviated notation, a matrix A is often written as $A = [a_{ij}]$.

The ith row of A is

$$(a_{i1}, \ldots, a_{in}),$$

and the jth column of A is

$$\begin{pmatrix} a_{1j} \\ \vdots \\ a_{mj} \end{pmatrix}.$$

The order of the matrix is the number of rows and columns that the matrix contains, and a matrix with m rows and n columns ($m \times n$ matrix) is described as having order m by n.

An $m \times n$ matrix all of whose elements are zero is called a zero or null matrix and denoted by $\mathbf{0}$.

A.2 Matrix

A square matrix has exactly the same number of rows as columns. In other words, an $m \times n$ matrix A is called square if $m = n$ and is said to be of order n.

A diagonal matrix is a square matrix whose off-diagonal elements, a_{ij} for $i \neq j$, are all equal to zeros.

A unit matrix or an identity matrix of order n, denoted by I and sometimes by I_n to denote the order, is a diagonal matrix with all diagonal elements equal to unities and other elements equal to zeros. Thus, the $n \times n$ unit matrix I is formed by arranging n unit vectors $e_1, \ldots, e_n$ in this order, i.e.,

$$I = [e_1, \ldots, e_n] = \begin{bmatrix} 1 & 0 & \ldots & 0 \\ 0 & 1 & \ldots & 0 \\ \vdots & \vdots & & \vdots \\ 0 & 0 & \ldots & 1 \end{bmatrix}.$$

Two $m \times n$ matrices $A = [a_{ij}]$ and $B = [b_{ij}]$ are said to be equal, denoted by $A = B$, if and only if their corresponding elements are equal, namely, $a_{ij} = b_{ij}$ for all $i = 1, \ldots, m; j = 1, \ldots, n$.

(i) Addition

For two $m \times n$ matrices $A = [a_{ij}]$ and $B = [b_{ij}]$, the addition, denoted by $A + B$, is defined by

$$A + B = [a_{ij} + b_{ij}].$$

(ii) Scalar Multiplication

For a scalar α and an $m \times n$ matrix $A = [a_{ij}]$, the scalar multiplication is defined by

$$\alpha A = [\alpha a_{ij}].$$

(iii) Product

If $A = [a_{ik}]$ is an $m \times l$ and $B = [b_{kj}]$ is an $l \times n$ matrix, then the product of A and B, denoted by AB, is the $m \times n$ matrix

$$C = [c_{ij}]$$

with elements

$$c_{ij} = \sum_{k=1}^{l} a_{ik} b_{kj}.$$

The product of an $m \times n$ matrix and an n-dimensional column vector, or that of an m-dimensional row vector and an $m \times n$ matrix, can be respectively performed by considering the column vector as an $n \times 1$ matrix or the row vector as a $1 \times m$ matrix.

The transpose, the rank, and the inverse of a matrix are defined as follows:

(iv) Transpose

The transpose of an $m \times n$ matrix A with a_{ij} as its (i, j) element, denoted by A^T, is the $n \times m$ matrix whose (j, i) element is a_{ij}. In other words, A^T is formed by letting the jth column of A be the jth row of A^T (similarly by letting the jth row of A be the jth column of A^T).

A matrix A is called symmetric if $A^T = A$. Such a matrix must, obviously, be square, and $a_{ij} = a_{ji}$.

(v) Rank of Matrix

The rank of an $m \times n$ matrix A, denoted by rank(A), is the maximum number of linearly independence columns or, equivalently, the maximum number of linearly independence rows of A. If the rank of A is equal to the minimum of $\{m, n\}$, then A is said to of full rank.

(vi) Inverse

For an $m \times m$ square matrix $A = [a_{ij}]$ and an $m \times m$ unit matrix I, if there exists an $m \times m$ matrix X such that

$$AX = XA = I,$$

then X is called the inverse of A. The inverse matrix, if exists, is unique and is denoted by A^{-1}. If A has an inverse, A is called nonsingular; otherwise, A is called singular. An $m \times m$ square matrix A is nonsingular if and only if rank(A) = m or, alternatively, its determinant, defined below, is not equal to zero.

Associated with each square $m \times m$ matrix is a real number called the determinant of the matrix.

(vii) Determinant

The determinant of an $m \times m$ square matrix can be calculated by expanding the matrix in the cofactors of its ith row, i.e.,

$$\det A = \sum_{j=1}^{m} a_{ij} A_{ij}$$

where A_{ij}, the ijth cofactor, is the determinant of the minor M_{ij} with the appropriate sign, i.e.,

$$A_{ij} = (-1)^{i+j} \det M_{ij}.$$

The minor M_{ij} is the matrix formed from A by deleting the ith row and jth column of A.

Note that the determinant of a 1×1 matrix is just the value of its element.

Partitioned Matrices

It is often convenient to partition a given matrix by drawing the dashed lines through the matrix. For an $m \times n$ matrix $A = [a_{ij}]$, by drawing $M - 1$ horizontal dashed lines and $N - 1$ vertical dashed lines, A is partitioned into MN submatrices A_{IJ}, $I = 1, \ldots, M; J = 1, \ldots, N$ which are

$$A = \begin{bmatrix} A_{11} & A_{12} & \cdots & A_{1N} \\ A_{21} & A_{22} & \cdots & A_{2N} \\ \vdots & \vdots & \ddots & \vdots \\ A_{m1} & A_{m2} & \cdots & A_{mn} \end{bmatrix}.$$

Frequently, partitioning is used to simplify operations with matrices, such as multiplication.

For such submatrices, the rules for the addition, the scalar multiplication, and the matrix multiplication can follow the usual rules.

(i) Addition

For two submatrices $A = [A_{IJ}]$, $I = 1, \ldots, M; J = 1, \ldots, L$, and $B = [B_{IJ}]$, $I = 1, \ldots, M; J = 1, \ldots, L$, the addition, denoted by $A + B$, is defined by

$$A + B = [A_{IJ} + B_{IJ}].$$

(ii) Scalar Multiplication

For a scalar α and a submatrix $A = [A_{IJ}]$, $I = 1, \ldots, M; J = 1, \ldots, L$, the scalar multiplication is defined by

$$\alpha A = [\alpha A_{IJ}].$$

(iii) Product

For two submatrices $A = [A_{IK}]$, $I = 1, \ldots, M$; $K = 1, \ldots, L$, and $B = [B_{KJ}]$, $K = 1, \ldots, L$; $J = 1, \ldots, N$, the product of A and B, denoted by AB, is defined by

$$AB = C = [C_{IJ}]$$

where

$$C_{IJ} = \sum_{k=1}^{L} A_{IK} B_{KJ}.$$

Note that the columns of $A = [A_{IK}]$ must be partitioned in the same way as the rows of $B = [B_{KJ}]$ so that the product AB of the submatrices is well defined.

For example, when $M = N = L = 2$, the product AB of the submatrices is

$$\begin{bmatrix} A_{11} & A_{12} \\ A_{21} & A_{22} \end{bmatrix} \begin{bmatrix} B_{11} & B_{12} \\ B_{21} & B_{22} \end{bmatrix} = \begin{bmatrix} A_{11}B_{11} + A_{12}B_{21} & A_{11}B_{12} + A_{12}B_{22} \\ A_{21}B_{11} + A_{22}B_{21} & A_{21}B_{12} + A_{22}B_{22} \end{bmatrix}.$$

In case of square matrices, it is important to draw the same number of horizontal and vertical dashed lines so that the resulting submatrices A_{II}, $I = 1, \ldots, M$ also become square. For the square submatrix

$$A = \begin{bmatrix} A_{11} & A_{12} \\ A_{21} & A_{22} \end{bmatrix}$$

if both A_{11} and $Z = A_{22} - A_{21} A_{11}^{-1} A_{12}$ are nonsingular, it can be shown the following formula holds

$$A^{-1} = \begin{bmatrix} A_{11}^{-1} + A_{11} A_{12} Z^{-1} A_{21} A_{11}^{-1} & A_{11} A_{12} Z^{-1} \\ Z^{-1} A_{21} A_{11} & Z^{-1} \end{bmatrix}.$$

Quadratic Form

(i) Eigenvalue and Eigenvector

For an $n \times n$ square matrix A, a scalar λ is called an eigenvalue of A, and a corresponding nonzero vector x is called an eigenvector of A if and only if

$$Ax = \lambda x$$

The eigenvalues of A are given as the roots of the eigen equation

$$\det(A - \lambda I) = 0.$$

A.2 Matrix

(ii) Quadratic Form

A function $Q(x)$ is called a quadratic form if

$$Q(x) = x^T Q x$$

where x is an n-dimensional column vector and Q is an $n \times n$ symmetric matrix.

(iii) Positive Definite

A symmetric matrix Q is called positive definite, denoted by $Q > 0$, if and only if the quadratic form $x^T Q x$ is positive for all nonzero vector x, i.e.,

$$x^T Q x > 0, \text{ for all } x \neq \mathbf{0}.$$

A symmetric matrix Q is called positive semidefinite, denoted by $Q \geq 0$, if and only if

$$x^T Q x \geq 0, \text{ for all } x \neq \mathbf{0}.$$

Similar definitions hold for negative definiteness and negative semidefiniteness.

Note that a symmetric matrix Q is positive definite if and only if all eigenvalues of Q are positive.

Appendix B
Nonlinear Programming

B.1 Problem Formulation

To introduce the basic notions of nonlinear programming, first recall the production planning problem discussed in Example 1.1.

Example B.1 (Production planning problem). A manufacturing company desires to maximize the total profit from producing two products P_1 and P_2 utilizing three different materials M_1, M_2, and M_3. The company knows that to produce 1 ton of product, P_1 requires 2 tons of material M_1, 3 tons of material M_2, and 4 tons of material M_3, while to produce 1 ton of product P_2 requires 6 tons of material M_1, 2 tons of material M_2, and 1 ton of material M_3. The total amounts of available materials are limited to 27 tons, 16 tons, and 18 tons for M_1, M_2, and M_3, respectively. It also knows that product P_1 yields a profit of 3 million yen per ton, while P_2 yields 8 million yen. Given these limited materials, the company is trying to figure out how many units of products P_1 and P_2 should be produced to maximize the total profit.

Denote the numbers of tons produced of products P_1 and P_2 by x_1 and x_2, respectively. This production planning problem can be formulated as the following linear programming problem:

$$\begin{aligned}
\text{minimize } & z = -3x_1 - 8x_2 \\
\text{subject to } & 2x_1 + 6x_2 \leq 27 \\
& 3x_1 + 2x_2 \leq 16 \\
& 4x_1 + x_2 \leq 18 \\
& x_1 \geq 0, \quad x_2 \geq 0.
\end{aligned}$$

◊

Example B.2 (Nonlinear production planning problem). In practice, however, it is found that the unit profit contributions of products P_1 and P_2 are not always

Fig. B.1 Graphical solution for Example B.1

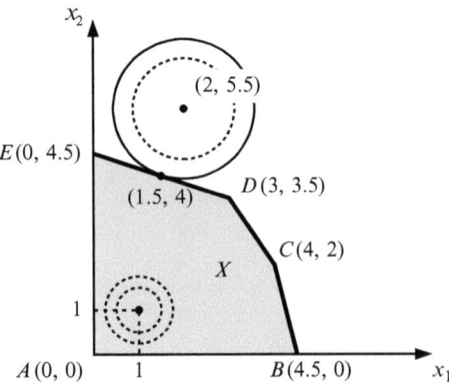

constants as in the linear profit function. They may depend on the amounts of products P_1 and P_2. Suppose that they are $4 - x_1$ and $11 - x_2$ for P_1 and P_2, respectively.

The resulting profit function is represented by

$$(4 - x_1)x_1 + (11 - x_2)x_2.$$

In such a situation, it is more appropriate to formulate the production planning problem as the following nonlinear programming problem than as a linear one:

$$\begin{aligned}
\text{minimize } z = &\ x_1^2 + x_2^2 - 4x_1 - 11x_2 \\
\text{subject to} \quad &\ 2x_1 + 6x_2 \leq 27 \\
&\ 3x_1 + 2x_2 \leq 16 \\
&\ 4x_1 + x_2 \leq 18 \\
&\ x_1 \geq 0,\ x_2 \geq 0.
\end{aligned}$$

Although the feasible region, having five extreme points, is the same as Example 1.1, the nonlinear objective function has contours of constant value which are not parallel lines, as in the linear case, but concentric circles. This simple nonlinear programming problem involving only two decision variables can be solved graphically in the x_1-x_2 plane as shown in Fig. B.1.

The minimum value of z corresponds to the contour of lowest value having at least one point in common with the feasible region, as can be seen in Fig. B.1. Therefore, the optimal solution occurs at the point $(1.5, 4)$ where the boundary of the feasible region is a tangent to the contour $z = -31.75$. Hence, the optimal solution is $x_1 = 1.5, x_2 = 4$ with $z = -31.75$. Observe that this is not an extreme point of the feasible region but it is a boundary point.

If the objective function of the problem is changed to

$$z = (x_1 - 1)^2 + (x_2 - 1)^2,$$

the optimal solution is obviously $x_1 = 1, x_2 = 1$, which is not even a boundary point but an interior point of the feasible region. ◇

B.1 Problem Formulation

It is significant to realize here that the optimal solution of a nonlinear programming problem does in general not occur at an extreme point of the feasible region and may not even be on the boundary. Therefore, nonlinear programming problems could not be solved by just examining only the finite set of extreme points as in the simplex method of linear programming. As a result, they are far more difficult to solve than linear programming problems.

Also, the possible existence of local minima which might not be optimal overall is another characteristic of nonlinear programming problems which can cause serious difficulty. For example, if the objective function of the previous problem has two minima and at least one of them is an interior point of the feasible region, then the problem would have two local minima.

These properties, discussed thus far via simple numerical examples, are direct consequences of the consideration of nonlinearities and take us into the field of nonlinear programming.

The preceding simple nonlinear production planning problem can be immediately generalized to the following nonlinear programming problem[1]:

Minimize the single-valued nonlinear objective function

$$f(x_1, x_2, \ldots, x_n)$$

subject to the m nonlinear inequality constraints

$$g_1(x_1, x_2, \ldots, x_n) \leq 0$$
$$g_2(x_1, x_2, \ldots, x_n) \leq 0$$
$$\ldots\ldots\ldots\ldots$$
$$g_m(x_1, x_2, \ldots, x_n) \leq 0,$$

or in more compact vector form,

$$\text{minimize } f(\boldsymbol{x})$$
$$\text{subject to } g_i(\boldsymbol{x}) \leq 0, \; i = 1, \ldots, m,$$

or

$$\text{minimize } f(\boldsymbol{x})$$
$$\text{subject to } \boldsymbol{x} \in X = \{\boldsymbol{x} \in \mathbb{R}^n \mid g_i(\boldsymbol{x}) \leq \boldsymbol{0}, \; i = 1, \ldots, m\},$$

where $\boldsymbol{x}$ is an n-dimensional vector of decision variables; $f(\boldsymbol{x}) = f(x_1, x_2, \ldots, x_n)$ and $g_i(\boldsymbol{x}) = g_i(x_1, x_2, \ldots, x_n)$, $i = 1, \ldots, m$ are given real-valued nonlinear functions of n real variables $x_1, x_2, \ldots, x_n$; and X is the feasible region of the nonlinearly constrained set.

Several basic notions and definitions involved in nonlinear programming follow.

---

[1] For convenience, we start from the problem with inequality constraints, but similar discussions can be made for the problems with both equality and inequality constraints.

B.2 Basic Notions and Optimality Conditions

Minimum Point and Convexity

(i) Local and Global Minimum

A point x^* is called a local minimum point of the nonlinear programming problem if there exists a real number $\delta > 0$ such that $f(x) \geq f(x^*)$ for all $x \in X$ satisfying $\|x - x^*\| < \delta$. If $f(x) > f(x^*)$ for all $x \in X$, $x \neq x^*$, within a distance δ of x^*, then x^* is called a strict local minimum point of the nonlinear programming problem. A point x^* is called a global minimum point of the nonlinear programming problem if $f(x) \geq f(x^*)$ for all $x \in X$. If $f(x) > f(x^*)$ for all $x \in X$, $x \neq x^*$, then x^* is called a strict global minimum point of the nonlinear programming problem.

(ii) Convex Set

A nonempty set S is called a convex set if the line segment joining any two points in the set also belongs to the set, i.e.,

$$\lambda x_1 + (1 - \lambda) x_2 \in S \quad \forall x_1, x_2 \in S \text{ and } \forall \lambda \in [0, 1].$$

(iii) Convex Function

A function $f(x)$ defined on a nonempty convex set S in $\mathbb{R}^n$ is called a convex function if

$$f(\lambda x_1 + (1 - \lambda) x_2) \leq \lambda f(x_1) + (1 - \lambda) f(x_2) \quad \forall x_1, x_2 \in S \text{ and } \forall \lambda \in [0, 1].$$

A function $f(x)$ is called a strictly convex function if the above inequality holds as a strict inequality for $x_1 \neq x_2$ and $\lambda \in (0, 1)$. A function $f(x)$ is called a concave (strictly concave) function if $-f(x)$ is a convex (strictly convex) function.

(iv) Convex Programming

A nonlinear programming problem to minimize a convex function (or maximize a concave function) is called convex programming.

Geometrically, a function is convex if the line joining two points on its graph lies nowhere below the graph. It is easy to see that a linear function is both convex and concave. Conversely, if a function is both convex and concave, it is linear.

The following properties follow directly from the definitions of convex sets and convex functions:

(v) Intersection of Convex Sets

The intersection of any number of convex sets is also a convex set.

(vi) Level Set of Convex Function

If $f(x)$ is a convex function defined on a convex set S, then the level set S_α defined by

$$S_\alpha = \{x \in S \mid f(x) \leq \alpha\}$$

is also convex for all real numbers α.

As an immediate consequence of these properties, if all the functions $f(x)$ and $g_i(x)$, $i = 1, \ldots, m$ are convex, the nonlinear programming problem becomes a convex programming problem. Observe that, since a linear function is convex, a linear programming problem is also a convex programming problem.

The desirable feature of convex programming follows.

Theorem B.1 (Convex programming). *Any local minimum of a convex programming problem is a global minimum.*

To introduce the most important theoretical results in the field of nonlinear programming, first introduced by Kuhn and Tucker (1951), define the following basic notions:

Basic Notions

(i) Active Constraint

An inequality constraint $g_i(x) \leq 0$ is called active at a point x^* if $g_i(x^*) = 0$.

(ii) Regular Point

A point x^* satisfying $g_i(x^*) \leq 0$ is called a regular point if the gradient vectors $\nabla g_i(x)$ of all active constraints (i.e., for all $g_i(x) \leq 0$ such that $g_i(x^*) = 0$) are linearly independent.

(iii) Lagrangian Function

The Lagrangian function associated with the nonlinear programming problem is defined as

$$L(x, \lambda) = f(x) + \sum_{i=1}^{m} \lambda_i g_i(x), \quad \lambda_i \geq 0,$$

where λ_i, $i = 1, \ldots, m$ are Lagrange multipliers for the inequality constraints $g_i(x) \leq 0$.

(iv) Nondegenerate

An active constraint is called nondegenerate if the corresponding Lagrange multiplier for the active constraint is strictly positive.

Optimality Conditions

By imposing the constraint qualification of a regular point assumption, the first-order Kuhn–Tucker necessary conditions for optimality can be expressed as follows:

Theorem B.2 (Kuhn–Tucker necessary conditions). *Let x^* be a regular point of the constraints of the nonlinear programming problem, and assume all the functions $f(x)$ and $g_i(x)$ of the nonlinear programming problem are differentiable. If x^* is a local minimum of the nonlinear programming problem, then there exist Lagrange multipliers λ_i, $i = 1, \ldots, m$ such that*

$$\nabla_x L(x, \lambda) = \nabla f(x) + \sum_{i=1}^{m} \lambda_i \nabla g_i(x) = 0, \quad \text{(B.1)}$$

$$\lambda_i g_i(x) = 0, \quad i = 1, \ldots, m, \quad \text{(B.2)}$$

$$g_i(x) \leq 0, \quad i = 1, \ldots, m, \quad \text{(B.3)}$$

$$\lambda_i \geq 0, \quad i = 1, \ldots, m. \quad \text{(B.4)}$$

Although the Kuhn–Tucker conditions have been shown to be necessary for both convex and nonconvex problems, they are also sufficient for convex programming as shown in the following theorem:

Theorem B.3 (Kuhn–Tucker sufficient conditions). *Let all the functions $f(x)$ and $g_i(x)$ of the nonlinear programming problem be convex and differentiable. Suppose that a point x^* satisfies the Kuhn–Tucker conditions. The point x^* is then a global minimum point of the nonlinear programming problem.*

By assuming that all the functions of the nonlinear programming problem are twice continuously differentiable and using the Hessian matrix of $L(x, \lambda)$, the second-order necessary or sufficient conditions for optimality can be expressed as follows.

Theorem B.4 (Second-order necessary conditions). *Let x^* be a regular point of the constraints of the nonlinear programming problem. If x^* is a local minimum point of the nonlinear programming problem, then there exist Lagrange multipliers λ_i^*, $i = 1, \ldots, m$ such that the Kuhn–Tucker conditions (B.1)–(B.3) hold and the Hessian matrix*

$$\nabla_x^2 L(x^*, \lambda^*) = \nabla^2 f(x^*) + \sum_{i=1}^{m} \lambda_i^* \nabla^2 g_i(x^*)$$

is positive semidefinite on the tangent subspace T of the active constraint at x^*, where

$$T = \{y \mid \nabla g_i(x^*) y = 0 \text{ for all } i \text{ with } \lambda_i > 0\}.$$

B.2 Basic Notions and Optimality Conditions

Theorem B.5 (Second-order sufficient conditions). *Let x^* be a regular point of the constraints of the nonlinear programming problem. Assume that there exist Lagrange multipliers λ_i^*, $i = 1, \ldots, m$ such that the Kuhn–Tucker conditions (B.1)–(B.3) hold and the Hessian matrix*

$$\nabla_x^2 L(x^*, \lambda^*) = \nabla^2 f(x^*) + \sum_{i=1}^{m} \lambda_i^* \nabla^2 g_i(x^*)$$

is positive definite on T. The point x^* is then a local minimum point of the nonlinear programming problem.

The following sensitivity theorem is very useful for developing the trade-off information in multiobjective programming problems.

Theorem B.6 (Sensitivity theorem). *Let $f(x)$ and $g_i(x)$ be twice continuously differentiable, and consider the following nonlinear programming problem:*

$$\text{minimize } f(x)$$
$$\text{subject to } g_i(x) \leq \varepsilon_i, \ i = 1, \ldots, m.$$

Let x^* be a local optimal solution of this problem for $\varepsilon_i = 0$, $i = 1, \ldots, m$, satisfying the following conditions:

(i) The solution x^* is a regular point of the constraints of the problem.
(ii) The second-order sufficiency conditions are satisfied at x^*.
(iii) There are no degenerate active constraints at x^*.

Then there exists a continuously differentiable vector-valued function $x(\cdot)$ defined in a neighborhood of $\mathbf{0}$ in $\mathbb{R}^m$ such that $x(\mathbf{0}) = x^*$ and such that for every ε near $\mathbf{0}$, $x(\varepsilon)$ is a strict local minimum point of the problem. Furthermore,

$$-\frac{\partial f(x(\mathbf{0}))}{\partial \varepsilon_i} = \lambda_i, \ i = 1, \ldots, m.$$

Further details of the theory and algorithms of nonlinear programming can be found in standard texts including Mangasarian (1969), Zangwill (1969), Avriel (1976), Bazarra and Shetty (1979, 1993, 2006), Gill et al. (1981), and Luenberger (1973, 1984, 2008).

Appendix C
Usage of Excel Solver

C.1 Setup for Solver

Microsoft Excel is one of calculation softwares so-called spreadsheets, and it is widely used and becomes an industry standard. In this book, we explain the usage of Excel Solver on Microsoft Excel 2010 which is its latest version at the present time, and even with an earlier version, one can use it in a similar procedure.

Simply installing Excel 2010 does not allow to solve a linear programming problem. By additionally installing Solver, one can finally solve it on Excel. Solver can be installed in the following steps: After launching Excel, click tag *File* on the upper left of the window shown in Fig. C.1. Click again *Options* in the lower right corner of the new screen, and then one finds the screen of options given in Fig. C.2. Clicking *Add-ins* in the left-hand side of the screen, the list of add-in options shown in Fig. C.3 appears.

From the screen in Fig. C.3, one finds that *Solver add-in* is not active, that is, it is not yet installed. To make *Solver add-in* available in Excel, click *Go...*, and then the add-in window shown in Fig. C.4 appears.

Solver can be installed by checking the box for *Solver add-in* and then clicking button *OK*. After installing, clicking tag *Data* in Excel, one can see the icon of *Solver* on the upper right of the window shown in Fig. C.5.

C.2 Solving a Production Planning Problem

We provide the illustration of solving the production planning problem in Example 1.1 given in Chap. 1 by using Solver. Selecting *Data* in the menu bar and clicking icon *Solver*, a window for specifying parameters of Solver appears, and then one can define a linear programming problem.

A linear programming problem can be solved by connecting data on Excel to parameters of Solver in the window of Solver parameters in Fig. C.6 such as

Fig. C.1 Initial screen of Excel

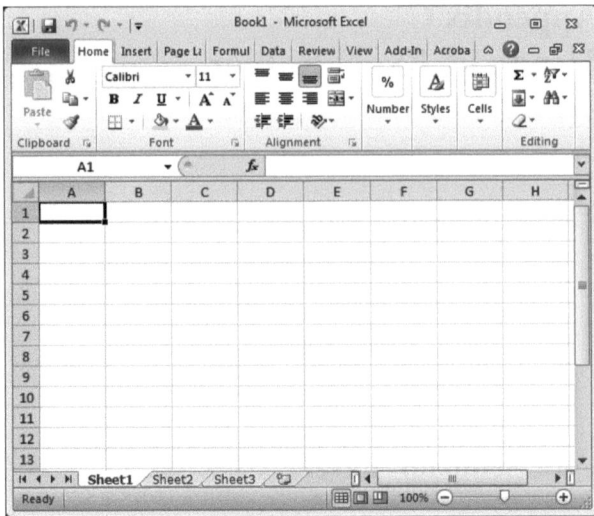

Fig. C.2 Options of Excel

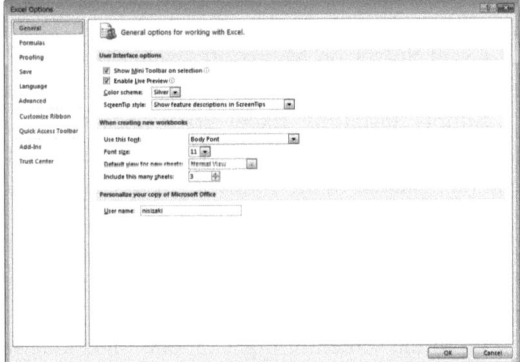

Set Objective for defining an objective function, ***To*** for specifying maximizing or minimizing the objective function, ***By Changing Variable Cells*** for specifying cells of decision variables, and ***Subject to the Constraints*** for specifying cells of constraints. By Solver, one can solve not only linear programming problems but also nonlinear programming problems in which objective and constraint functions are not always linear. When solving linear programming problems, ***Simplex LP*** should be selected in ***Select a Solving Method***.

Before clicking icon ***Solver***, the following procedure is needed in order to prepare the data for a linear programming problem:

1. Allocate a cell to each of the decision variables, and input the coefficients into cells.
2. After allocating a cell to the objective function, define it by reference to the cells of the decision variables and those of the coefficients.

Fig. C.3 List of add-in options

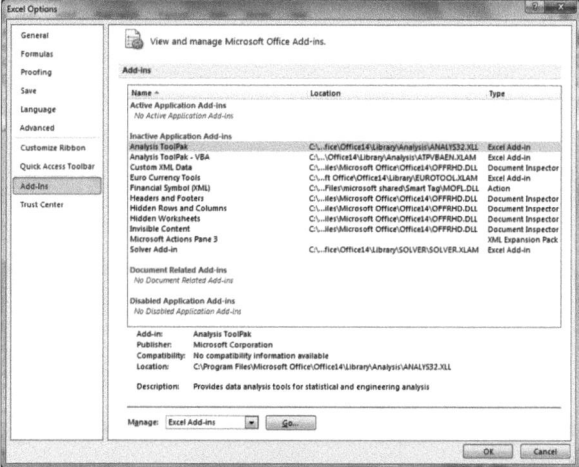

Fig. C.4 Add-in for Solver

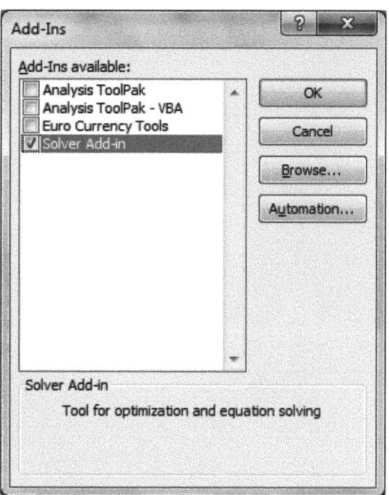

3. After allocating a cell to each of the left-hand sides of the constraints, define it by reference to the cells of the decision variables and those of the coefficients.
4. Input the right-hand-side constants of the constraints into the corresponding cells.

Now, solve the following production planning problem given in Example 1.1:

$$\begin{aligned}
\text{minimize } z = & -3x_1 - 8x_2 \\
\text{subject to } & 2x_1 + 6x_2 \leq 27 \\
& 3x_1 + 2x_2 \leq 16 \\
& 4x_1 + x_2 \leq 18 \\
& x_1 \geq 0, \ x_2 \geq 0.
\end{aligned}$$

Fig. C.5 Icon *Solver*

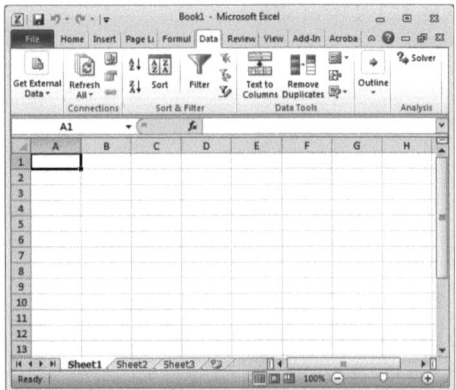

Fig. C.6 Parameters of Solver

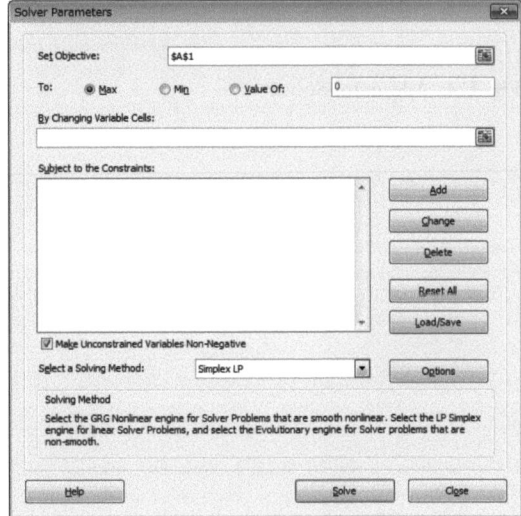

When solving a linear programming problem, it is not required to transform from an inequality function to an equality one as in the simplex method. Therefore, one does not need to introduce slack variables, surplus variables, and artificial variables, and then it is only necessary to input the data of the formulated problem in a sheet of Excel directly. It is also not necessary to force the formulation either in maximization problems or in minimization ones. There is no one way to specifying a problem in an Excel sheet, and therefore, one has only to keep in mind that the formulation is easy to understand oneself. The specification in an Excel sheet shown in Fig. C.7 is illustrative only, and it is given as an example. Needless to say, one can employ another way of specification of the problem.

Although in this sheet the decision variables are arranged in a horizontal direction, one can also arrange them in a vertical direction. In column A, to make it easy for readers to understand, captions such as "decision variable," "coefficients

C.2 Solving a Production Planning Problem

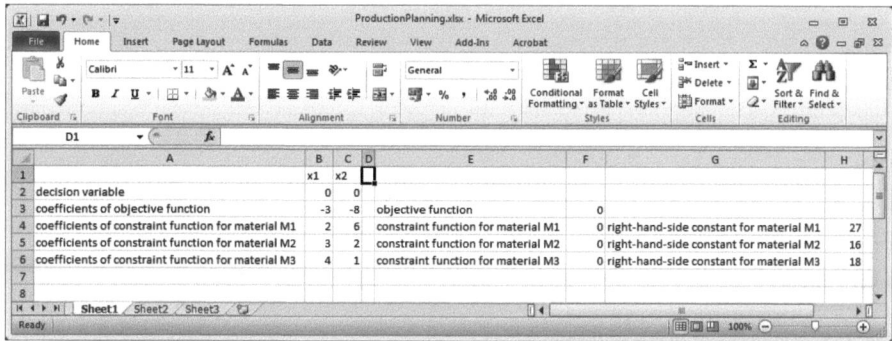

Fig. C.7 Specification of a linear programming problem in an Excel sheet

of objective function," and "coefficients of constraint function for material M1" are given. Columns E and G are also used as displaying captions such as "constraint function for material M1" and "right-hand-side constant for material M3." "x1" in cell B1 and "x2" in cell C1 are also captions, and it should be noted that they have no relation to calculation. A procedure for solving the linear programming problem by using Solver is summarized as follows:

Step 1. Relate a decision variable to a cell.

The decision variables x_1 and x_2 in the production planning problem are related to cells B2 and C2, whose initial values are set at 0s. It is noted that "x1" and "x2" in cells B1 and C1 are just captions and the actual values of x_1 and x_2 are in cells B2 and C2.

Step 2. Input the coefficients of the objective function and the constraint functions into cells.

The coefficients of x_1 and x_2 in the objective function are input into cells B3 and C3, respectively, and similarly, those of the constraint functions for materials M1, M2, and M3 are input into cells B4, C4, B5, C5, B6, and C6, respectively. Although this step is not necessary if the objective function and the constraint functions are defined in only single cells without cell references, all the coefficients are input below the cells of the decision variables in order to define the functions by using SUMPRODUCT function of Excel. Giving the coefficients separately facilitates calculation of optimal solutions to the reformulated problems for sensitivity analysis or small changes of the parameters.

Step 3. Define the objective function and the constraint functions.

In order to define the objective function, the following function is input into cell F3:

$$= \text{SUMPRODUCT}(\$B\$2:\$C\$2, B3:C3)$$

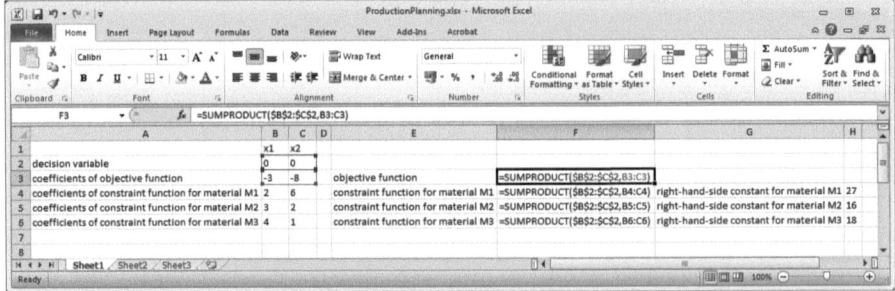

Fig. C.8 Presentment of Excel formulas

By using SUMPRODUCT function, one can calculate the sum of products. In this case, "B2 × B3 + C2 × C3" is calculated, and it follows that the objective function $-3x_1 - 8x_2$ is defined. While a cell reference with symbol $ such as $B$2 in SUMPRODUCT function is called to be absolute, a cell reference without $ such as B3 is called to be relative. Employing this expression, the constraint functions for materials M1, M2, and M3 are appropriately copied as shown in Fig. C.8 when the formula SUMPRODUCT($B$2:$C$2, B3:C3) in cell F3 is copied into from cell F4 to cell F6. Namely, the values of x_1 and x_2 are fixed at cells B2 and C2, and the coefficients of x_1 and x_2 in the constraint functions are adjusted downward. Actually, as shown in Fig. C.8, the contents in cells F4, F5, and F6 are as follows:

$$= \text{SUMPRODUCT}(\$B\$2:\$C\$2, B4:C4)$$

$$= \text{SUMPRODUCT}(\$B\$2:\$C\$2, B5:C5)$$

$$= \text{SUMPRODUCT}(\$B\$2:\$C\$2, B6:C6)$$

Step 4. Input the right-hand-side constants of the constraints into cells.

The right-hand-side constants of the constraints are input into from cell H4 to cell H6.

The values and the formulas specified in from step 1 to step 4 can be verified in Fig. C.8, which can be displayed by pressing keys [ctrl], [shift], and [@] simultaneously.

Step 5. Define the linear programming problem in the window for specifying Solver parameters.

Selecting **Data** in the menu bar and clicking icon **Solver**, a window for specifying parameters of Solver shown in Fig. C.9 appears. Cell F3 in which the objective function is defined is specified in **Set Objective**, and cells B2 and C2 in which the decision variables x_1 and x_2 are allocated are specified in **By Changing Variable Cells** in the window. It is noted that

Fig. C.9 Solver parameters

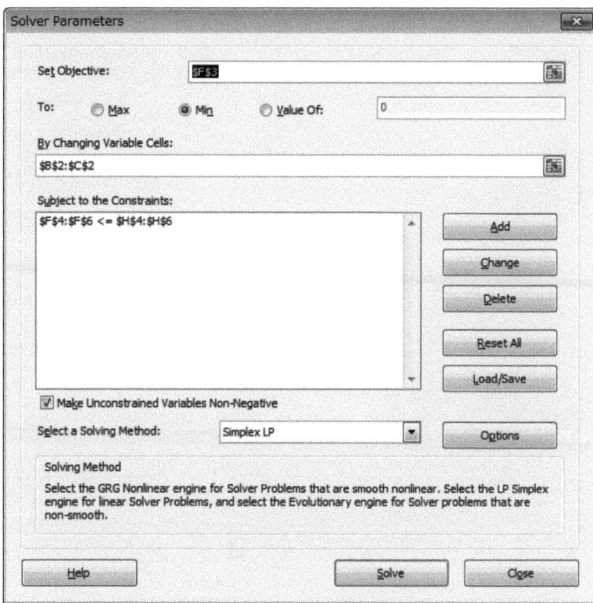

the cell references are represented to be absolute such as $F$3 when one specifies cells in an Excel sheet by using the cursor instead of by typing them and the notation of $B$2:$C$2 means from $B$2 to $C$2 in an Excel sheet. To specify the constraints of the problem, after clicking button *Add*, one can specify cells F4:F6 for the three constraint functions, cells H4:H6 for the right-hand-side constants of the constraints, and the inequalities "$\leq$" in *Subject to the Constraints*. Moreover, since the decision variables are nonnegative and the production planning problem is linear, the box of *Make Unconstrained Variables Non-Negative* is checked and *Simplex LP* is selected in *Select a Solving Method*.

Step 6. Specify options for the computational result, and activate Solver.

Clicking button *Solve*, the window for showing the computational result and specifying report options shown in Fig. C.10 appears.

After selecting *Answer*, *Sensitivity*, and *Limits* in *Reports*, by clicking *OK*, the answer, the sensitivity, and the limits reports are made in three separate sheets, and the optimal solution is incorporated in the original Excel sheet. As shown in Fig. C.11, 3 and 3.5 appear in cells B2 and C2 corresponding to the decision variables x_1 and x_2, respectively. This means that the optimal solution is $(x_1^*, x_2^*) = (3, 3.5)$ and it is coincided with the optimal solution obtained by the simplex method.

Step 7. Check reports of the computational result.

Clicking the tab of worksheet *Answer Report 1* in the bottom edge of the window, one can see the answer report of Solver shown in Fig. C.12. In the

Fig. C.10 Report options

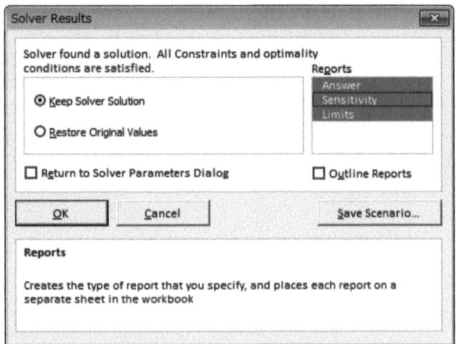

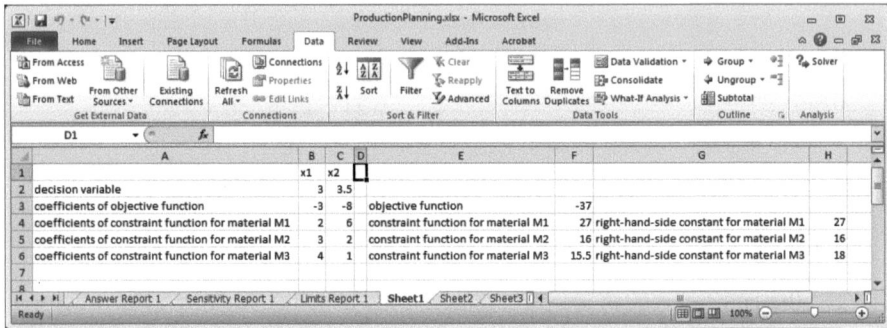

Fig. C.11 Computational result

answer report, the original and the final values of the objective function and the decision variables are given together with the slacks of the constraints.

The sensitivity report is given in Fig. C.13. The reduced cost for each of the decision variables is given in the report, and it represents the variation of the objective function when the decision variable is increased by one unit. In the text, we call it the *relative cost coefficient*, and it is noted that these values are coincident with the relative cost coefficients for x_1 and x_2 in Table 2.4 (at the row of $-z$ and the columns of x_1 and x_2 in cycle 2). The shadow price in the report also corresponds to the relative cost coefficient, and those of the constraints for materials M_1, M_2, and M_3 are coincident with the relative cost coefficients for x_3, x_4, and x_5 in Table 2.4 (at the row of $-z$ and the columns of x_3, x_4, and x_5 in cycle 2) times -1. For example, while the relative cost coefficient for x_3 is $\frac{9}{7}$ in Table 2.4, the shadow price in the report of Fig. C.13 is -1.285714286. The difference is due to the fact that the relative cost coefficient for x_3 is the variation of the

Fig. C.12 Answer report

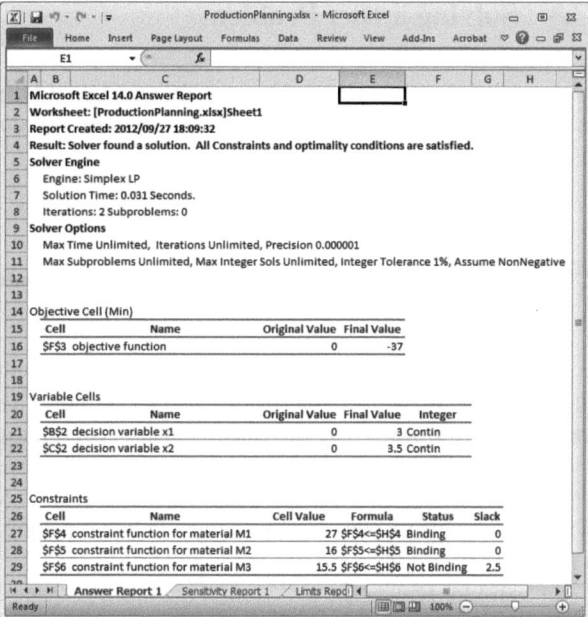

Fig. C.13 Sensitivity report

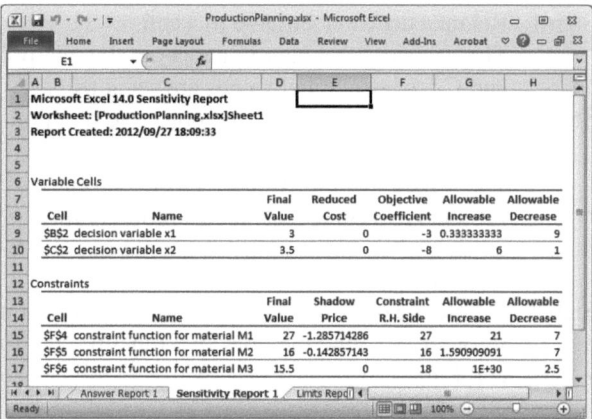

objective function when the slack variable x_3 is increased by one unit, and the shadow price for the constraint for material M_1 is the variation of the objective function when material M_1 is increased by one unit. Namely, this is because the increase of the slack variable has the opposite effect of the increase of the material.

C.3　Solving a Diet Problem

We provide one more illustration of solving a linear programming problem by using Solver and deal with the following diet problem in Example 2.3:

$$\begin{aligned}
\text{minimize } z &= 4x_1 + 3x_2 \\
\text{subject to } \quad x_1 + 3x_2 &\le 12 \\
x_1 + 2x_2 &\le 10 \\
2x_1 + x_2 &\le 15 \\
x_1 \ge 0, \ x_2 &\ge 0.
\end{aligned}$$

As we described in Chap. 2, when the simplex method is employed for solving the diet problem, by introducing the surplus variables x_3, x_4, and x_5 and the artificial variables x_6, x_7, and x_8, the above linear programming problem is transformed into the standard form of linear programming. However, when one solves the diet problem by using Solver, such transformation is not required, and what one has to do is only to specify the data of the above linear programming problem in an Excel sheet.

In a similar way to the production planning problem, a procedure for solving the diet problem by using Solver is summarized in the following seven-step procedure:

Step 1. Relate a decision variable to a cell.

The decision variables x_1 and x_2 in the diet problem are related to cells B2 and C2, whose initial values are set at 0s.

Step 2. Input the coefficients of the objective function and the constraint functions into cells.

The coefficients of x_1 and x_2 of the objective function are input in cells B3 and C3, respectively, and similarly, those of the constraint functions for nutrients N1, N2, and N3 are input in cells B4, C4, B5, C5, B6, and C6, respectively.

Step 3. Define the objective function and the constraint functions.

In order to define the objective function, the following function is input in cell F3:

$$= \text{SUMPRODUCT}(\$B\$2{:}\$C\$2, B3{:}C3)$$

By using SUMPRODUCT function, one can calculate the sum of products. In this case, "B2 × B3 + C2 × C3" is calculated, and it follows that the objective function $4x_1 + 3x_2$ is defined. By copying the formula SUMPRODUCT($B$2:$C$2, B3:C3) in cell F3 into from cell F4 to cell F6, the constraint functions are defined as follows:

$$= \text{SUMPRODUCT}(\$B\$2{:}\$C\$2, B4{:}C4)$$

$$= \text{SUMPRODUCT}(\$B\$2{:}\$C\$2, B5{:}C5)$$

$$= \text{SUMPRODUCT}(\$B\$2{:}\$C\$2, B6{:}C6)$$

C.3 Solving a Diet Problem

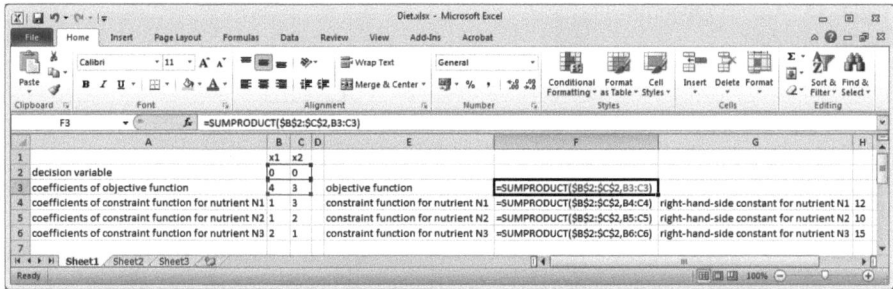

Fig. C.14 Specification of the diet problem in an Excel sheet

Step 4. Input the right-hand-side constants of the constraints into cells.

The right-hand-side constants of the constraints are input into from cell H4 to cell H6. The values and the formulas specified in from step 1 to step 4 can be verified in Fig. C.14.

Step 5. Define the linear programming problem in the window for specifying Solver parameters.

In a similar way to the production planning problem, the diet problem can be defined in the window of Solver parameters. Cell F3 in which the objective function is defined is specified in *Set Objective*, and cells B2 and C2 in which the decision variables x_1 and x_2 are allocated are specified in *By Changing Variable Cells* in the window. To specify the constraints of the problem, after clicking button *Add*, one can specify cells F4:F6 for the three constraint functions, cells H4:H6 for the right-hand-side constants of the constraints, and the inequalities "$\geq$" in *Subject to the Constraints*. It is noted that the inequalities in the diet problem are reverse directions to those of the production planning problem.

Step 6. Specify options for the computational result, and activate Solver.

By clicking button *Solve* and selecting *Answer*, *Sensitivity*, and *Limits* in *Reports*, Solver is activated, and then the optimal solution to the diet problem is shown in the Excel sheet. It is found that the optimal solution is $(x_1^*, x_2^*) = (6.6, 1.8)$, and it is coincident with the solution obtained through the two-phase method given in Chap. 2.

Step 7. Check reports of the computational result.

Clicking the tab of worksheet *Answer Report 1* in the bottom edge of the window, one can see the answer report of Solver shown in Fig. C.15. In the answer report, the original and the final values of the objective function and the decision variables are given together with the slacks of the constraints. Compare the answer report given in Fig. C.15 to the simplex tableau in Table 2.7. The final value of the objective cell is 31.8 in the answer report, and the value at the row of $-z$ and the column of *constants* is -31.8 in the simplex tableau. It follows that the same optimal value is obtained by the two methods. Moreover, one finds that the optimal basic solution is

Fig. C.15 Answer report for the diet problem

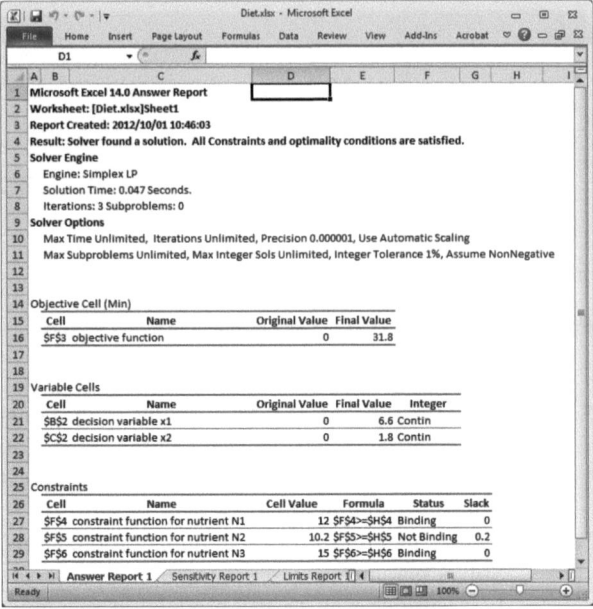

Fig. C.16 Sensitivity report for the diet problem

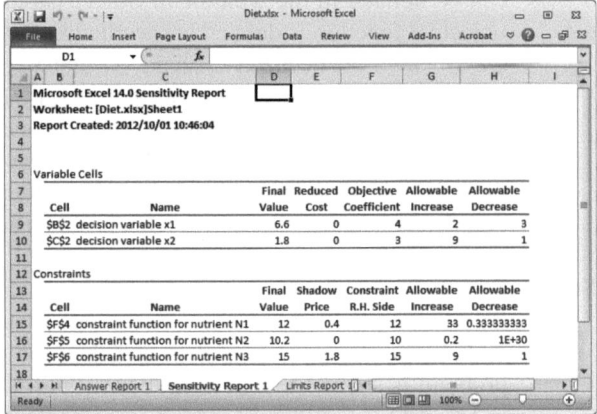

$(x_1, x_2, x_4) = (6.6, 1.8, 0.2)$ in the simplex tableau, and these values are given in the 21st, 22nd, and 28th rows in the answer report. In this solution, the surplus variable of $x_4 = 0.2$ means that the amount of nutrient N_2 is larger than 10 mg of the minimum nutritional requirement by 0.2.

The sensitivity report is given in Fig. C.16. The relative cost coefficients for x_1 and x_2 at the row of $-z$ and the columns of x_1 and x_2 in cycle 3 in Table 2.7 are 0s, and the reduced costs for the decision variables x_1 and x_2 are 0s given in the 9th and 10th rows of the sensitivity report. Moreover, the relative cost coefficients for x_3, x_4, and x_5 in the row of $-z$ are 0.4, 0, and 1.8, and the shadow prices for nutrients N_1, N_2, and N_3 in the 15th,

C.4 Solving a Nonlinear Programming Problem

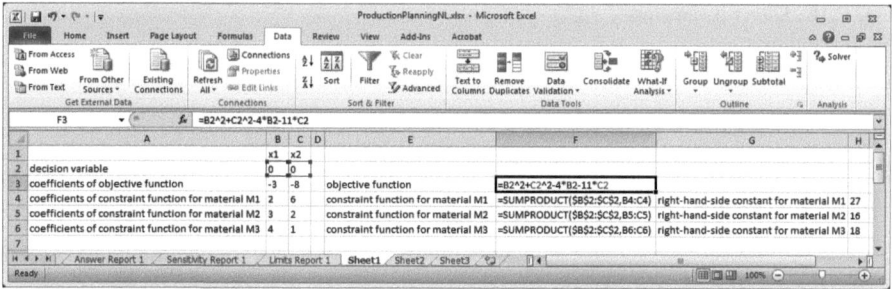

Fig. C.17 Specification of a nonlinear programming problem in an Excel sheet

16th, and 17th rows are the same values. Since the increase of the surplus variable has the same effect of the increase of the nutrient, the signs of them are the same unlike the result of the production planning problem.

C.4 Solving a Nonlinear Programming Problem

We briefly illustrate how to solve a nonlinear programming problem through Solver by the following production planning problem with the quadratic profit function which is given in Appendix B:

$$\begin{aligned}
\text{minimize } z = &\ x_1^2 + x_2^2 - 4x_1 - 11x_2 \\
\text{subject to} &\ 2x_1 + 6x_2 \leq 27 \\
&\ 3x_1 + 2x_2 \leq 16 \\
&\ 4x_1 + x_2 \leq 18 \\
&\ x_1 \geq 0,\ x_2 \geq 0.
\end{aligned}$$

We clarify the difference between the procedures for solving the linear production planning problem and the nonlinear one. Although in the linear production planning problem the objective function is defined as follows

$$= \text{SUMPRODUCT}(\$B\$2{:}\$C\$2, B3{:}C3),$$

in the nonlinear production planning problem the quadratic profit function $x_1^2 + x_2^2 - 4x_1 - 11x_2$ is employed as an objective function, and it is defined as

$$= \text{B2\textasciicircum{}2+C2\textasciicircum{}2-4*B2-11*C2}$$

in Solver, which is input into cell F3 without any cell reference as seen in Fig. C.17. The other specification is the same as that of the linear case.

Fig. C.18 Solver parameters for the nonlinear programming problem

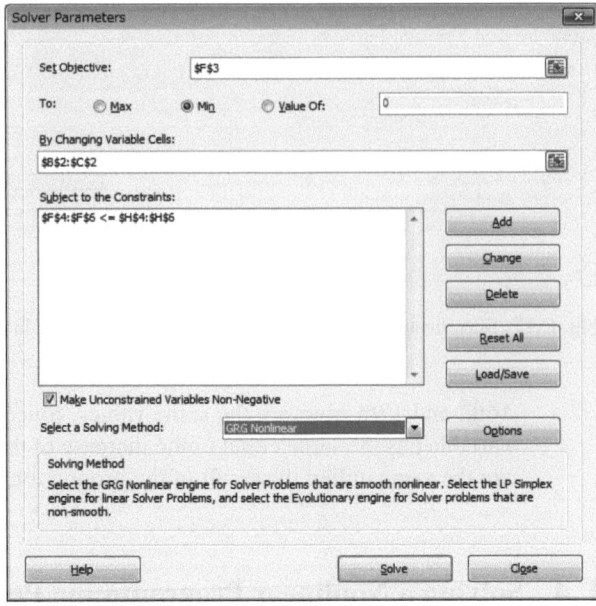

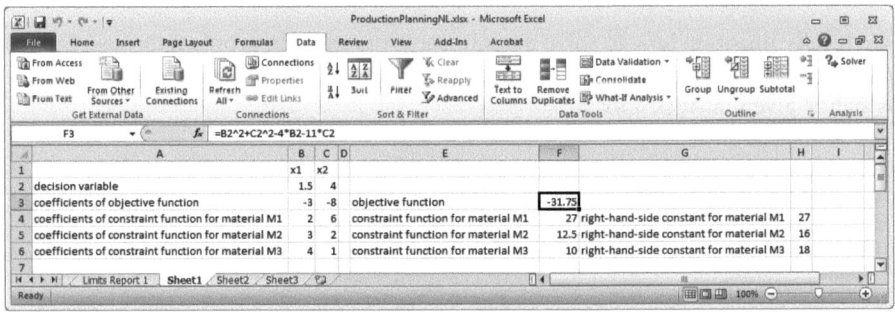

Fig. C.19 Computational result of the nonlinear programming problem

The nonlinear production planning problem cannot solve by the simplex algorithm evidently, and therefore, one should not select **Simplex LP** in **Select a Solving Method** in the window of Solver parameters. Since the production planning problem with quadratic profit function is a convex programming problem, one should select **GRG Nonlinear** as shown in Fig. C.18.

The computational result is given in Fig. C.19, and the optimal solution is $(x_1^*, x_2^*) = (1.5, 4)$ and $z^* = -31.75$, which is coincided with the graphical solution in Fig. B.1. The answer and the sensitivity reports, which are given in Figs. C.20 and C.21, can be obtained in a similar way.

C.4 Solving a Nonlinear Programming Problem

Fig. C.20 Answer report for the nonlinear programming problem

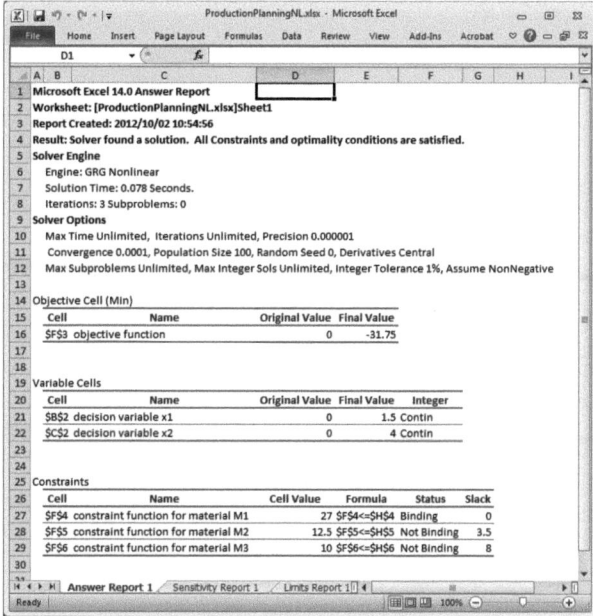

Fig. C.21 Sensitivity report for the nonlinear programming problem

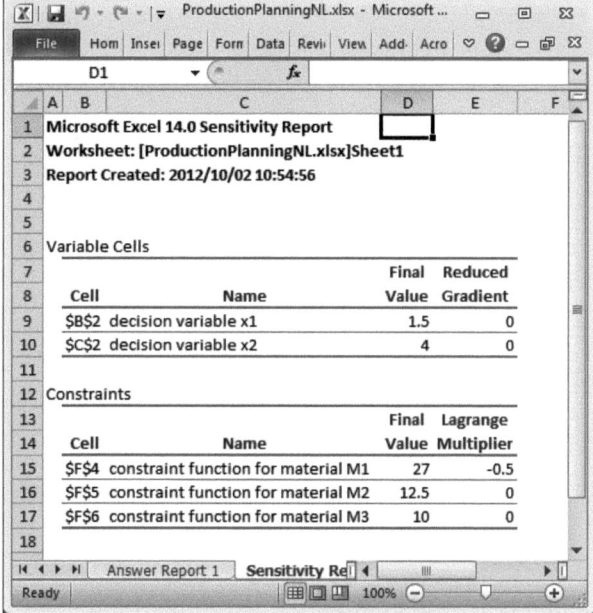

Solutions

Chapter 1

1.1 (1) Maximize the total profit $2x_1 + 5x_2$ subject to $2x_1 + 6x_2 \leq 27, 8x_1 + 6x_2 \leq 45, 3x_1 + x_2 \leq 15$, and $x_1 \geq 0, x_2 \geq 0$.
(2) From the following figure, an optimal solution is $x_1 = 3, x_2 = 3.5, z = 23.5$.

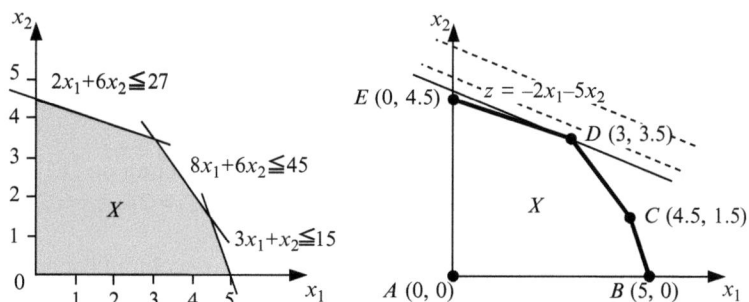

1.2 Minimize the total transportation cost $\sum_{i=1}^{m}\sum_{j=1}^{n} c_{ij} x_{ij}$ subject to $\sum_{j=1}^{n} x_{ij} = a_i, i = 1, \ldots, m$, $\sum_{i=1}^{m} x_{ij} = b_j, j = 1, \ldots, n$, and $x_{ij} \geq 0$.

1.3 Maximize overall effectiveness $\sum_{i=1}^{n}\sum_{j=1}^{n} c_{ij} x_{ij}$ subject to $\sum_{j=1}^{n} x_{ij} = 1, i = 1, \ldots, n$, $\sum_{i=1}^{n} x_{ij} = 1, j = 1, \ldots, n$, and $x_{ij} = 0$ or 1.

Chapter 2

2.1 (i) Letting $x_j = x_j^+ - x_j^-$, $x_j^+ \geq 0$, $x_j^- \geq 0$ yields $|x_j| = x_j^+ + x_j^-$, and the problem is to minimize $\sum_{j=1}^{n} c_j(x_j^+ + x_j^-)$ subject to $\sum_{j=1}^{n} a_{ij}(x_j^+ - x_j^-) = b_i, i = 1, \ldots, m, x_j^+ \geq 0, x_j^- \geq 0, j = 1, \ldots, n$.

(ii) Introducing the auxiliary variable $t = 1/(\sum_{j=1}^{n} d_j x_j + d_0)$, $y_j = x_j t$, $j = 1, 2, \ldots, n$, the problem id to minimize $\sum_{j=1}^{n} c_j y_j + c_0 t$, subject to $\sum_{j=1}^{n} d_j y_j + d_0 t = 1$, $\sum_{j=1}^{n} a_{ij} y_j - b_i t = 0$, $i = 1, \ldots, m$, $y_j \geq 0$, $j = 1, \ldots, n$, $t > 0$.

(iii) Introducing the auxiliary variable λ and converting $\sum_{j=1}^{n} c_j^l x_j$, $l = 1, 2, \ldots, L$ into the constraints, the problem is to minimize λ subject to $\sum_{j=1}^{n} c_j^l x_j \leq \lambda$, $l = 1, \ldots, L$, $\sum_{j=1}^{n} a_{ij} x_j = b_i$, $i = 1, \ldots, m$, $x_j \geq 0$, $j = 1, \ldots, n$.

2.2 (i) Maximize $z = 3x_1 + 2x_2$ subject to $2x_1 + 5x_2 \leq 40$, $3x_1 + 1x_2 \leq 30$, $3x_1 + 4x_2 \leq 39$, $x_1 \geq 0$, $x_2 \geq 0$.

(ii) Minimize $z = 9x_1 + 15x_2$ subject to $9x_1 + 2x_2 \geq 54$, $1x_1 + 5x_2 \geq 25$, $1x_1 + 1x_2 \geq 13$, $x_1 \geq 0$, $x_2 \geq 0$.

2.3 Optimality of all $z^* = cx^l$, $l = 1, \ldots, L$ yields $cx^* = \sum_{l=1}^{L} \lambda_l cx^l = \sum_{l=1}^{L} \lambda_l z^* = z^*$. Also, $Ax^l = b$, $l = 1, \ldots, L$ yields $\sum_{l=1}^{L} \lambda_l Ax^l = \sum_{l=1}^{L} \lambda_l b$ which implies $Ax^* = b$. Obviously, $x^* \geq 0$.

2.4 From the linearly independence of the column vectors corresponding to x_k^+ and x_k^-, the statement follows.

2.5 Converting the second problem into minimize $(z/\mu\lambda)$ subject to $A(x/\lambda) = b$, $(x/\lambda) \geq 0$, the relationships are obvious.

2.6 (i) Introducing the slack variables x_3, x_4, x_5 and pivoting on [6] and [6], cycle 2 yields an optimal solution $x_2 = 3.5$, $x_1 = 3$ ($x_3 = x_4 = 0, x_5 = 2.5$), $\min z = -23.5$.

(ii) Introducing the slack variables x_3, x_4 and pivoting on [6] and [4], cycle 2 yields an optimal solution $x_2 = 70/3$, $x_1 = 20/3$ ($x_3 = x_4 = 0$), $\min z = -200/3$.

(iii) Introducing the slack variables x_3, x_4, x_5 and pivoting on [12] and [25/4], cycle 2 yields an optimal solution $x_2 = 32/3$, $x_4 = 24$, $x_1 = 272/3$ ($x_3 = x_5 = 0$), $\min z = -944/3$.

(iv) Pivoting on [10], at cycle 1, all the values of x_1 column become negative, the solution is unbounded.

(v) Letting the free variable be $x_1 = x_1^+ - x_1^-$ and pivoting on [20], [5], and [0.2], cycle 3 yields an optimal solution $x_2 = 11$, $x_1^+ = 9.75$ ($x_1^- = x_2 = x_3 = x_4 = 0$), $\min z = -315$.

(vi) Introducing the artificial variables x_5, x_6 and performing phase 1 with the pivot elements [3] and [5/3], cycle 2 yields $w = 0$. In phase 2, pivoting on [4/5], cycle 1 yields an optimal solution $x_2 = 5/4$, $x_3 = 1/2$ ($x_1 = x_4 = 0$), $\min z = -17/4$. Noting $\bar{c}_1 = 0$, when x_1 becomes the basic variables, another optimal solution is obtained.

2.7 (i) Pivoting on [2], [1.5], [2/3], cycle 3 yields an optimal solution $x_1 = 1.5$, $x_2 = 0$, $x_3 = 3.5$, $\min z = 8.5$.

(ii) Pivoting on [2], [3], [2.5], cycle 3 yields an optimal solution $x_1 = 5/6$, $x_2 = 1/3$, $x_3 = 0$, $\min z = 0.6$.

(iii) Pivoting on [1], [2], [6], cycle 3 yields an optimal solution $x_1 = 5/6$, $x_2 = 0$, $x_3 = 5/3$, $\min z = -2.5$.

Solutions   307

2.8 (i) In the canonical form for I' and J', setting only x_t be -1 and the remaining nonbasic variables $x_j (j \in J' - \{t\})$ zeros yields $x_t = -1$, $x_j = 0$ $j \in J' - \{t\}$, $x_i = \bar{b}_i + \bar{a}'_{it}$, $i \in I'$, $-z = -\bar{z} + \bar{c}'_t$. Noting that these solutions naturally satisfy the canonical form for I and J, substituting them into the equation for z yields $-\bar{z} + \bar{c}'_t + \sum_{j \in J} \bar{c}_j x_j = -\bar{z}$, which gives $-\bar{c}'_t = \sum_{j \in J} \bar{c}_j x_j$.

(ii) From $\bar{c}'_t < 0$, the right hand must have a positive term. Let it be $\bar{c}_r x_r > 0$, $r \in J$, and consider r. From $r \in J$, $r \notin I$, and from $x_j = 0$ $j \in J' - \{t\}$, $r \notin J' - \{t\}$. Thus, $r \in I' \cup \{t\}$. If $r = t$, $t \in J$, and from $x_t = -1$, $\bar{c}_t < 0$, which implies x_t becomes a candidate for entering the basis in the canonical form for I and J. However, at this stage, x_q is assumed to enter the basis and $t < q$, which contradict the step 1B. Hence, $r \in I'$. From $r \notin I$, $r \in I'$, x_r must enter the basis during cycling, which means $r \in T$. Also, from $\bar{c}_q < 0$, $\bar{c}'_t < 0$, $\bar{a}'_{qt} > 0$, $t \in T$, $t < q$, and $x_i = \bar{b}_i + \bar{a}'_{it}$, $i \in I'$, $\bar{c}_q x_q = \bar{c}_q (\bar{b}_q + \bar{a}'_{qt}) = \bar{c}'_q \bar{a}'_{qt} < 0$ ($\because \bar{b}_q = 0$). Thus, from $\bar{c}_r x_r > 0$, $r \in J$, $r \neq q$. From $q = \max\{j \mid j \in T\}$ and $r \in T$, $r < q$.

(iii) In the canonical form for I and J, since x_q is assumed to enter the basis, from step 1, $\bar{c}_r \geq 0$ ($\because r \in T$). From $\bar{c}_r x_r > 0$, $r \in J$, $\bar{c}_r > 0$, $x_r > 0$. Noting $r \in I'$, $x_i = \bar{b}_i + \bar{a}'_{it}$, $i \in I'$, from which $x_r = \bar{b}_t + \bar{a}'_{rt}$. Here, $\bar{b}_r = 0$. Because, from $r \notin I$, $r \in I'$, during cycling from the canonical form for I and J to that for I' and J', although r enters the basis, the value of $\bar{b}$ for x_r is zero, and after the pivot operations as long as the rows having the zero value of $\bar{b}$ are selected, the value of $\bar{b}$ for x_r is zero. Hence, from $\bar{c}_r > 0$, $x_r > 0$, $x_r = \bar{a}'_{rt} > 0$. Noting $\bar{c}'_t < 0$, $\bar{a}'_{rt} > 0$, $\bar{b}_r = 0$, $r < q$, the assumption that q leaves the basis in the canonical forms for I' and J' contradicts step 3B.

2.9 Using the standard simplex method with the pivot elements $[1/4]$, $[30]$, $[8/25]$, $[1/40]$, $[50]$, and $[1/3]$, cycle 6 yields cycle 1, which implies the cycling occurs. Employing the simplex method incorporating the Bland's rule, pivoting on $[1/4]$, $[30]$, $[8/25]$, $[1/40]$, $[125/2]$, and $[2/15]$, cycle 6 yields an optimal solution $x_3 = 1$, $x_5 = 3/100$, $x_1 = 1/25$ ($x_2 = x_4 = x_6 = x_7 = 0$), $\min z = -1/20$.

2.10 Multiplying both sides of m constrains by $\pi_1, \ldots, \pi_m$ and subtracting them from the z equation, the coefficient of x_j becomes $c_j - \pi p_j$. Setting $c_j - \pi p_j = 0$, $j = 1, 2, \ldots, m$ yields $\pi B = c_B$ whose solution is the simplex multiplier vector $\pi = c_B B^{-1}$.

2.11 Referring to the obtained results for problem 2.6 by the simplex method, do it by yourself.

2.12 Conditions for self-dual are $c = -b^T$, $A^T = -A$.

2.13 For any primal feasible $\bar{x}$, $A\bar{x} = b$ and $c\bar{x} = \bar{z}$. Multiplying both sides of the first equation by some dual feasible $\bar{\pi}$ and subtracting the result from the second equation yield $c\bar{x} - \bar{\pi} A\bar{x} = \bar{z} - \bar{\pi} b$. Setting $\bar{\pi} b = \bar{v}$ gives $(c - \bar{\pi} A)\bar{x} = \bar{z} - \bar{v}$. If x^o and π^o are primal and dual optimal with optimal values z^o and v^o, respectively, then, by the strong duality theorem, $(c - \pi^o A)x^o = z^o - v^o = 0$.

Conversely, if x^o and π^o satisfy $(c - \pi^o A)x^o = cx^o - \pi^o b = 0$, then, from the weak duality theorem, it immediately follows that x^o and π^o are primal and dual optimal.

2.14 The alternative (ii) of the Gordon's theorem can be equivalently expressed as $\pi A \leq -I\varepsilon$, $\varepsilon > 0$ for some π and $\varepsilon > 0$, where $-I = (-1, -1, \ldots, -1)$ is an n-dimensional row vector. Rewriting this in the form of the alternative (2) of the Farkas' theorem gives $(\pi, \varepsilon) \left(\dfrac{A}{I} \right) \leq \mathbf{0}^T$, $(\pi, \varepsilon) \left(\dfrac{0}{1} \right) > 0$ for some (π, ε). From the Farkas' theorem, the associated alternative (1) states that $\left(\dfrac{A}{I} \right) x = \left(\dfrac{0}{1} \right)$ for some $x \geq \mathbf{0}$. This implies $Ax = \mathbf{0}$, $Ix = 1$ for some $x \geq \mathbf{0}, x \neq \mathbf{0}$, and hence, the alternative (1) of the Gordon's theorem follows.

2.15 (i) Introducing the surplus variables x_3, x_4, x_5 and multiplying both sides by (-1), then performing the dual simplex method with the pivot elements $[-3], [-5/3], [-3/5]$, cycle 3 yields an optimal solution $x_2 = 11/3$, $x_3 = 5/3$, $x_1 = 8/3$ ($x_4 = x_5 = 0$), $\min z = 65/3$.

(ii) Introducing the surplus variables x_3, x_4, x_5 and multiplying both sides by (-1), then performing the dual simplex method with the pivot elements $[-3], [-16/3]$, cycle 2 yields an optimal solution $x_5 = 20.5$, $x_2 = 1$, $x_1 = 8.5$ ($x_3 = x_4 = 0$), $\min z = 30.5$.

(iii) Introducing the surplus variables x_4, x_5 and multiplying both sides by (-1), then performing the dual simplex method with the pivot elements $[-5], [-19/5], [-4/19]$, cycle 3 yields an optimal solution $x_2 = 3.5$, $x_3 = 0.25$ ($x_1 = x_4 = x_5 = 0$), $\min z = 7.75$.

(iv) Introducing the surplus variables x_4, x_5, x_6 and multiplying both sides by (-1), then performing the dual simplex method with the pivot elements $[-8], [-2]$, cycle 2 yields an optimal solution $x_3 = 17.5$, $x_5 = 183.75$, $x_1 = 66.25$ ($x_2 = x_4 = x_6 = 0$), $\min z = 317.5$.

2.16 (i) $x_B^* = (67/14, 15/7, 65/14)^T \geq \mathbf{0}$ itself is an optimal basic feasible solution and $\min z = -313/7$.

(ii) From the calculated result of $x_B^* = (39/14, 36/7, -75/14)^T$ $\bar{z}^* = -264/7$, noting x_B^* involves the negative value of $-75/14$, performing the dual simplex method with the pivot elements $[-11/7]$, cycle 1 yields an optimal solution $x_2 = 36/11$, $x_1 = 81/22$, $x_4 = 75/22$ ($x_3 = x_5 = 0$), $\min z = -819/22$.

2.17 (i) $x_B^* = (25/6, 2.5, 10/3)^T \geq \mathbf{0}$ itself is an optimal basic feasible solution and $\min z = -155/6$.

(ii) From the calculated result of $x_B^* = (19/6, 4, -1/6)^T$ $\bar{z}^* = -143/6$, noting x_B^* involves the negative value of $-1/6$, performing the dual simplex method with the pivot elements $[-4/9]$, cycle 1 yields an optimal solution $x_2 = 51/16$, $x_1 = 63/16$, $x_4 = 3/8$ ($x_3 = x_5 = 0$), $\min z = -381/16$.

Chapter 3

3.1 The Pareto optimal solutions of the original problem, the weak Pareto optimal solutions for $z_1 = -x_1 - 8x_2$ and $z_2 = -x_1 - x_2$, and the complete optimal solution for $z_1 = -2x_1 - 3x_2$ and $z_2 = -x_1 - x_2$ are shown in the following figures:

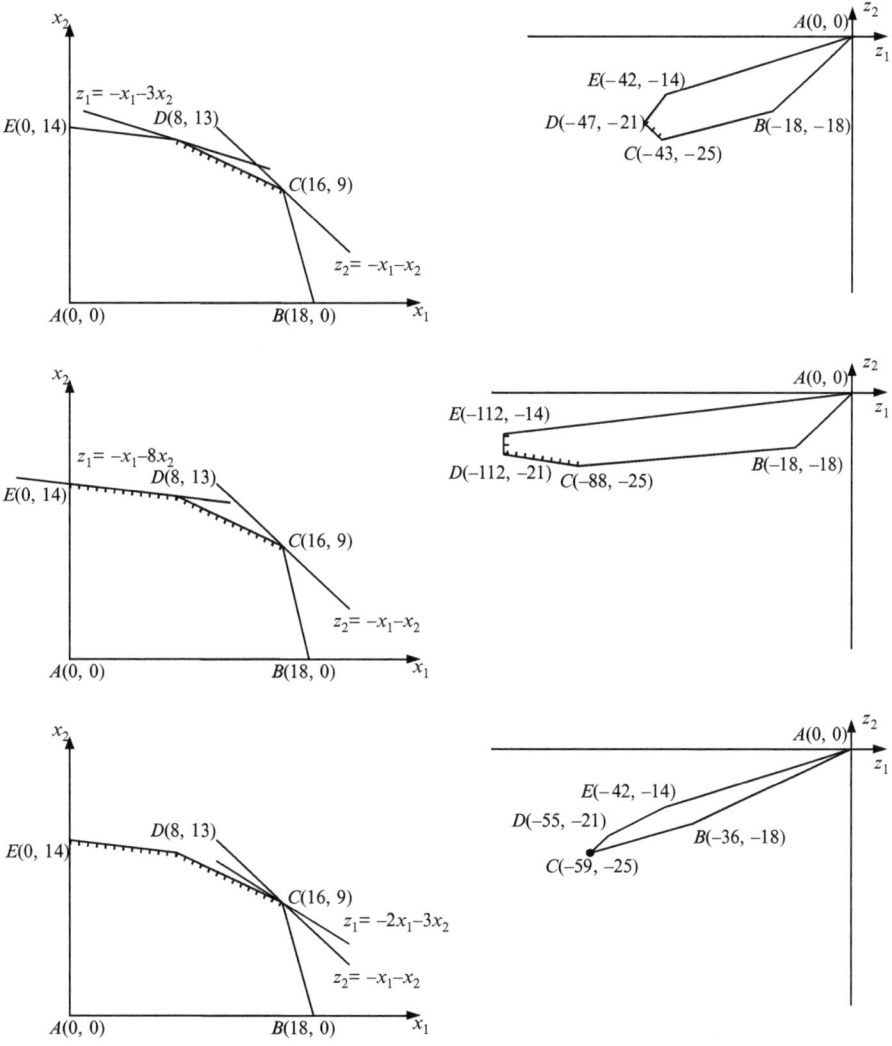

3.2 If an optimal solution x^* of the weighting problem is not a Pareto optimal solution of the multiobjective linear programming, then there exists $x \in X$ such that $z_j(x) < z_j(x^*)$ for some j and $z_i(x) \leq z_i(x^*), i = 1, \ldots, k; i \neq j$. Noting

$w = (w_1, \ldots, w_k) > \mathbf{0}$, this implies $\sum_i w_i z_i(x) < \sum_i w_i z_i(x^*)$. However, this contradicts the assumption that x^* is an optimal solution of the weighting problem for some $w > \mathbf{0}$.

3.3 Pareto optimal solutions in the x_1-x_2 plane and the z_1-z_2 plane are shown in the following figures.

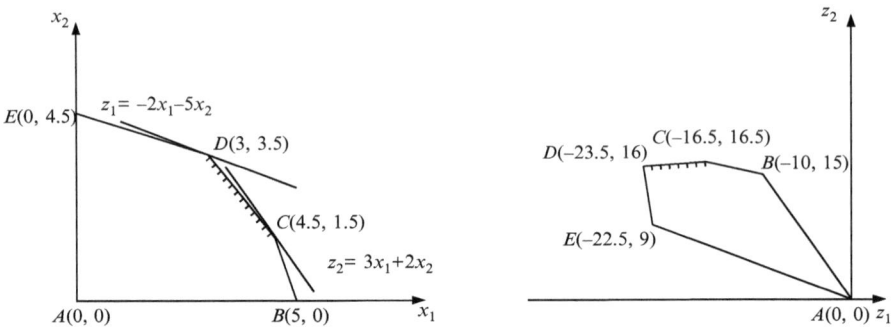

3.4 (i) The Pareto optimal solution is $(z_1, z_2) = (-22.5, 9)$ and $(x_1, x_2) = (0, 4.5)$. (See the figure.)

(ii) The Pareto optimal solution is $(z_1, z_2) = (-20, 8)$ and $(x_1, x_2) = (0, 4)$. (See the figure.)

(iii) The Pareto optimal solution is $(z_1, z_2) = (-16.8, 6.72)$ and $(x_1, x_2) = (0, 3.36)$. (See the figure.)

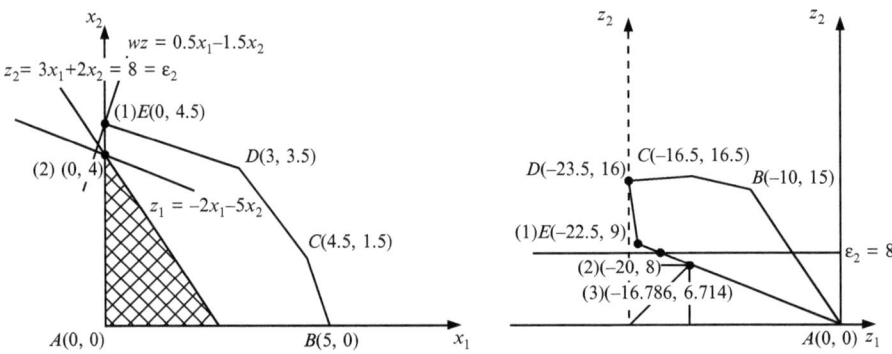

3.5 To graphically obtain an optimal solution for the problem in the x_1-x_2 plane, the two priority goals are depicted as straight lines together with the original feasible region in the figure. The effect of increasing either d_i^+ or d_i^- is reflected by the arrow signs. The region which satisfies both the original constraints and the first priority goal, i.e., $d_1^+ \geq 0$ and $d_1^- = 0$, is shown as the cross-hatched region. To achieve the second priority goal without degrading the achievement of the first priority goal, the area of feasible solution should be limited to the crisscross-hatched area in the figure. However, as can been seen, concerning the third priority goals, d_3^+ cannot increase to be positive. As just

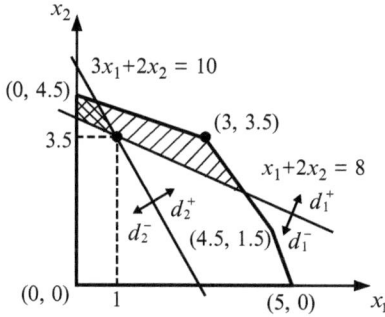

described, the final solution of this problem occurs at the point $(x_1, x_2) = (1, 3.5)$ in which only the first and second priority goals are satisfied.

3.6 From the definition, we have $z_1^{\min(1)} = -10$, $z_2^{\min(1)} = 0$. In step 3, the following two-objective programming problem is formulated: minimize$_{x \in C_w^{(0)}}$ ($\tilde{d}_1(z(x), z^{\min(1)})$, $\tilde{d}_\infty(z(x), z^{\min(1)})$), where $\tilde{d}_1(z(x), z^{\min(1)}) = \min_{x \in C_w^{(0)}} (-5x_1 - 5x_2 - (-10)) + (5x_1 + 1x_2 - 0)$, $\tilde{d}_\infty(z(x), z^{\min(1)}) = \max_{x \in C_w^{(0)}} \{(-5x_1 - 5x_2 - (-10)), (5x_1 + 1x_2 - 0)\}$. From this problem, we have the compromise set $C_w^{(1)}$ which is a straight-line segment between points $A(-9, 3)$ and $B(-7.667, 2.333)$ shown in the figure, where point A corresponding to a solution $x = (0.3, 1.5)$ minimizes $\tilde{d}_1(z(x), z^{\min(1)})$ and point B corresponding to a solution $x^* = (0.2, 1.333)$ minimizes $\tilde{d}_\infty(z(x), z^{\min(1)})$. Suppose that the DM cannot select the final solution. In step 2, we have $z_1^{\min(2)} = \min_{x \in C_w^{(1)}} z_1(x) = -9$, $z_2^{\min(2)} = \min_{x \in C_w^{(1)}} z_2(x) = 2.333$. In step 3, the following two-objective programming problem is reformulated: minimize$_{x \in C_w^{(1)}}$ ($\tilde{d}_1(z(x), z^{\min(2)})$, $\tilde{d}_\infty(z(x), z^{\min(2)})$), where $\tilde{d}_1(z(x), z^{\min(2)}) = \min_{x \in C_w^{(1)}} (-5x_1 - 5x_2 - (-9)) + (5x_1 + 1x_2 - 2.333)$, $\tilde{d}_\infty(z(x), z^{\min(2)}) = \max_{x \in C_w^{(1)}} \{(-5x_1 - 5x_2 - (-9)), (5x_1 + 1x_2 - 2.333)\}$. From

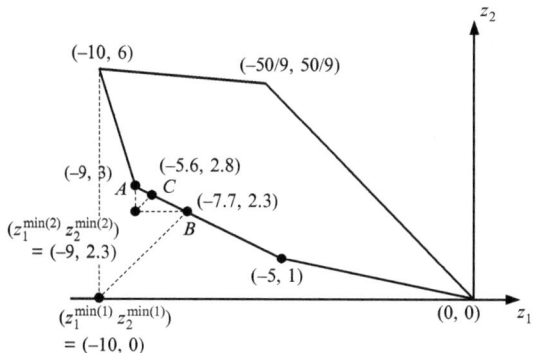

this problem, we have the revised compromise set $C_w^{(2)}$ which is a straight-line segment between points $A(-9, 3)$ and $-C(-8.556, 2.778)$ shown in Fig. 3.8, where point A minimizes $\tilde{d}_1(z(x), z^{\min(2)})$ and point C corresponding to a solution $x^* = (0.267, 1.444)$ minimizes $\tilde{d}_\infty(z(x), z^{\min(2)})$. One finds that the compromise set diminishes from $C_w^{(1)}$ to $C_w^{(2)}$. If the DM still cannot select the final solution in $C_w^{(2)}$, it follows that the procedure continues.

3.7 Using the Excel solver, do it by yourself.

3.8 Considering $z_1^{\min} = -23.5$, $z_1^{\max} = 0$, $z_2^{\min} = 0$, $z_2^{\max} = 16.5$, assume the DM specifies the reference point as $\hat{z}_1 = -23.5$, $\hat{z}_2 = 10$ and solving the corresponding minimax problem yields the Pareto optimal solution $z_1 = -22.75$, $z_2 = 10.75$ ($x_1 = 0.75$, $x_2 = 4.25$). On the basis of such information, assume that the DM updates the reference point to $\hat{z}_1 = -23.5$, $\hat{z}_2 = 8$ and solving the corresponding minimax problem yields the Pareto optimal solution $z_1 = -22.5$, $z_2 = 9$ ($x_1 = 0$, $x_2 = 4.5$). If the DM is satisfied with the current values of $z_1 = -22.5$, $z_2 = 9$, this Pareto optimal solution becomes a satisficing solution of the DM.

Chapter 4

4.1 $(A \cap B)_\alpha = \{x \mid \min\{\mu_A(x), \mu_B(x)\} \geq \alpha\} = \{x \mid \mu_A(x) \geq \alpha \text{ and } \mu_B(x) \geq \alpha\} = \{x \mid \mu_A(x) \geq \alpha\} \cap \{x \mid \mu_B(x) \geq \alpha\} = A_\alpha \cap B_\alpha$.

4.2 If $\mu_G(x) \leq \mu_C(x)$, then $\min\{\mu_G(x), \mu_C(x)\} = 1 \cdot \mu_G(x) + 0 \cdot \mu_C(x) \leq \alpha\mu_G(x) + (1-\alpha)\mu_C(x)$, and if $\mu_G(x) > \mu_C(x)$, then $\min\{\mu_G(x), \mu_C(x)\} = 0 \cdot \mu_G(x) + 1 \cdot \mu_C(x) \leq \alpha\mu_G(x) + (1-\alpha)\mu_C(x)$.

If $\mu_G(x) \leq \mu_C(x)$, then $\min\{\mu_G(x), \mu_C(x)\} = \mu_G(x) \geq \mu_G(x)\mu_C(x)$, and if $\mu_G(x) > \mu_C(x)$, then $\min\{\mu_G(x), \mu_C(x)\} = \mu_C(x) \geq \mu_G(x)\mu_C(x)$.

4.3 The problem to be solved becomes maximize$\{\lambda \mid \frac{2}{3}x_1 + 2x_2 + \lambda \leq 10, 1.6x_1 + 1.2x_2 + \lambda \leq 10, 1.5x_1 + 0.5x_2 + \lambda \leq 8.5, x_1 + 2x_2 - \lambda \geq 9.5, x_1 \geq 0, x_2 \geq 0\}$, and an optimal solution is $(x_1, x_2) = (3.0789, 3.5921)$, $\lambda = 0.76316$. Thus, compared with the nonfuzzy solution $x_1 = 3$, $x_2 = 3.5$, $z = -10$, the fuzzy solution $x_1 = 3.0789$, $x_2 = 3.5921$, $z = -10.2631$ gives about 2.7% additional value.

4.4 The problem to be solved becomes maximize$\{\lambda \mid 5x_1 + 7x_2 + 3\lambda \leq 15, 9x_1 + x_2 + 2\lambda \leq 12, -5x_1 + 3x_2 + \lambda \leq 8, 5x_1 + 5x_2 - 4\lambda \geq 8, x_1 \geq 0, x_2 \geq 0\}$, and an optimal solution is $(x_1, x_2) = (1.059140, 1.091398)$, $\lambda = 0.688172$. Thus, compared with the nonfuzzy solution $x_1 = 1$, $x_2 = 1$, $z = -10$, the fuzzy solution $x_1 = 1.059140$, $x_2 = 1.091398$, $z = -10.75269$ gives about 7.5% additional value.

4.5 From $z_1^{\min} = -37$, $z_2^{\min} = 0$, $z_1^m = 0$, $z_1^m = 29$, the problem to be solved becomes maximize $\{\lambda \mid 3x_1 + 8x_2 - 37\lambda \geq 0, 5x_1 + 4x_2 + 29\lambda \leq 29, 2x_1 + 6x_2 \leq 27, 3x_1 + 2x_2 \leq 16, 4x_1 + x_2 \leq 18, , x_1 \geq 0, x_2 \geq 0\}$, which yields an optimal solution $(x_1, x_2) = (0, 2.82368)$, $(z_1, z_2) = (-22.58947, 11.29474)$, $\lambda = 0.61053$.

Solutions 313

4.6 The problem to be solved becomes maximize$\{\lambda \mid x_1 + 2x_2 - 2\lambda \geq 8, 3x_1 + 2x_2 + 5\lambda \leq 14, 2x_1 + 6x_2 \leq 27, 8x_1 + 6x_2 \leq 45, 3x_1 + x_2 \leq 15, x_1 \geq 0, x_2 \geq 0\}$, and an optimal solution is $(x_1, x_2) = (0.78947, 4.2368)$, $(z_1, z_2) = (-9.26316, 10.84201)$, $\lambda = 0.63158$.

4.7 The problem to be solved becomes maximize$\{\lambda \mid x_1 + x_2 - \lambda \geq 1, 5x_1 + x_2 + 4\lambda \leq 5, 5x_1 + 7x_2 \leq 12, 9x_1 + x_2 \leq 10, -5x_1 + 3x_2 \leq 3, x_1 \geq 0, x_2 \geq 0\}$, and an optimal solution is $(x_1, x_2) = (0.230769, 1.384615)$, $(z_1, z_2) = (-8.07692, 2.53846)$, $\lambda = 0.615385$.

4.8 Proof of Theorem 4.2: Suppose that $x^* \in X$ is not an M-Pareto optimal solution to the generalized multiobjective linear programming problem. Then, there exists some $x \in X$ such that $\mu_i(z_i(x)) \geq \mu_i(z_i(x^*))$, $i = 1, \ldots, k$ and $\mu_j(z_j(x)) \neq \mu_j(z_j(x^*))$ for at least one j. Moreover, for any $\hat{\mu}_i$, $i = 1, \ldots, k$, there exists some $x \in X$ such that $\hat{\mu}_i - \mu_i(z_i(x)) \leq \hat{\mu}_i - \mu_i(z_i(x^*))$, $i = 1, \ldots, k$. Thus, $x^* \in X$ is not a unique optimal solution to (4.65).

Proof of Theorem 4.3: Suppose that $x^* \in X$ and v^* is not an optimal solution to (4.65) with any $\hat{\mu}$. We can specify $\hat{\mu}$ such that $\hat{\mu}_i - \mu_i(z_i(x^*)) = v^*$, $i = 1, \ldots, k$ hold. Since (x^*, v^*) is not an optimal solution to (4.65), there exists some $x \in X$ such that $\mu_i(z_i(x)) > \mu_i(z_i(x^*))$, $i = 1, \ldots, k$ hold. Thus, $x^* \in X$ is not an M-Pareto optimal solution to the generalized multiobjective linear programming problem.

4.9 Solving the minimax problem for the initial reference membership levels $\hat{\mu}_1 = \hat{\mu}_2 = \hat{\mu}_3 = 1$ yields the M-Pareto optimal solution $z_1 = 6.833$, $z_2 = 10.94$, $z_3 = -2.722$, $(x_1 = 2.056, x_2 = 2.389)$ and $\mu_1 = 0.6111$, $\mu_2 = 0.6111$, $\mu_3 = 0.6111$, $-\partial\mu_1/\partial\mu_2 = 0.8333$, $-\partial\mu_1/\partial\mu_3 = 1.1667$. On the basis of such information, assume that the DM updates the reference membership levels to $\hat{\mu}_1 = 0.7$, $\hat{\mu}_2 = 0.8$, $\hat{\mu}_3 = 0.5$. For the updated reference membership levels, the corresponding minimax problem yields $z_1 = 6.983$, $z_2 = 10.19$, $z_3 = -3.772$, $(x_1 = 1.606, x_2 = 2.689)$, $\mu_1 = 0.6611$, $\mu_2 = 0.7611$, $\mu_3 = 0.4611$, $-\partial\mu_1/\partial\mu_2 = 0.8333$, $-\partial\mu_1/\partial\mu_3 = 1.1667$. If the DM is satisfied with the current values of the membership functions, this M-Pareto optimal solution becomes a satisficing solution of the DM.

4.10 Solving the minimax problem for the initial reference membership levels $\hat{\mu}_1 = \hat{\mu}_2 = \hat{\mu}_3 = 1$ yields the M-Pareto optimal solution $z_1 = 5.856354$, $z_2 = 3.071823$, $z_3 = -4.14365$, $(x_1 = 0.475138, x_2 = 0.696133)$ and $\mu_1 = 0.585635$, $\mu_2 = 0.585635$, $\mu_3 = 0.585635$, $-\partial\mu_1/\partial\mu_2 = 0.6395$, $-\partial\mu_1/\partial\mu_3 = 0.4651$. On the basis of such information, assume that the DM updates the reference membership levels to $\hat{\mu}_1 = 0.63$, $\hat{\mu}_2 = 0.63$, $\hat{\mu}_3 = 0.52$. For the updated reference membership levels, the corresponding minimax problem yields $z_1 = 6.099448$, $z_2 = 2.950276$, $z_3 = -5.00055$, $(x_1 = 0.432597, x_2 = 0.787293)$, $\mu_1 = 0.609945$, $\mu_2 = 0.609945$, $\mu_3 = 0.499945$, $-\partial\mu_1/\partial\mu_2 = 0.6395$, $-\partial\mu_1/\partial\mu_3 = 0.4651$. If the DM is satisfied with the current values of the membership functions, this M-Pareto optimal solution becomes a satisficing solution of the DM.

4.11 (i) From $x \geq 0$ and $c_i^1 \leq c_i^2$, $c_i^1 x \leq c_i^2 x \leq d_{iR}^{-1}(\hat{\mu}_i - v)$ and $c_i^2 x \geq c_i^1 x \geq c_i x \geq d_{iL}^{-1}(\hat{\mu}_i - v)$, and then $S_{iR}(c_i^1) \supseteq S_{iR}(c_i^2)$ and $S_{iL}(c_i^1) \subseteq S_{iL}(c_i^2)$. (ii) Similarly, since $a_j^1 x \leq a_j^2 x \leq b_j$, $T_j(a_j^1, b_j) \supseteq T_j(a_j^2, b_j)$. (iii) Similarly, since $a_j x \leq b_j^1 \leq b_j^2$, $T_j(a_j, b_j^1) \subseteq T_j(a_j, b_j^2)$.

4.12 Proof of Theorem 4.4: Suppose that $x^* \in X(a_\alpha^L, b_\alpha^R)$ is not an M-α-Pareto optimal solution to the generalized α-multiobjective linear programming problem. Then, for the α-level optimal parameters $(a_\alpha^L, b_\alpha^R, c_{i\alpha}^L, i \in I_1 \cup I_{3R}, c_{i\alpha}^R, i \in I_2 \cup I_{3L})$, there exists some $x \in X(a_\alpha^L, b_\alpha^R)$ such that $\mu_i(c_{i\alpha}^L x) \geq \mu_i(c_{i\alpha}^L x^*)$, $i \in I_1 \cup I_{3R}$ and $\mu_i(c_{i\alpha}^R x) \geq \mu_i(c_{i\alpha}^R x^*)$, $i \in I_2 \cup I_{3L}$ hold. From this fact, for any $\hat{\mu}_i$, $i = 1, \ldots, k$, there exists some $x \in X(a_\alpha^L, b_\alpha^R)$ such that $\hat{\mu}_i - \mu_i(c_{i\alpha}^L x) \leq \hat{\mu}_i - \mu_i(c_{i\alpha}^L x^*)$, $i \in I_1 \cup I_{3R}$ and $\hat{\mu}_i - \mu_i(c_{i\alpha}^R x) \leq \hat{\mu}_i - \mu_i(c_{i\alpha}^R x^*)$, $i \in I_2 \cup I_{3L}$ hold. Thus, it follows that $x^* \in X(a_\alpha^L, b_\alpha^R)$ is not a unique optimal solution to (4.83). From the strictly monotone decreasing or increasing property of $d_{iR}(\cdot)$ or $d_{iL}(\cdot)$ together with the fact that $\mu_i(\cdot) = d_{iR}(\cdot)$, $i \in I_1 \cup I_{3R}$ and $\mu_i(\cdot) = d_{iL}(\cdot)$, $i \in I_2 \cup I_{3L}$, the optimal solution of (4.83) coincides with that of (4.87). Thus, $x^* \in X(a_\alpha^L, b_\alpha^R)$ is not a unique optimal solution to (4.87).

Proof of Theorem 4.5: Suppose that $x^* \in X(a_\alpha^L, b_\alpha^R)$ and v^* is not an optimal solution to (4.87) with any $\hat{\mu}$. From the strictly monotone decreasing or increasing property of $d_{iR}(\cdot)$ or $d_{iL}(\cdot)$, we can specify $\hat{\mu}$ such that $c_{i\alpha}^L x^* = d_{iR}^{-1}(\hat{\mu}_i - v^*)$, $i \in I_1 \cup I_{3R}$ and $c_{i\alpha}^R x^* = d_{iL}^{-1}(\hat{\mu}_i - v^*)$, $i \in I_2 \cup I_{3L}$ hold. Since (x^*, v^*) is not an optimal solution to (4.87), there exists some $x \in X(a_\alpha^L, b_\alpha^R)$ such that $c_{i\alpha}^L x < c_{i\alpha}^L x^* = d_{iR}^{-1}(\hat{\mu}_i - v^*)$, $i \in I_1 \cup I_{3R}$ and $c_{i\alpha}^R x > c_{i\alpha}^R x^* = d_{iL}^{-1}(\hat{\mu}_i - v^*)$, $i \in I_2 \cup I_{3L}$ hold. From this fact, from the strictly monotone decreasing or increasing property of $d_{iR}(\cdot)$ or $d_{iL}(\cdot)$, $\mu_i(c_{i\alpha}^L x) > \mu_i(c_{i\alpha}^L x^*) = \hat{\mu}_i - v^*$, $i \in I_1 \cup I_{3R}$ and $\mu_i(c_{i\alpha}^R x) > \mu_i(c_{i\alpha}^R x^*) = \hat{\mu}_i - v^*$, $i \in I_2 \cup I_{3L}$ hold. Thus, $x^* \in X(a_\alpha^L, b_\alpha^R)$ is not an M-α-Pareto optimal solution to the generalized α-multiobjective linear programming problem.

4.13 Solving the minimax problem for the initial reference membership levels $\hat{\mu}_1 = \hat{\mu}_2 = \hat{\mu}_3 = 1$ and the α-level $\alpha = 0.5$ yields the M-α-Pareto optimal solution $z_1 = 6.194741$, $z_2 = 2.90263$, $z_3 = -3.80526$, $(x_1 = 0.4946695, x_2 = 0.676617)$ and $\mu_1 = 0.619474$, $\mu_2 = 0.619474$, $\mu_3 = 0.619474$, $-\partial \mu_1/\partial \mu_2 = 0.737261$, $-\partial \mu_1/\partial \mu_3 = 0.503185$, $-\partial \mu_1/\alpha = -0.19038$. On the basis of such information, assume that the DM updates the reference membership levels to $\hat{\mu}_1 = 0.65$, $\hat{\mu}_2 = 0.65$, $\hat{\mu}_3 = 0.58$. For the updated reference membership levels, the corresponding minimax problem yields $z_1 = 6.351955$, $z_2 = 2.824023$, $z_3 = -4.348045$, $(x_1 = 0.464819, x_2 = 0.732338)$, $\mu_1 = 0.6355195$, $\mu_2 = 0.635195$, $\mu_3 = 0.565195$, $-\partial \mu_1/\partial \mu_2 = 0.737261$, $-\partial \mu_1/\partial \mu_3 = 0.503185$, $-\partial \mu_1/\partial \alpha = -0.18855$. If the DM is satisfied with the current values of the membership functions, this M-Pareto optimal solution becomes a satisficing solution of the DM.

Chapter 5

5.1 The two-stage programming problem is formulated as follows: minimize $-5x_1 - 5x_2 + E[5y_1^+ + 7y_2^+ + 10y_1^- + 9y_2^-]$ subject to $x_1 + y_1^+ - y_1^- = \bar{b}_1$, $x_2 + y_2^+ - y_2^- = \bar{b}_2$, $5x_1 + 7x_2 \leq 12$, $9x_1 + 1x_2 \leq 10$, $-5x_1 + 3x_2 \leq 3$, $x_1 \geq 0$, $x_2 \geq 0$. The optimal solution is $(x_1^*, x_2^*) = (0.969163, 1.022026)$, and the optimal value is -2.255507.

5.2 It immediately follows from $P\left(\sum_{j=1}^n a_{ij} x_j \geq \bar{b}_i\right) = F_i\left(\sum_{j=1}^n a_{ij} x_j\right)$, $i = 1, \ldots, m$.

5.3 The corresponding equivalent deterministic constraint is $\prod_{i=1}^m G_i(\sum_{j=1}^n \bar{a}_{ij} x_j) \geq \beta$, where $G_i = 1 - F_i$.

5.4 The chance constraints can be expressed by the following inequalities: $5x_1 + 7x_2 \leq 12 + 2\Phi^{-1}(0.2) = 10.31676$, $9x_1 + 1x_2 \leq 10 + \Phi^{-1}(0.2) = 9.15838$, $-5x_1 + 3x_2 \leq 3 + 0.5\Phi^{-1}(0.2) = 2.57919$, where Φ is the distribution function of the standard normal distribution $N(0, 1)$.

5.5 The chance constraints can be expressed by the following inequalities:

$$5x_1 + 7x_2 - \Phi^{-1}(0.2)\sqrt{2^2 + (x_1, x_2)\begin{bmatrix} 0.5 & 0.2 \\ 0.2 & 1 \end{bmatrix}\begin{pmatrix} x_1 \\ x_2 \end{pmatrix}} \leq 12$$

$$9x_1 + 1x_2 - \Phi^{-1}(0.2)\sqrt{1^2 + (x_1, x_2)\begin{bmatrix} 1 & 0.1 \\ 0.1 & 0.25 \end{bmatrix}\begin{pmatrix} x_1 \\ x_2 \end{pmatrix}} \leq 10$$

$$-5x_1 + 3x_2 - \Phi^{-1}(0.2)\sqrt{0.5^2 + (x_1, x_2)\begin{bmatrix} 0.5 & 0.1 \\ 0.1 & 0.25 \end{bmatrix}\begin{pmatrix} x_1 \\ x_2 \end{pmatrix}} \leq 3.$$

5.6 Minimize $x^T V_{\bar{c}} x$ subject to $m_c x \leq \gamma$, $P\left(\sum_{j=1}^n \bar{a}_{ij} x_j \leq \bar{b}_i\right) \geq \beta_i$, $i = 1, \ldots, m$, $x \geq 0$.

5.7 Minimize $x V_r x$ subject to $m_r x \geq a$, $\sum_{j=1}^n x_j = 1$, $x_j \geq 0$, $j = 1, \ldots, n$.

5.8 The expectation model with the chance constraints is formulated as follows:

minimize $-5x_1 - 5x_2$

subject to $5x_1 + 7x_2 - \Phi^{-1}(0.2)\sqrt{2^2 + (x_1, x_2)\begin{bmatrix} 0.5 & 0.2 \\ 0.2 & 1 \end{bmatrix}\begin{pmatrix} x_1 \\ x_2 \end{pmatrix}} \leq 12$

$9x_1 + 1x_2 - \Phi^{-1}(0.2)\sqrt{1^2 + (x_1, x_2)\begin{bmatrix} 1 & 0.1 \\ 0.1 & 0.25 \end{bmatrix}\begin{pmatrix} x_1 \\ x_2 \end{pmatrix}} \leq 10$

$-5x_1 + 3x_2 - \Phi^{-1}(0.2)\sqrt{0.5^2 + (x_1, x_2)\begin{bmatrix} 0.5 & 0.1 \\ 0.1 & 0.25 \end{bmatrix}\begin{pmatrix} x_1 \\ x_2 \end{pmatrix}} \leq 3$

$x_1 \geq 0$, $x_2 \geq 0$.

The optimal solution is $(x_1^*, x_2^*) = (0.886634, 0.803562)$, and the optimal value is -8.450981.

5.9 The expectation model with the chance constraints is formulated as follows: minimize $-5x_1 - 5x_2$ subject to $5x_1 + 7x_2 \leq 12 + 2\Phi^{-1}(0.2) = 10.316758$, $9x_1 + 1x_2 \leq 10 + 1\Phi^{-1}(0.2) = 9.158379$, $-5x_1 + 3x_2 \leq 3 + 0.5\Phi^{-1}(0.2) = 2.579189$, $x_1 \geq 0$, $x_2 \geq 0$. The optimal solution is $(x_1^*, x_2^*) = (0.889521, 0.712756)$, and the optimal value is -8.011386.

5.10 The variance model with the chance constraints and the expectation level is formulated as follows:

$$\text{minimize } (x_1, x_2) \begin{bmatrix} 1 & 0.5 \\ 0.5 & 1 \end{bmatrix} \begin{pmatrix} x_1 \\ x_2 \end{pmatrix}$$

$$\text{subject to } 5x_1 + 7x_2 - \Phi^{-1}(0.2)\sqrt{2^2 + (x_1, x_2)\begin{bmatrix} 0.5 & 0.2 \\ 0.2 & 1 \end{bmatrix}\begin{pmatrix} x_1 \\ x_2 \end{pmatrix}} \leq 12$$

$$9x_1 + 1x_2 - \Phi^{-1}(0.2)\sqrt{1^2 + (x_1, x_2)\begin{bmatrix} 1 & 0.1 \\ 0.1 & 0.25 \end{bmatrix}\begin{pmatrix} x_1 \\ x_2 \end{pmatrix}} \leq 10$$

$$-5x_1 + 3x_2 - \Phi^{-1}(0.2)\sqrt{0.5^2 + (x_1, x_2)\begin{bmatrix} 0.5 & 0.1 \\ 0.1 & 0.25 \end{bmatrix}\begin{pmatrix} x_1 \\ x_2 \end{pmatrix}} \leq 3$$

$$-5x_1 - 5x_2 \leq -8, \ x_1 \geq 0, \ x_2 \geq 0.$$

The optimal solution is $(x_1^*, x_2^*) = (0.8, 0.8)$, and the optimal value is 1.92.

5.11 The variance model with the chance constraints and the expectation level is formulated as follows: minimize $(x_1, x_2) \begin{bmatrix} 1 & 0.5 \\ 0.5 & 1 \end{bmatrix} \begin{pmatrix} x_1 \\ x_2 \end{pmatrix}$ subject to $5x_1 + 7x_2 \leq 12 + 2\Phi^{-1}(0.2) = 10.316758$, $9x_1 + 1x_2 \leq 10 + 1\Phi^{-1}(0.2) = 9.158379$, $-5x_1 + 3x_2 \leq 3 + 0.5\Phi^{-1}(0.2) = 2.579189$, $-5x_1 - 5x_2 \leq -8$, $x_1 \geq 0$, $x_2 \geq 0$. The optimal solution is $(x_1^*, x_2^*) = (0.8, 0.8)$, and the optimal value is 1.92.

5.12 It directly follows from $P(\bar{c}x \leq f_0) = P\left(\frac{\bar{c}x - c^0 x}{c^1 x} \leq \frac{f_0 - c^0 x}{c^1 x}\right) = \Phi\left(\frac{f_0 - c^0 x}{c^1 x}\right)$.

5.13 The probability model with the chance constraints is formulated as follows:

$$\text{minimize } \frac{(-8) - (-5x_1 - 5x_2)}{\sqrt{(x_1, x_2)\begin{bmatrix} 1 & 0.5 \\ 0.5 & 1 \end{bmatrix}\begin{pmatrix} x_1 \\ x_2 \end{pmatrix}}}$$

$$\text{subject to } 5x_1 + 7x_2 - \Phi^{-1}(0.2)\sqrt{2^2 + (x_1, x_2)\begin{bmatrix} 0.5 & 0.2 \\ 0.2 & 1 \end{bmatrix}\begin{pmatrix} x_1 \\ x_2 \end{pmatrix}} \leq 12$$

$$9x_1 + 1x_2 - \Phi^{-1}(0.2)\sqrt{1^2 + (x_1, x_2)\begin{bmatrix} 1 & 0.1 \\ 0.1 & 0.25 \end{bmatrix}\begin{pmatrix} x_1 \\ x_2 \end{pmatrix}} \le 10$$

$$-5x_1 + 3x_2 - \Phi^{-1}(0.2)\sqrt{0.5^2 + (x_1, x_2)\begin{bmatrix} 0.5 & 0.1 \\ 0.1 & 0.25 \end{bmatrix}\begin{pmatrix} x_1 \\ x_2 \end{pmatrix}} \le 3$$

$x_1 \ge 0, x_2 \ge 0.$

The optimal solution is $(x_1^*, x_2^*) = (0.886634, 0.803562)$, and the optimal value of the original objective function (probability) is 0.620949.

5.14 The probability model with the chance constraints is formulated as follows: minimize $\dfrac{(-8)-(-5x_1-5x_2)}{\sqrt{(x_1,x_2)\begin{bmatrix} 1 & 0.5 \\ 0.5 & 1 \end{bmatrix}\begin{pmatrix} x_1 \\ x_2 \end{pmatrix}}}$ subject to $5x_1 + 7x_2 \le 12 + 2\Phi^{-1}(0.2) = 10.316758$, $9x_1 + 1x_2 \le 10 + 1\Phi^{-1}(0.2) = 9.158379$, $-5x_1 + 3x_2 \le 3 + 0.5\Phi^{-1}(0.2) = 2.579189$, $x_1 \ge 0$, $x_2 \ge 0$. The optimal solution is $(x_1^*, x_2^*) = (0.927446, 0.811361)$, and the optimal value of the original objective function (probability) is 0.694036.

5.15 The fractile model with the chance constraints is formulated as follows:

$$\text{minimize } (-5x_1 - 5x_2) + \Phi^{-1}(0.6)(x_1, x_2)\begin{bmatrix} 1 & 0.5 \\ 0.5 & 1 \end{bmatrix}\begin{pmatrix} x_1 \\ x_2 \end{pmatrix}$$

$$\text{subject to } 5x_1 + 7x_2 - \Phi^{-1}(0.2)\sqrt{2^2 + (x_1, x_2)\begin{bmatrix} 0.5 & 0.2 \\ 0.2 & 1 \end{bmatrix}\begin{pmatrix} x_1 \\ x_2 \end{pmatrix}} \le 12$$

$$9x_1 + 1x_2 - \Phi^{-1}(0.2)\sqrt{1^2 + (x_1, x_2)\begin{bmatrix} 1 & 0.1 \\ 0.1 & 0.25 \end{bmatrix}\begin{pmatrix} x_1 \\ x_2 \end{pmatrix}} \le 10$$

$$-5x_1 + 3x_2 - \Phi^{-1}(0.2)\sqrt{0.5^2 + (x_1, x_2)\begin{bmatrix} 0.5 & 0.1 \\ 0.1 & 0.25 \end{bmatrix}\begin{pmatrix} x_1 \\ x_2 \end{pmatrix}} \le 3$$

$x_1 \ge 0, x_2 \ge 0.$

The optimal solution is $(x_1^*, x_2^*) = (0.886634, 0.803562)$, and the optimal value is -8.079994.

5.16 The fractile model with the chance constraints is formulated as follows: minimize $(-5x_1 - 5x_2) + \Phi^{-1}(0.6)(x_1, x_2)\begin{bmatrix} 1 & 0.5 \\ 0.5 & 1 \end{bmatrix}\begin{pmatrix} x_1 \\ x_2 \end{pmatrix}$ subject to $5x_1 + 7x_2 \le 12 + 2\Phi^{-1}(0.2) = 10.316758$, $9x_1 + 1x_2 \le 10 + 1\Phi^{-1}(0.2) = 9.158379$, $-5x_1 + 3x_2 \le 3 + 0.5\Phi^{-1}(0.2) = 2.579189$, $x_1 \ge 0$, $x_2 \ge 0$. The optimal solution is $(x_1^*, x_2^*) = (0.927446, 0.811361)$, and the optimal value is -8.312250.

5.17 Referring to the obtained results by using the Excel solver, do it by yourself.

Chapter 6

6.1 The expectation model with the chance constraints is formulated as follows:

$$\text{minimize } z_1^E(x) = -5x_1 - 5x_2$$
$$\text{minimize } z_2^E(x) = 5x_1 + x_2$$
$$\text{subject to } 5x_1 + 7x_2 \leq 12 + 2\Phi^{-1}(0.2) = 10.317$$
$$9x_1 + x_2 \leq 10 + \Phi^{-1}(0.2) = 9.158$$
$$-5x_1 + 3x_2 \leq 3 + 0.5\Phi^{-1}(0.2) = 2.579$$
$$x_1 \geq 0, \ x_2 \geq 0.$$

An example of the interactive process is shown in the following table.

Iteration	First	Second	Third
$\hat{\mu}_1$	1.00	0.72	0.71
$\hat{\mu}_2$	1.00	0.67	0.67
$\mu_1(z_1^E(x))$	0.687878	0.71569	0.710127
$\mu_2(z_2^E(x))$	0.687878	0.66569	0.670127
$z_1^E(x)$	−5.980438	−6.22223	−6.17387
$z_2^E(x)$	1.700624	1.821521	1.797342

6.2 The variance model with the chance constraints and the expectation levels is formulated as follows:

$$\text{minimize } z_1^V(x) = (x_1, x_2) \begin{bmatrix} 1 & 0.5 \\ 0.5 & 1 \end{bmatrix} \begin{pmatrix} x_1 \\ x_2 \end{pmatrix}$$
$$\text{minimize } z_2^V(x) = (x_1, x_2) \begin{bmatrix} 1 & 0.1 \\ 0.1 & 0.2 \end{bmatrix} \begin{pmatrix} x_1 \\ x_2 \end{pmatrix}$$
$$\text{subject to } 5x_1 + 7x_2 \leq 12 + 2\Phi^{-1}(0.2) = 10.317$$
$$9x_1 + x_2 \leq 10 + \Phi^{-1}(0.2) = 9.158$$
$$-5x_1 + 3x_2 \leq 3 + 0.5\Phi^{-1}(0.2) = 2.579$$
$$z_1^E(x) = -5x_1 - 5x_2 \leq -5$$
$$z_2^E(x) = 5x_1 + x_2 \leq 3$$
$$x_1 \geq 0, \ x_2 \geq 0.$$

An example of the interactive process is shown in the following table.

Iteration	First	Second	Third
$\hat{\mu}_1$	1.00	0.77	0.77
$\hat{\mu}_2$	1.00	0.74	0.73
$\mu_1(z_1^V(x))$	0.75	0.764775	0.76960
$\mu_2(z_2^V(x))$	0.75	0.734775	0.72960
$z_1^V(x)$	0.79	0.787636	0.786864
$z_2^V(x)$	0.23	0.232436	0.233264

6.3 The probability model with the chance constraints is formulated as follows:

$$\text{minimize } z_1^P(x) = \Phi\left(\{(-5) - (-5x_1 - 5x_2)\} \Big/ \sqrt{(x_1, x_2)\begin{bmatrix} 1 & 0.5 \\ 0.5 & 1 \end{bmatrix}\begin{pmatrix} x_1 \\ x_2 \end{pmatrix}}\right)$$

$$\text{minimize } z_2^P(x) = \Phi\left(\{(3) - (5x_1 + 1x_2)\} \Big/ \sqrt{(x_1, x_2)\begin{bmatrix} 1 & 0.1 \\ 0.1 & 0.2 \end{bmatrix}\begin{pmatrix} x_1 \\ x_2 \end{pmatrix}}\right)$$

subject to $5x_1 + 7x_2 \leq 12 + 2\Phi^{-1}(0.2) = 10.317$
$9x_1 + x_2 \leq 10 + \Phi^{-1}(0.2) = 9.158$
$-5x_1 + 3x_2 \leq 3 + 0.5\Phi^{-1}(0.2) = 2.579$
$x_1 \geq 0, x_2 \geq 0.$

An example of the interactive process is shown in the following table.

Iteration	First	Second	Third
$\hat{\mu}_1$	1.00	0.75	0.75
$\hat{\mu}_2$	1.00	0.72	0.71
$z_1^P(x^*)$	0.868748	0.871942	0.872990
$z_2^P(x^*)$	0.818748	0.812942	0.810990
$\mu_1(z_1^P(x^*))$	0.739805	0.707839	0.743299
$\mu_2(z_2^P(x^*))$	0.709805	0.667839	0.703299

6.4 The probability model with the chance constraints is formulated as follows:

$$\text{minimize } z_1^P(x) = \Phi\left(\{(-5) - (-5x_1 - 5x_2)\} \Big/ \{x_1 + x_2\}\right)$$

$$\text{minimize } z_2^P(x) = \Phi\left(\{(3) - (5x_1 + x_2)\} \Big/ \{x_1 + \sqrt{0.2}x_2\}\right)$$

subject to $5x_1 + 7x_2 \leq 12 + 2\Phi^{-1}(0.2) = 9.475$
$9x_1 + x_2 \leq 10 + \Phi^{-1}(0.2) = 8.317$
$-5x_1 + 3x_2 \leq 3 + 0.5\Phi^{-1}(0.2) = 1.317$
$x_1 \geq 0, x_2 \geq 0.$

An example of the interactive process is shown in the following table.

Iteration	First	Second	Third
$\hat{\mu}_1$	1.000	0.91	0.91
$\hat{\mu}_2$	1.000	0.89	0.87
$\mu_1(z_1^P(x))$	0.896568	0.902343	0.907969
$\mu_2(z_2^P(x))$	0.896568	0.882343	0.867969
$z_1^P(x)$	0.918970	0.920703	0.922391
$z_2^P(x)$	0.868970	0.864703	0.860391
$-\partial\mu_1/\partial\mu_2$	0.413524	0.398576	0.384416

6.5 The fractile model with the chance constraints is formulated as follows:

$$\text{minimize } z_1^F(x) = (-5x_1 - 5x_2) + \Phi^{-1}(0.6)\sqrt{(x_1, x_2)\begin{bmatrix} 1 & 0.5 \\ 0.5 & 1 \end{bmatrix}\begin{pmatrix} x_1 \\ x_2 \end{pmatrix}}$$

$$\text{minimize } z_2^F(x) = (5x_1 + x_2) + \Phi^{-1}(0.6)\sqrt{(x_1, x_2)\begin{bmatrix} 1 & 0.1 \\ 0.1 & 0.2 \end{bmatrix}\begin{pmatrix} x_1 \\ x_2 \end{pmatrix}}$$

subject to $5x_1 + 7x_2 \leq 12 + 2\Phi^{-1}(0.2) = 10.317$
$9x_1 + x_2 \leq 10 + \Phi^{-1}(0.2) = 9.158$
$-5x_1 + 3x_2 \leq 3 + 0.5\Phi^{-1}(0.2) = 2.579$
$x_1 \geq 0, \ x_2 \geq 0.$

An example of the interactive process is shown in the following table.

Iteration	First	Second	Third
$\hat{\mu}_1$	1.00	0.71	0.70
$\hat{\mu}_2$	1.00	0.68	0.68
$z_1^F(x^*)$	−5.66785	−5.80694	−5.76058
$z_2^F(x^*)$	1.819517	1.895397	1.870101
$\mu_1(z_1^F(x^*))$	0.681868	0.69860	0.693023
$\mu_2(z_2^F(x^*))$	0.681868	0.66860	0.673023

6.6 The simple recourse model is formulated as follows:

$$\text{minimize } z_1^R(x) = -5x_1 - 5x_2 + E[5y_1^+ + 7y_2^+ + 10y_1^- + 9y_2^-]$$
$$\text{minimize } z_2^R(x) = 5x_1 + x_2$$
subject to $x_1 + y_1^+ - y_1^- = \bar{b}_1$
$x_2 + y_2^+ - y_2^- = \bar{b}_2$
$5x_1 + 7x_2 \leq 12$
$9x_1 + 1x_2 \leq 10$
$-5x_1 + 3x_2 \leq 3$
$x_1 \geq 0, \ x_2 \geq 0.$

An example of the interactive process is shown in the following table.

Iteration	First	Second	Third
$\hat{\mu}_1$	1.00	0.71	0.72
$\hat{\mu}_2$	1.00	0.68	0.67
$\mu_1(z_1^R(x^*))$	0.689493	0.702726	0.711405
$\mu_2(z_2^R(x^*))$	0.689493	0.672726	0.661405
$z_1^R(x^*)$	2.170922	1.982280	1.858562
$z_2^R(x^*)$	1.822005	1.920390	1.986824

Solutions   321

Chapter 7

7.1 $z = 90x_1 + 111x_2 - (50y_{11} + 57y_{12} + 60y_{21} + 50y_{22}) - (20y_{11} + 8y_{12} + 8y_{21} + 10y_{22})$

7.2 $4000 \leq x_1 \leq 5000, 4000 \leq x_2 \leq 5000$

7.3 $x_1 = y_{11} + y_{21}, x_2 = y_{12} + y_{22}$

7.4 $50y_{11} + 57y_{12} \leq 220000, 60y_{21} + 50y_{22} \leq 200000$

7.5 maximize $z = 90x_1 + 111x_2 - (50y_{11} + 57y_{12} + 60y_{21} + 50y_{22}) - (20y_{11} + 8y_{12} + 8y_{21} + 10y_{22})$ subject to $4000 \leq x_1 \leq 5000, 4000 \leq x_2 \leq 5000, x_1 = y_{11} + y_{21}, x_2 = y_{12} + y_{22}, 50y_{11} + 57y_{12} \leq 220000, 60y_{21} + 50y_{22} \leq 200000$

   An optimal solution is $x_1 = 4000, x_2 = 4351, y_{11} = 4000, y_{12} = 351, y_{21} = 0, y_{22} = 4000, z = 300140$

7.6 $z_1 = 90x_1 + 111x_2 - (50y_{11} + 57y_{12} + 60y_{21} + 50y_{22})$
   $z_2 = 20y_{11} + 8y_{12} + 8y_{21} + 10y_{22}$
   minimize $-z_1 + 4z_2$ subject to $4000 \leq x_1 \leq 5000, 4000 \leq x_2 \leq 5000, x_1 = y_{11} + y_{21}, x_2 = y_{12} + y_{22}, 50y_{11} + 57y_{12} \leq 220000, 60y_{21} + 50y_{22} \leq 200000$

   An optimal solution is $x_1 = 4000, x_2 = 4000, y_{11} = 2913, y_{12} = 1304, y_{21} = 1087, y_{22} = 2696, z_1 = 384000, z_2 = 104348$

7.7 maximize λ subject to $\frac{z_1 - 384000}{422947 - 384000} - \lambda \geq 0, \frac{z_2 - 122807}{104348 - 122807} - \lambda \geq 0, 4000 \leq x_1 \leq 5000, 4000 \leq x_2 \leq 5000, x_1 = y_{11} + y_{21}, x_2 = y_{12} + y_{22}, 50y_{11} + 57y_{12} \leq 220000, 60y_{21} + 50y_{22} \leq 200000$

   An optimal solution is $x_1 = 4000, x_2 = 4175, y_{11} = 3457, y_{12} = 828, y_{21} = 543, y_{22} = 3348, \lambda = 0.5, z_1 = 403474, z_2 = 113577$

7.8 minimize v subject to $-390000 - (-90x_1 - 111x_2 + (50y_{11} + 57y_{12} + 60y_{21} + 50y_{22})) \geq \Phi^{-1}((\hat{\mu}_1 - v)(422947 - 384000) + 384000)(5y_{11} + 6y_{12} + 5y_{21} + 5y_{22}), 110000 - (20y_{11} + 8y_{12} + 8y_{21} + 10y_{22}) \geq \Phi^{-1}((\hat{\mu}_2 - v)(104348 - 122807) + 122807)(3y_{11} + 2y_{12} + 2y_{21} + 2y_{22}), 4000 \leq x_1 \leq 5000, 4000 \leq x_2 \leq 5000, x_1 = y_{11} + y_{21}, x_2 = y_{12} + y_{22}, 50y_{11} + 57y_{12} - 220000 + \Phi^{-1}(0.8)(5y_{11} + 6y_{12}) \leq 0, 60y_{21} + 50y_{22} - 200000 + \Phi^{-1}(0.8)(5y_{21} + 5y_{12}) \leq 0$

References

Alves, M. J., & Clìmaco, J. (2007). A review of interactive methods for multiobjective integer and mixed-integer programming. *European Journal of Operational Research, 180*, 99–115.

Aouadni, S., Allouche, M. A., & Rebai, A. (2013). Supplier selection: an analytic network process and imprecise goal programming model integrating the decision-maker's preferences. *International Journal of Operational Research, 16*, 137–154.

Arrow, K. J. (1963). *Social choice and individual values* (2nd ed.). New York: Wiley.

Avriel, M. (1976). *Nonlinear programming: Analysis and method*. Englewood: Prentice-Hall.

Babbar, M. M. (1955). Distributions of solutions of a set of linear equation with an application to linear programming. *Journal of American Statistical Association, 50*, 854–869.

Bard, J. F. (1983). An efficient point algorithm for a linear two-stage optimization problem. *Operations Research, 38*, 556–560.

Baucells, M., & Shapley, L. S. (2008). Multiperson utility. *Games and Economic Behavior, 62*, 329–347.

Baushke, H. H., & Borwein, J. M. (1996). On projection algorithms for solving convex feasibility problems. *SIAM Review, 38*, 367–426.

Bazarra, M. S., & Shetty, C. M. (1979). *Nonlinear programming: Theory and algorithms* (2nd ed. (1993), 3rd ed. (2006)). New York: Wiley.

Beale, E. M. L. (1955). On minimizing a convex function subject to linear inequalities. *Journal of the Royal Statistical Society, B17*, 173–184.

Beale, E. M. L., Forrest, J. J. H., & Taylor, C. J. (1980). Multi-time-period stochastic programming. In M. A. H. Dempster (Ed.), *Stochastic programming* (pp. 387–402). New York: Academic.

Bellman, R. E., & Zadeh, L. A. (1970). Decision making in a fuzzy environment. *Management Science, 17*, 141–164.

Belton, V., & Stewart, T. J. (2001). *Multiple criteria decision analysis: An integrated approach*. Norwell: Kluwer.

Ben Abdelaziz, F. (2012). Solution approaches for the multiobjective stochastic programming. *European Journal of Operational Research, 216*, 1–16.

Benayoun, R., de Montgolfier, J., Tergny, J., & Larichev, O. I. (1971). Linear programming with multiple objective functions: STEP method (STEM). *Mathematical Programming, 1*, 366–375.

Bereanu, B. (1967). On stochastic linear programming. Distribution problems, stochastic technology matrix. *Zeitschrift für Wahrscheinlichkeitstheorie und verwandte Gebiete, 8*, 148–152.

Bereanu, B. (1980). Some numerical methods in stochastic linear programming under risk and uncertainty. In M. A. H. Dempster (Ed.), *Stochastic programming* (pp. 196–205). London: Academic.

Birge, J. R., & Louveaux, F. (1997). *Introduction to stochastic programming*. London: Springer.

Bitran, G. R., & Novaes, A. G. (1973). Linear programming with a fractional objective function. *Operations Research, 21*, 22–29.

Bland, R. G. (1977). New finite pivoting methods for the simplex method. *Mathematics of Operations Research, 2*, 103–107.

Borde, J., & Crouzeix, J. P. (1987). Convergence of a Dinkelbach-type algorithm in generalized fractional programming. *Zeitschrift fur Operations Research, 31*, 31–54.

Bowman, V. J. (1976). On the relationship of the Tchebycheff norm and the efficient frontier of multi-criteria objectives. In H. Thiriez, & S. Zionts (Eds.), *Multiple criteria decision making* (pp. 76–86). Berlin: Springer.

Branke, J., Deb, K., Miettinen, K., & Słowiński, R. (Eds.) (2008). *Multiobjective optimization—interactive and evolutionary approach*. Berlin, Heidelberg: Springer.

Buchanan, J. T., & Daellenbach, H. G. (1987). A comparative evaluation of interactive solution methods for multiple objective decision models. *European Journal of Operational Research, 29*, 353–359.

Caballero, R., Cerdá, E., Muñoz, M. M., Rey, L., & Stancu-Minasian, I. M. (2001). Efficient solution concepts and their relations in stochastic multiobjective programming. *Journal of Optimization Theory and Applications, 110*, 53–74.

Carlsson, C., & Fullér, R. (2002). *Fuzzy reasoning in decision making and optimization*. Heidelberg: Physica-Verlag.

Changkong, V., & Haimes, Y. Y. (1983). *Multiobjective decision making: Theory and methodology*. Amsterdam: North-Holland.

Charnes, A., & Cooper, W. W. (1959). Chance constrained programming. *Management Science, 6*, 73–79.

Charnes, A., & Cooper, W. W. (1961). *Management models and industrial applications of linear programming* (Vols. I and II). New York: Wiley.

Charnes, A., & Cooper, W. W. (1962). Programming with linear fractional functions. *Naval Research Logistic Quarterly, 9*, 181–186.

Charnes, A., & Cooper, W. W. (1963). Deterministic equivalents for optimizing and satisficing under chance constraints. *Operations Research, 11*, 18–39.

Charnes, A., & Cooper, W. W. (1977). Goal programming and multiple objective optimizations. *European Journal of Operational Research, 1*, 39–54.

Chen, L. -H., & Chen, H. -H. (2013). Considering decision decentralizations to solve bi-level multiobjective decision-making problems: A fuzzy approach. *Applied Mathematical Modelling, 37*, 6884–6898.

Cheng, H., Huang, W., Zhou, Q., & Cai, J. (2013). Solving fuzzy multi-objective linear programming problems using deviation degree measures and weighted max-min method. *Applied Mathematical Modelling, 37*, 6855–6869.

Choo, E. U., & Atkins, D. R. (1980). An interactive algorithm for multicriteria programming. *Computer and Operations Research, 7*, 81–87.

Chung, K. L. (1974). *Elementary probability theory with stochastic processes*. Berlin: Springer.

Chvatal, V. (1983). *Linear programming*. New York: W.H. Freeman and Company.

Cochrane, J. J., & Zeleny, M. (Eds.) (1973). *Multiple criteria decision making*. Columbia: University South Carolina Press.

Contini, B. (1968). A stochastic approach to goal programming. *Operations Research, 16*, 576–586.

Cramér, H. (1999). *Mathematical methods of statistics*. Princeton Landmarks in Mathematics and Physics. Princeton: Princeton University Press.

Dantzig, G. B. (1955). Linear programming under uncertainty. *Management Science, 1*, 197–206.

Dantzig, G. B. (1963). *Linear programming and extensions*. Princeton: Princeton University Press.

Dantzig, G. B., & Thapa, M. N. (1997). *Linear programming, 1: Introduction*. New York: Springer.

Davis, E., Freedman, M., Lane, J., McCall, B., Nestoriak, N., & Park, T. (2009). Product market competition and human resource practices in the retail food sector. *Industrial Relations, 48*, 350–371.

Delgado, M., Kacprzyk, J., Verdegay, J. L., & Vila, M. A. (Eds.) (1994). *Fuzzy optimization: Recent advances*. Heidelberg: Physica-Verlag.
Dempster, M. A. H. (Eds.) (1980). *Stochastic programming*. New York: Academic.
Dinkelbach, W. (1967). On nonlinear fractional programming. *Management Science Ser A, 13*, 492–498.
Doumpos, M., & Grigoroudis, E. (Eds.) (2013). *Multicriteria decision aid and artificial intelligence: Links, theory, and applications*. New York: Wiley.
Dubois, D., & Prade, H. (1978). Operations on fuzzy numbers. *International Journal of Systems Science, 9*, 613–626.
Dubois, D., & Prade, H. (1980). *Fuzzy sets and systems: Theory and application*. New York: Academic.
Ehrgott, M., & Gandibleux, X. (Eds.) (2002). *Multiple criteria optimization –state of the art annotated bibliographic surveys*. Boston: Kluwer.
Eskelinen, P., & Miettinen, K. (2012). Trade-off analysis approach for interactive nonlinear multiobjective optimization. *OR Spectrum, 34*, 803–816.
Everitt, R., & Ziemba, W. T. (1978). Twoperiod stochastic programs with simple recourse. *Operations Research, 27*, 485–502.
Faccio, M., Ferrari, E., Persona, A., & Vecchiato, P. (2013). Lean distribution principles to food logistics: a product category approach. *International Journal of Operational Research, 16*, 214–240.
Feller, W. (1968). *An introduction to probability theory and its applications* (3rd ed., Vol. 1). New York: Wiley.
Feller, W. (1978). *An introduction to probability theory and its applications* (2nd ed., Vol. 2). New York: Wiley.
Fiacco, A. V. (1983). *Introduction to sensitivity and stability analysis in nonlinear programming*. New York: Academic.
Fichefet, J. (1976). GPSTEM: an interactive multiobjective optimization method. *Progress in operations research* (Vol. 1, pp. 317–332). Amsterdam: North-Holland.
Fletcher, R. (1980). *Practical methods of optimization* (Vol. 2). New York: Wiley.
Fuji-Keizai (2003). *Map of regional supermarkets 2003*. Japan: Fuji-Keizai (in Japanese).
Gartska, S. J. (1980a). The economic equivalence of several stochastic programming models. In M. A. H. Dempster (Ed.), *Stochastic programming* (pp. 83–91). New York: Academic.
Gartska, S. J. (1980b). An economic interpretation of stochastic programs. *Mathematical Programming, 18*, 62–67.
Gartska, S. J., & Wets, R. J-B. (1974). On decision rules in stochastic programming. *Mathematical Programming, 7*, 117–143.
Garvin, W. W. (1960). *Introduction to linear programming*. New York: McGraw-Hill.
Gass, S. I. (1958). *Linear programming* (5th ed.). New York: McGraw-Hill (1985).
Geem, Z. W., Kim, J. H., & Loganathan, G. V. (2001). A new heuristic optimization algorithm: harmony search. *Simulation, 76*, 60–68.
Geoffrion, A. M. (1967). Stochastic programming with aspiration or fractile criteria. *Management Science, 13*, 672–679.
Geoffrion, A. M., Dyer, J. S., & Feinberg, A. (1972). An interactive approach for multicriterion optimization, with an application to the operation of an academic department. *Management Science, 19*, 357–368.
Gill, P. E., Murray, W., & Wright, M. H. (1981). *Practical optimization*. London: Academic.
Goicoecha, A., Hansen, D. R., & Duckstein, L. (1982). *Multiobjective decision analysis with engineering and business applications*. New York: Wiley.
Gorton, M., Sauer, J., & Supatpongkul, P. (2011). Wet markets, supermarkets and the "Big Middle" for food retailing in developing countries: Evidence from Thailand. *World Development, 39*, 1624–1637.
Guddat, J., Vasquez, F. G., Tammer, K., & Wendler K. (1985). *Multiobjective and stochastic optimization based on parametric optimization*. Berlin: Akademie-Verlag.

Hamdouch, Y. (2011). Multi-period supply chain network equilibrium with capacity constraints and purchasing strategies. *Transportation Research Part C: Emerging Technologies, 19*, 803–820.

Haimes, Y. Y., & Chankong V. (1979). Kuhn–Tucker multipliers as trade-offs in multiobjective decision-making analysis. *Automatica, 15*, 59–72.

Haimes, Y. Y., & Hall, W. A. (1974). Multiobjectives in water resources systems analysis: the surrogate worth trade-off method. *Water Resources Research, 10*, 614–624.

Haimes, Y. Y., Lasdon, L., & Wismer, D. (1971). On a bicriteria formulation of the problems of integrated system identification and system optimization. *IEEE Transactions on Systems, Man, and Cybernetics, SMC-1*, 296–297.

Hannan, E.L. (1980). Linear programming with multiple fuzzy goals. *Fuzzy Sets and Systems, 6*, 235–248.

Hadley, G. (1962). *Linear programming*. Massachusetts: Addison-Wesley.

Hillier, F. S., & Lieberman, G. J. (1990). *Introduction to mathematical programming*. New York: McGraw-Hill.

Hulsurkar, S., Biswal, M. P., & Sinha, S. B. (1997). Fuzzy programming approach to multiobjective stochastic linear programming problems. *Fuzzy Sets and Systems, 88*, 173–181.

Hwang, C. L., & Masud, A. S. M. (1979). *Multiple objective decision making: Methods and applications*. Berlin: Springer.

Ignizio, J. P. (1976). *Goal programming and extensions*. Lexington: Lexington Books, D. C. Heath and Company.

Ingnizio, J. P. (1982). *Linear programming in single and multiple objective systems*. Englewood Cliffs, New Jersey: Prentice-Hall.

Ignizio, J. P. (1983). Generalized goal programming: an overview. *Computer and Operations Research, 10*, 277–289.

Ingnizio, J. P., & Cavalier, T. M. (1994). *Linear programming*. Englewood Cliffs, New Jersey: Prentice-Hall.

Ijiri, Y. (1965). *Management goals and accounting for control*. Amsterdam: North-Holland.

Johnson, N. L., & Kotz, S. (1972). *Distributions in statistics: Continuous multivariate distributions*. New York: Wiley.

Kacprzyk, J., & Orlovski S. A. (Eds.) (1987). *Optimization models using fuzzy sets and possibility theory*. Dordrecht: D. Reidel Publishing Company.

Kahraman, C. (Ed.) (2008). *Fuzzy multi-criteria decision making: Theory and applications with recent developments*. New York: Springer.

Kaliszewski, I., Miroforidis, J., & Podkopaev, D. (2012). Interactive multiple criteria decision making based on preference driven evolutionary multiobjective optimization with controllable accuracy. *European Journal of Operational Research, 216*, 188–199.

Kall, P. (1976). *Stochastic linear programming*. Berlin: Springer.

Kall, P., & Mayer, J. (2005). *Stochastic linear programming: Models, theory, and computation*. New York: Springer.

Kataoka, S. (1963). A stochastic programming model. *Econometrica, 31*, 181–196.

Kaufmann, A., & Gupta, M. (1991). *Introduction to fuzzy arithmetic*. New York: Van Nostrand Reinhold.

Kickert, W. J. M. (1978). *Fuzzy theories on decision-making*. Leiden: Martinus Nijhoff.

Kidachi, M. (2006). Evolutionand development of retailing-oriented distribution systems. In Kidate, Tatsuma (Eds.), *Theory, history and analysis on distributive trades* (pp. 133–174). Tokyo: Chuo University Press (in Japanese).

Klein, G., Moskowitz, H., & Ravindran, A. (1990). Interactive multiobjective optimization under uncertainty. *Management Science, 36*, 58–75.

Köksalan, M., Wallenius, J., & Zionts, S. (2011). *Multiple criteria decision making: From early history to the 21st century*. New Jersey, World Scientific.

Kornbluth, J. S. H., & Steuer, R. E. (1981). Goal programming with linear fractional criteria. *European Journal of Operational Research, 8*, 58–65.

Kotz, S., Balakrishnan, N., & Johnson, N. L. (2000). *Continuous multivariate distributions, volume 1, models and applications* (2nd ed.). New York: Wiley.
Kruse, R., & Meyer, K. D. (1987). *Statistics with vague data*. Dordrecht: D. Reidel Publishing Company.
Kuhn, H. W., & Tucker, A. W. (1951). Nonlinear programming. In J. Neyman (Ed.), *Proceedings of the second berkeley symposium on mathematical statistics and probability* (pp. 481–492). California: University of California Press.
Kumano-Hongu (2005). http://www2.w-shokokai.or.jp/hongu/shokokai-seinenbu-kabu.html
Lai, Y. J., & Hwang, C. L. (1992). *Fuzzy mathematical programming*. Berlin: Springer.
Lai, Y. J., & Hwang, C. L. (1994). *Fuzzy multiple objective decision making: Methods and applications*. Berlin: Springer.
Lasdon, L. S. (1970). *Optimization theory for large systems*. New York: Macmillan.
Leberling, H. (1980). On finding compromise solution in multicriteria problems using the fuzzy min-operator. *Fuzzy Sets and Systems, 6*, 105–118.
Lee, S. M. (1972). *Goal programming for decision analysis*. Philadelphia: Auerbach.
Leclercq, J. -P. (1982). Stochastic programming: an interactive multicriteria approach. *European Journal of Operational Research, 10*, 33–41.
Lemke, C. E. (1965). The dual method of solving the linear programming problem. *Naval Research Logistics Quarterly, 1*, 36–47.
Liu, B., & Iwamura, K. (1998). Chance constrained programming with fuzzy parameters. *Fuzzy Sets and Systems, 94*, 227–237.
Louveaux, F. V. (1980). A solution method for multistage stochastic programs with recourse with applications to an energy investment problem. *Operations Research, 27*, 889–902.
Luenberger, D. G. (1973). *Linear and nonlinear programming* (2nd ed. (1984), 3rd ed. (2008)). California: Addison-Wesley.
Luque, M., Ruiz, F., & Cabello, J. M. (2012). A synchronous reference point-based interactive method for stochastic multiobjective programming. *OR Spectrum, 34*, 763–784.
Luque, M., Ruiz, F., & Miettinen, K. (2011). Global formulation for interactive multiobjective optimization. *OR Spectrum, 33*, 27–48.
Luhandjula, M. K. (1984). Fuzzy approaches for multiple objective linear fractional optimization. *Fuzzy Sets and Systems, 13*, 11–23.
Luhandjula, M. K. (1987). Multiple objective programming problems with possibilistic coefficients. *Fuzzy Sets and Systems, 21*, 135–145.
Luhandjula, M. K. (1996). Fuzziness and randomness in an optimization framework. *Fuzzy Sets and Systems, 77*, 291–297.
Luhandjula, M. K. (2006). Fuzzystochastic linear programming: Survey and future research directions *European Journal of Operational Research, 174*, 1353–1367.
Luhandjula, M. K., & Gupta, M. M. (1996). On fuzzy stochastic optimization. *Fuzzy Sets and Systems, 81*, 47–55.
Mangasarian, O. L. (1969). *Nonlinear programming*. New York: McGraw-Hill.
March, J. G., & Simon H. A. (1958). *Organizations*. New York: Wiley.
Massa, S., & Testa, S. (2011). Beyond the conventional-specialty dichotomy in food retailing business models: an Italian case study. *Journal of Retailing and Consumer Services, 18*, 476–482.
Miettinen, K. (1999). *Nonlinear multiobjective optimization*. Boston: Kluwer.
Miettinen, K., Ruiz, F., & Wierzbicki, A. P. (2008). Introduction to multiobjective optimization: interactive approaches. In J. Branke, K. Deb, K. Miettinen, & R. Słowiński (Eds.), *Multiobjective optimization: Interactive and evolutionary approach* (pp. 25–57). Berlin, Heidelberg: Springer.
Miller, B. L., & Wagner, H. M. (1965). Chance constrained programming with joint constraints. *Operations Research, 3*, 930–945.
Ministry of Agriculture, Forestry and Fisheries of Japan (2008). *Summary of report on price formation in each stage of food distribution 2008*, Statistics of agriculture, forestry and fisheries, Ministry of Agriculture, Forestry and Fisheries of Japan.

Mohan, C., & Nguyen, H. T. (2001). An interactive satisficing method for solving multiobjective mixed fuzzy-stochastic programming problems. *Fuzzy Sets and Systems, 117*, 61–79.

Nering, E. D., & Tucker, A. W. (1993). *Linear programs and related problems*. Boston: Academic.

Nikulin, Y., Miettinen, K., & Mäkelä, M. M. (2012). A new achievement scalarizing function based on parameterization in multiobjective optimization. *OR Spectrum, 34*, 69–87.

Nishizaki, I., & Sakawa, M. (2001). *Fuzzy and multiobjective games for conflict resolution*. Heidelberg: Physica-Verlag.

Nowak, M. (2007). Aspiration level approach in stochastic MCDM problems. *European Journal of Operational Research, 177*, 1626–1640.

Orlovski, S. A. (1984). Multiobjective programming problems with fuzzy parameters. *Control and Cybernetics, 13*, 175–183.

Paksoy, T., & Chang, C. -T. (2010). Revised multi-choice goal programming for multi-period, multi-stage inventory controlled supply chain model with popup stores in Guerrilla marketing. *Applied Mathematical Modelling, 34*, 3586–3898.

Pierre, D. A., & Lowe, M. J. (1975). *Mathematical programming via augmented lagrangians*. London: Addison-Wesley Publishing Company.

Pishvaee, M. S., & Razmi, J. (2011). Environmental supply chain network design using multi-objective fuzzy mathematical programming. *Applied Mathematical Modelling, 36*, 3433–3446.

Perkgoz, C., Kato, K., Katagiri, H., & Sakawa, M. (2004). An interactive fuzzy satisficing method for multiobjective stochastic integer programming problems through variance minimization model. *Scientiae Mathematicae Japonicae, 60*, 327–336.

Perkgoz, C., Sakawa, M., Kato, K., & Katagiri, H. (2005). An interactive fuzzy satisficing method for multiobjective stochastic integer programming problems through a probability maximization model. *Asia Pacific Management Review: An International Journal, 10*, 29–35.

Powell, M. J. D. (1983). Variable metric methods for constrained optimization. In A. Bachem, M. Grotschel, & B. Korte (Eds.), *Mathematical programming: The state of the art* (pp. 288–311). New York: Springer.

Prekopa, A. (1995). *Stochastic programming*. Dordrecht: Kluwer.

Prekopa, A. (1970). Onprobabilistic constrained programming. *Mathematical Programming Study, 28*, 113–138.

Rommelfanger, H. (1990). FULPAL: an interactive method for solving multiobjective fuzzy linear programming problems. In R. Słowìnski, & J. Teghem (Eds.), *Stochastic versus fuzzy approaches to multiobjective mathematical programming under uncertainty* (pp. 279–299). Dordrecht: D. Reidel Publishing Company.

Rommelfanger, H. (1996). Fuzzy linear programming and applications. *European Journal of Operational Research, 92*, 512–527.

Rommelfanger, H. (2007). A general concept for solving linear multicriteria programming problems with crisp, fuzzy or stochastic variables. *Fuzzy Sets and Systems, 158*, 1892–1904.

Ruiz, F., Luque, M., & Miettinen, K. (2012). Improving the computational efficiency in a global formulation (GLIDE) for interactive multiobjective optimization. *Annals of Operations Research, 197*, 47–70.

Ruiz, F., Luque, M., Miguel, F., & del Mar Muñoz, M. (2008). An additive achievement scalarizing function for multiobjective programming problems. *European Journal of Operational Research, 188*, 683–694.

Sakawa, M. (1981). An interactive computer program for multiobjective decision making by the sequential proxy optimization technique. *International Journal of Man-Machine Studies, 14*, 193–213.

Sakawa, M. (1983). Interactive computer programs for fuzzy linear programming with multiple objectives. *International Journal of Man-Machine Studies, 18*, 489–503.

Sakawa, M. (1984b). Interactive fuzzy decision making for multiobjective nonlinear programming problems. In M. Grauer, & A. P. Wierzbicki (Eds.), *Interactive decision analysis* (pp. 105–112). New York: Springer.

Sakawa, M. (1986). Interactive multiobjective decision-making using fuzzy satisficing intervals and its application to an urban water resources system. *Large Scale Systems, 10*, 203–213.

Sakawa, M. (1993). *Fuzzy sets and interactive multiobjective optimization*. New York: Plenum Press.
Sakawa, M. (2000). *Large scale interactive fuzzy multiobjective programming*. Heidelberg: Physica-Verlag.
Sakawa, M. (2001). *Genetic algorithms and fuzzy multiobjective optimization*. Boston: Kluwer.
Sakawa, M. (2002). Fuzzy multiobjective and multilevel optimization. In M. Ehrgott, & X. Gandibleux (Eds.), *Multiple criteria optimization: State of the art annotated bibliographic surveys* (pp. 171–226). Boston: Kluwer.
Sakawa, M. (2013). Fuzzy multiobjective optimization. In M. Doumpos, & E. Grigoroudis (Eds.), *Multicriteria decision aid and artificial intelligence: Links, theory, and applications* (pp. 235–271). New York: Wiley.
Sakawa, M., & Katagiri, H. (2010). Interactive fuzzy programming based on fractile criterion optimization model for two-level stochastic linear programming problems. *Cybernetics and Systems, 41*, 508–521.
Sakawa, M., & Katagiri, H. (2012). Stackelberg solutions for fuzzy random two-level linear programming through level sets and fractile criterion optimization. *Central European Journal of Operations Research, 20*, 101–117.
Sakawa, M., Katagiri, H., & Kato, K. (2001). An interactive fuzzy satisficing method for multiobjective stochastic linear programming problems using fractile criterion model. *The 10th IEEE international conference on fuzzy systems* (Vol. 3, pp. 25–31), Melbourne.
Sakawa, M., Katagiri, H., & Matsui, T. (2011). Interactive fuzzy random two-level linear programming through fractile criterion optimization. *Mathematical and Computer Modelling, 54*, 3153–3163.
Sakawa, M., Katagiri, H., & Matsui, T. (2012a). Stackelberg solutions for fuzzy random two-level linear programming through probability maximization with possibility. *Fuzzy Sets and Systems, 188*, 45–57.
Sakawa, M., Katagiri, H., & Matsui, T. (2012b). Interactive fuzzy stochastic two-level integer programming through fractile criterion optimization. *Operational Research: An International Journal, 12*, 209–227.
Sakawa, M., Katagiri, H., & Matsui, T. (2012c). Fuzzy random bilevel linear programming through expectation optimization using possibility and necessity. *International Journal of Machine Learning and Cybernetics, 3*, 183–192.
Sakawa, M., Katagiri, H., & Matsui, T. (2012d). Stackelberg solutions for fuzzy random bilevel linear programming through level sets and probability maximization. *Operational Research: An International Journal, 12*, 271–286.
Sakawa, M., & Kato, K. (2000). Integer programming through genetic algorithms with double strings based on reference solution updating. *Proceedings of 2000 IEEE international conference on industrial electronics, control and instrumentation* (pp. 2744–2749), Nagoya.
Sakawa, M., & Kato, K. (2002). An interactive fuzzy satisficing method for multiobjective stochastic linear programming problems using chance constrained conditions. *Journal of Multi-Criteria Decision Analysis, 11*, 125–137.
Sakawa, M., & Kato, K. (2003). Genetic algorithms with double strings for 0–1 programming problems. *European Journal of Operational Research, 144*, 581–597.
Sakawa, M., & Kato, K. (2008). Interactive fuzzy multi-objective stochastic linear programming. In C. Kahraman (Ed.), *Fuzzy multi-criteria decision making: Theory and applications with recent developments* (pp. 375–408). New York: Springer.
Sakawa, M., & Kato, K. (2009a). Interactive fuzzy programming for stochastic two-level linear programming problems through probability maximization. *International institute for applied systems analysis (IIASA), Interim report*, IR-09-013.
Sakawa, M., & Kato, K. (2009b). Interactive fuzzy random two-level linear programming through fractile criterion optimization. *International institute for applied systems analysis (IIASA), Interim report*, IR-09-020.

Sakawa, M., & Kato, K. (2009c). Fuzzy random noncooperative two-level linear programming through absolute deviation minimization using possibility and necessity. *International institute for applied systems analysis (IIASA), Interim report*, IR-09-021.

Sakawa, M., Kato, K., Azad, M. A. K., & Watanabe, R. (2005). A genetic algorithm with double string for nonlinear integer programming problems. *Proceedings of the 2005 IEEE international conference on systems, man and cybernetics* (pp. 3281–3286), Hawaii.

Sakawa, M., Kato, K., & Katagiri, H. (2002). An interactive fuzzy satisficing method through a variance minimization model for multiobjective linear programming problems involving random variables. *Knowledge-based intelligent information engineering systems and allied technologies KES2002* (pp. 1222–1226), Crema.

Sakawa, M., Kato, K., & Katagiri, H. (2004). An interactive fuzzy satisficing method for multiobjective linear programming problems with random variable coefficients through a probability maximization model. *Fuzzy Sets and Systems, 146*, 205–220.

Sakawa, M., Kato, K., Katagiri, H., & Wang, J. (2003). Interactive fuzzy programming for two-level linear programming problems involving random variable coefficients through a probability maximization model. In T. Bilgic, B. Baets, & O. Kaynak (Eds.), *Fuzzy sets and systems, IFSA 2003: 10th international fuzzy systems association world congress (Lecture notes in computer science)* (pp. 555–558). Istanbul: Springer.

Sakawa, M., Kato, K., & Nishizaki, I. (2003). An interactive fuzzy satisficing method for multiobjective stochastic linear programming problems through an expectation model. *European Journal of Operational Research, 145*, 665–672.

Sakawa, M., Kato, K., Nishizaki, I., & Wasada, K. (2001). An interactive fuzzy satisficing method for multiobjective stochastic linear programs through simple recourse model. *Joint 9th IFSA world congress and 20th NAFIPS international conference* (pp. 59–64), Vancouver.

Sakawa, M., Kato, K., Shibano, T., & Hirose, K. (2000). Genetic algorithms with double strings for multidimensional integer knapsack problems. *Journal of Japan Society for Fuzzy Theory and Systems, 12*, 562–569 (in Japanese).

Sakawa, M., Kato, K., Sunada, H., & Shibano, T. (1997). Fuzzy programming for multiobjective 0-1 programming problems through revised genetic algorithms. *European Journal of Operational Research, 97*, 149–158.

Sakawa, M., & Matsui, T. (2012a). Stackelberg solutions for random fuzzy two-level linear programming through possibility-based probability model. *Expert Systems with Applications, 39*, 10898–10903.

Sakawa, M., & Matsui, T. (2012b). Interactivefuzzy programming for random fuzzy two-level programming problems through possibility-based fractile model. *Expert Systems with Applications, 39*, 12599–12604.

Sakawa, M., & Matsui, T. (2012c). An interactivefuzzy satisficing method for multiobjective stochastic integer programming with simple recourse. *Applied Mathematics, 3*, 1245–1251.

Sakawa, M., & Matsui, T. (2013a). Interactive fuzzy programming for stochastic two-level linear programming problems through probability maximization. *Artificial Intelligence Research, 2*, 109–124.

Sakawa, M., & Matsui, T. (2013b). Interactive fuzzy random cooperative two-level linear programming through level sets based probability maximization. *Expert Systems with Applications, 40*, 1400–1406.

Sakawa, M., & Matsui, T. (2013c). Interactive fuzzy programming for fuzzy random two-level linear programming problems through probability maximization with possibility. *Expert Systems with Applications, 40*, 2487–2492.

Sakawa, M., & Matsui, T. (2013d). Interactive fuzzy random two-level linear programming based on level sets and fractile criterion optimization. *Information Sciences, 238*, 163–175.

Sakawa, M., & Matsui, T. (2013e). Interactive random fuzzy two-level programming through possibility-based probability model. *Information Sciences, 239*, 191–200.

Sakawa, M., & Matsui, T. (2013f). Fuzzy multiobjective nonlinear operation planning in district heating and cooling plants. *Fuzzy Sets and Systems, 231*, 58–69.

Sakawa, M., & Matsui, T. (2013g). Fuzzy random noncooperative two-level linear programming through fractile models with possibility and necessity. *Engineering Optimization, 45*, 811–833.

Sakawa, M., & Matsui, T. (2013h). Interactive fuzzy multiobjective stochastic programming with simple recourse. *International Journal of Multicriteria Decision Making* (in press).

Sakawa, M., & Matsui, T. (2013i). Interactive fuzzy programming for random fuzzy two-level integer programming problems through fractile criteria with possibility. *Applied Mathematics, 4*, 34–43.

Sakawa, M., & Matsui, T. (2013j). Random fuzzy bilevel linear programming through possibility-based fractile model. *International Journal of Machine Learning and Cybernetics* (in press).

Sakawa, M., & Nishizaki, I. (2002a). Interactivefuzzy programming for decentralized two-level linear programming problems. *Fuzzy Sets and Systems, 125*, 301–315.

Sakawa, M., & Nishizaki, I. (2002b). Interactive fuzzy programming for two-level nonconvex programming problems with fuzzy parameters through genetic algorithms. *Fuzzy Sets and Systems, 127*, 185–197.

Sakawa, M., & Nishizaki, I. (2009). *Cooperative and noncooperative multi-level programming*. New York: Springer.

Sakawa, M., & Nishizaki, I. (2012). Interactive fuzzy programming for multi-level programming problems: a review. *International Journal of Multicriteria Decision Making, 2*, 241–266.

Sakawa, M., Nishizaki, I., & Katagiri, H. (2011). *Fuzzy stochastic multiobjective programming*. New York: Springer.

Sakawa, M., Nishizaki, I., & Matsui, T. (2013). Fuzzy and multiobjective purchase and transportation planning for food retailing: case study in Japan. *International Journal of Multicriteria Decision Making, 3*, 277–300.

Sakawa, M., Nishizaki, I., Matsui, T., & Hayashida, T. (2012). A two-level purchase problem for food retailing in Japan. *American Journal of Operations Research, 2*, 482–494.

Sakawa, M., Nishizaki, I., Matsui, T., & Hayashida, T. (2013a). Purchase and transportation planning for food retailing in Japan. *Asia Pacific Management Review, 18*, 79–92.

Sakawa, M., Nishizaki, I., Matsui, T., & Hayashida, T. (2013b). Multi-store food retailing problem with outsourcing purchase operation: a case study in Japan. *International Journal of Operational Research* (in press).

Sakawa, M., Nishizaki, I., & Uemura, Y. (1998). Interactive fuzzy programming for multi-level linear programming problems. *Computers and Mathematics with Applications, 36*, 71–86.

Sakawa, M., Nishizaki, I., & Uemura, Y. (2000). Interactive fuzzy programming for multi-level linear programming problems with fuzzy parameters. *Fuzzy Sets and Systems, 109*, 3–19.

Sakawa, M., Nishizaki, I., & Uemura, Y. (2002). A decentralized two-level transportation problem in a housing material manufacturer –Interactive fuzzy programming approach. *European Journal of Operational Research, 141*, 167–185.

Sakawa, M., & Seo, F. (1980). Interactive multiobjective decision making for large-scale systems and its application to environmental systems. *IEEE Transactions on Systems, Man and Cybernetics, SMC-10*, 796–806.

Sakawa, M., & Shibano, T. (1996). Interactive fuzzy programming for multiobjective 0–1 programming problems through genetic algorithms with double strings. In Da Ruan (Ed.), *Fuzzy logic foundations and industrial applications* (pp. 111–128). Boston: Kluwer.

Sakawa, M., & Yano, H. (1985a). Interactive fuzzy decision-making for multi-objective nonlinear programming using reference membership intervals. *International Journal of Man-Machine Studies, 23*, 407–421.

Sakawa, M., & Yano, H. (1985b). Interactive decision making for multiobjective linear fractional programming problems with fuzzy parameters. *Cybernetics and Systems: An International Journal, 16*, 377–394.

Sakawa, M., & Yano, H. (1985c). Interactive fuzzy satisficing method using augmented minimax problems and its application to environmental systems. *IEEE Transactions on Systems, Man and Cybernetics, SMC-15*, 720–729.

Sakawa, M., & Yano, H. (1986a). An interactive fuzzy decisionmaking method using constraint problems. *IEEE Transactions on Systems, Man, and Cybernetics, SMC-16*, 179–182.

Sakawa, M., & Yano, H. (1986b). *An interactive fuzzy satisficing method for multiobjective nonlinear programming problems with fuzzy parameters CP-86-15*. Laxenburg, Austria: International Institute for Applied Systems Analysis.

Sakawa, M., & Yano, H. (1986c). Interactive fuzzy decision making for multiobjective nonlinear programming using augmented minimax problems. *Fuzzy Sets and Systems, 20*, 31–43.

Sakawa, M., & Yano, H. (1986d). Interactive decision making for multiobjective linear programming problems with fuzzy parameters. In G. Fandel, M. Grauer, A. Kurzhanski, & A. P. Wierzbicki (Eds.), *Large-scale modeling and interactive decision analysis* (pp. 88–96). New York: Springer.

Sakawa, M., & Yano, H. (1987). An interactive satisficing method for multiobjective nonlinear programming problems with fuzzy parameters. In M. Kacprzyk, & S. A. Orlovski (Eds.), *Optimization models using fuzzy sets and possibility theory* (pp. 258–271). Dordrecht: D. Reidel Publishing Company.

Sakawa, M., & Yano, H. (1988). An interactive fuzzy satisficing method for multiobjective linear fractional programming problems. *Fuzzy Sets and Systems, 28*, 129–144.

Sakawa, M., & Yano, H. (1989). Interactive decision making for multiobjective nonlinear programming problems with fuzzy parameters. *Fuzzy Sets and Systems, 29*, 315–326.

Sakawa, M., & Yano, H. (1990). An interactive fuzzy satisficing method for generalized multiobjective linear programming problems with fuzzy parameters. *Fuzzy Sets and Systems, 35*, 125–142.

Sakawa, M., & Yano, H. (1991). Feasibility and Pareto optimality for multiobjective nonlinear programming problems with fuzzy parameters. *Fuzzy Sets and Systems, 43*, 1–15.

Sakawa, M., Yano, H., & Takahashi, J. (1992). Pareto optimality for multiobjective linear fractional programming problems with fuzzy parameters. *Information Sciences, 63*, 33–53.

Sakawa, M., Yano, H., & Yumine, T. (1987). An interactive fuzzy satisficing method for multiobjective linear-programming problems and its application. *IEEE Transactions on Systems, Man, and Cybernetics, SMC-17*, 654–661.

Sakawa, M., & Yauchi, K. (1998). Coevolutionary genetic algorithms for nonconvex nonlinear programming problems: Revised GENOCOP III. *Cybernetics and Systems: An International Journal, 29*, 885–899.

Sakawa, M., & Yauchi, K. (1999). An interactive fuzzy satisficing method for multiobjective nonconvex programming through floating point genetic algorithms. *European Journal of Operational Research, 117*, 113–124.

Sakawa, M., & Yauchi, K. (2000). Interactive decision making for multiobjective nonconvex programming problems with fuzzy parameters through coevolutionary genetic algorithms. *Fuzzy Sets and Systems, 114*, 151–165.

Sakawa, M., & Yumine, T. (1983). Interactive fuzzy decision-making for multiobjective linear fractional programming problems. *Large Scale Systems, 5*, 105–113.

Salam, A., Bandaly, D., & Defersha, F. M. (2011). Optimising the design of a supply chain network with economies of scale using mixed integer programming. *International Journal of Operational Research, 10*, 398–415.

Sengupta, J. (1972). *Stochastic programming: Methods and applications*. Amsterdam: North-Holland.

Seo, F., & Sakawa, M. (1988). *Multiple criteria decision analysis in regional planning: Concepts, methods and applications*. Dordrecht: D. ReidelPublishing Company.

Słowìnski, R. (Ed.) (1998). *Fuzzy sets in decision analysis, operations research and statistics*. Dordrecht, Boston, London: Kluwer.

Słowìnski, R., & Teghem, J. (1988). Fuzzy vs. stochastic approaches to multicriteria linear programming under uncertainty. *Naval Research Logistics, 35*, 673–695.

Słowìnski, R., & Teghem, J. (Eds.) (1990). *Stochastic versus fuzzy approaches to multiobjective mathematical programming under uncertainty*. Dordrecht, Boston, London: Kluwer.

Sommer, G., & Pollastschek, M. A. (1978). A fuzzy programming approach to an air pollution regulation problem. In R. Trappl, G. J. Klir, & L. Ricciardi (Eds.), *Progress in cybernetics and systems research*, (pp. 303–323). John Wiley & Sons, New York.

Spronk, J. (1981). *Interactive multiple goal programming: Applications to financial planning*. Boston: Martinus Nijhoff Publishing.
Stancu-Minasian, I. M. (1984). *Stochastic programming with multiple objective functions*. Dordrecht: D. Reidel Publishing Company.
Stancu-Minasian, I. M. (1990). Overview of different approaches for solving stochastic programming problems with multiple objective functions. In R. Słowinski, & J. Teghem (Eds.), *Stochastic versus fuzzy approaches to multiobjective mathematical programming under uncertainty* (pp. 71–101). Dordrecht, Boston, London: Kluwer.
Stancu-Minasian, I. M. (1992). *Fractional programming*, Dordrecht, Boston, London: Kluwer.
Stancu-Minasian, I. M., & Wets, M. J. (1976). A research bibliography in stochastic programming, 1955–1975. *Operations Research, 24*, 1078–1119.
Stadler, W. (1979). A survey of multicriteria optimization or the vector maximum problem, Part 1: 1776–1960. *Journal of Optimization Theory and Applications, 29*, 1–52.
Steuer, R. E. (1986). *Multiple criteria optimization: Theory, computation, and application*. New York: Wiley.
Steuer, R. E., & Choo, E. U. (1983). An interactive weighted Tchebycheff procedure for multiple objective programming. *Mathematical Programming, 26*, 326–344.
Symonds, G. H. (1968). Chance-constrained equivalents of stochastic programming problems. *Operations Research, 16*, 1152–1159.
Taleizadeh, A. A., Niaki, S. T. A., & Barzinpour, F. (2011). Multiple-buyer multiple-vendor multi-product multi-constraint supply chain problem with stochastic demand and variable lead-time: a harmony search algorithm. *Applied Mathematics and Computation, 217*, 9234–9253.
Taleizadeh, A. A., Niaki, S. T. A., & Makui, A. (2012). Multiproduct multiple-buyer single-vendor supply chain problem with stochastic demand, variable lead-time, and multi-chance constraint. *Expert Systems with Applications, 39*, 5338–5348.
Teghem, J., Dufrane, D., Thauvoye, M., & Kunsch, P. (1986). STRANGE: an interactive method for multi-objective linear programming under uncertainty. *European Journal of Operational Research, 26*, 65–82.
Thie, P. R. (1988). *An introduction to linear programming and game theory*. New York: Wiley.
Tintner, G. (1955). Stochastic linear programming with applications to agricultural economics. *Proceedings of 2nd symposium on linear programming* (pp. 192–228), Washington.
Urli, B., & Nadeau, R., (1990). Stochastic MOLP with incomplete information: an interactive approach with recourse. *Journal of the Operational Research Society, 41*, 1143–1152.
Urli, B., & Nadeau, R. (2004). PROMISE/scenarios: an interactive method for multiobjective stochastic linear programming under partial uncertainty. *European Journal of Operational Research, 155*, 361–372.
Vajda, S. (1972). *Probabilistic programming*. New York: Academic.
Vanderpooten, D., & Vincke, P. (1989). Description and analysis of some representative interactive multicriteria procedures. *Mathematical and Computer Modelling, 12*, 1221–1238.
Verdegay, J. L., & Delgado, M. (Eds.) (1989). *The interface between artificial intelligence and operations research in fuzzy environment*. Köln: Verlag TÜV Rheinland.
Wagner, H. M. (1955). On the distribution of solutions in linear programming problems. *Journal of the American Statistical Association, 53*, 161–163.
Walkup, D. W., & Wets, R. (1967). Stochastic programs with recourse. *SIAM Journal on Applied Mathematics, 15*, 139–162.
Wets, R. (1966). Programming under uncertainty: the complete problem. *Zeitschrift für Wahrscheinlichkeitstheorie und Verwandte Gebiete, 4*, 316–339.
Wets, R. (1974). Stochastic programs with fixed recourse: the equivalent deterministic program. *SIAM Review, 16*, 309–339.
Wets, R. (1996). Challenges in stochastic programming. *Mathematical Programming, 75*, 115–135.
White, D. J. (1982). *Optimality and efficiency*. New York: Wiley.
Wierzbicki, A. P. (1977). Basic properties of scalarizing functionals for multiobjective optimization. *Mathematische Operations-Forshung und Statistik, Ser. Optimization, 8*, 55–60.

Wierzbicki, A. P. (1979). *A methodological guide to multiobjective optimization, WP-79-122.* Laxenburg, Austria: International Institutefor Applied Systems Analysis.

Wierzbicki, A. P. (1980). The use of reference objectives in multiobjective optimization. In G. Fandel, & T. Gal (Eds.), *Multiple criteria decision making: Theory and application* (pp. 468–486). Berlin: Springer.

Wierzbicki, A. P. (1982). A mathematical basis for satisficing decision making. *Mathematical Modeling, 3*, 391–405.

Williams, A. C. (1965). On stochastic linear programming. *SIAM Journal Applied Mathematics, 13*, 927–940.

Wolfe, P. (1959). The simplex method for quadratic programming. *Econometrica, 27*, 382–398.

Yano, H. (2011). Fuzzy approaches for multiobjective stochastic linear programming problems considering both probability maximization and fractile optimization. *Proceedings of 2011 IEEE international conference on fuzzy systems* (pp. 1866–1873).

Yano, H. (2012). Interactive fuzzy decision making for multiobjective stochastic linear programming problems with variance-covariance matrices. *Proceedings of 2012 IEEE international conference on systems, man, and cybernetics* (pp. 97–102).

Yano, H. (2012b). An interactive fuzzy satisficing method for multiobjective stochastic linear programming problems considering both probability maximization and fractile optimization. In J. Watada, T. Watanabe, G. P-Wren, R. J. Howlett, & L. C. Jain (Eds.), *Intelligent decision technologies, proceedings of the 4th international conference on intelligent decision technologies* (pp. 119–128). New York: Springer.

Yano, H., & Sakawa, M. (1989). Interactive fuzzy decision making for generalized multiobjective linear fractional programming problems with fuzzy parameters. *Fuzzy Sets and Systems, 32*, 245–261.

Yano, H., & Sakawa, M. (2009). A fuzzy approach to hierarchical multiobjective programming problems and its application to an industrial pollution control problem. *Fuzzy Sets and Systems, 160*, 3309–3322.

Yano, H., & Sakawa, M. (2012). Interactive multiobjective fuzzy random linear programming through fractile criteria. *Advanced in Fuzzy Systems, 2012*, 1–9.

Yu, P. L. (1985). *Multiple-criteria decision making: Concepts, techniques, and extensions.* New York: Plenum Press.

Yu, P. L. (1973). A class of solutions for group decision problems. *Management Science, 19*, 936–946.

Zadeh, L. A. (1963). Optimality and non-scalar valued performance criteria. *IEEE Transactions on Automatic Control, AC-8*, 59–60.

Zadeh, L. A. (1965). Fuzzy sets. *Information and Control, 8*, 338–353.

Zadeh, L. A. (1968). Probability measure of fuzzy events. *Journal of Mathematical Analysis and Applications, 23*, 421–427.

Zadeh, L. A. (1978). Fuzzy sets as a basis for a theory of possibility. *Fuzzy Sets and Systems, 1*, 3–28.

Zangwill, W. I. (1969). *Nonlinear programming: A unified approach.* Englewood Cliffs: Prentice-Hall.

Zeleny, M. (1973). Compromise programming. In J. L. Chochrane, & M. Zeleny (Eds.), *Multiple criteria decision making* (pp. 262–301). Columbia, SC: University of South Carolina Press.

Zeleny, M. (1974). *Linear multiobjective programming*, New York: Springer.

Zeleny, M. (1976). The theory of displaced ideal. In M. Zeleny (Ed.), *Multiple criteria decision making, Kyoto, 1975* (pp. 153–206). New York: Springer.

Zeleny, M. (1982). *Multiple criteria decision making.* New York: McGraw-Hill.

Zhang, J. -L., & Zhang, M. -Y. (2011). Supplier selection and purchase problem with fixed cost and constrained order quantities under stochastic demand. *International Journal of Production Economics, 129*, 1–7.

Zimmermann, H. -J. (1976). Description and optimization of fuzzy systems. *International Journal of General Systems, 2*, 209–215.

Zimmermann, H. -J. (1978). Fuzzy programming and linear programming with several objective functions. *Fuzzy Sets and Systems, 1*, 45–55.
Zimmermann, H. -J. (1985). *Fuzzy set theory and its application* (2nd ed. (1991)). Dordrecht: Kluwer.
Zimmermann, H. -J. (1987). *Fuzzy sets, decision-making and expert systems*. Boston: Kluwer.
Zionts, S., & Wallenius, J. (1976). An interactive programming method for solving the multiple criteria problem. *Management Science, 22*, 652–663.
Zionts, S., & Wallenius, J. (1983). An interactive multiple objective linear programming method for a class of underlying nonlinear utility functions. *Management Science, 29*, 519–529.

Index

A
α-level set, 109, 132
α-multiobjective linear programming problem, 133
α-Pareto optimal solution, 133
anti-ideal point, 93
artificial variable, 28
assured probability level, 182
augmented minimax problem, 207, 225
augmented system of equations, 14, 18
axioms of probability, 150

B
basic feasible solution, 17
basic form, 18
basic solution, 17
basic variable, 17
basis, 17, 40
basis matrix, 17, 40, 41
Bayes' theorem, 152
bell-shaped membership function, 112
bisection method, 127, 175, 202, 212
Bland's rule, 37

C
canonical form, 18, 55
central limit theorem, 159
chance constrained condition, 198
chance constrained problem, 188
chance constrained programming, 164
chance constraint, 259
characteristic function, 106
circling, 36
complement, 108, 149
complementary slackness, 55
complete optimal solution, 75
compromise solution, 92
conditional distribution function, 156
conditional expectation, 156
conditional probability, 151
constraint method, 81
continuous random variable, 153
convex fuzzy decision, 113
convex fuzzy set, 110
convex programming, 284
convex programming problem, 162
correlation coefficient, 157
correlation matrix, 158
cost coefficient, 13
covariance, 157
cycling, 36

D
decomposition theorem, 110
density function, 152
deviational variable, 89
diet problem, 10, 32, 53, 58, 61
Dinkelbach algorithm, 178
discrete random variable, 152
displaced ideal, 93
distribution function, 152
dual feasible canonical form, 56
dual problem, 49
dual simplex method, 55
dual variable, 49
duality, 49

E
economic interpretation, 53
enlarged basis inverse matrix, 41
enlarged basis matrix, 41
equality constraint, 13

equally likely, 149
event, 149
expectation, 153
expectation level, 206
expectation model, 168, 199, 259

F
Farkas's theorem, 54
feasible region, 206
feasible solution, 17
fractile model, 168, 181, 217, 262
fuzzy constraint, 113, 251
fuzzy decision, 113, 251
fuzzy equal, 124
fuzzy goal, 113, 123, 200, 206, 251
fuzzy inequality, 116
fuzzy linear programming, 115, 251
fuzzy max, 124
fuzzy min, 124
fuzzy multiobjective linear programming, 119
fuzzy multiobjective stochastic programming, 257
fuzzy number, 111
fuzzy parameter, 131
fuzzy set, 105

G
Gauss distribution, 153
general fuzzy multiobjective problem, 125
general linear goal programming, 90
generalized α-multiobjective linear programming problem, 134
global minimum point, 284
goal programming, 87
Gordon's theorem, 55

I
ideal point, 92
independent, 151
interactive fuzzy multiobjective linear programming, 123
interactive fuzzy satisficing method, 201, 203, 208, 213, 220, 255
interactive multiobjective linear programming, 96, 249
intersection, 108, 149

J
joint density function, 155
joint distribution, 154
joint distribution function, 155

K
Kuhn-Tucker conditions, 186, 187, 286

L
ℓ_1 norm, 88
Lagrange multiplier, 186, 285
Lagrangian function, 140, 186, 187, 285
linear programming problem, 1
linearly independent, 16
local minimum point, 284

M
M-α-Pareto optimal solution, 135
M-Pareto optimal solution, 125, 201
M-Pareto optimality, 200
M-Pareto optimality test, 202
marginal distribution, 154
marginal price, 54
marginal probability mass function, 155
maximizing decision, 113
mean, 153
membership function, 105
minimax problem, 97, 137, 249, 255, 260, 263, 267
minimum operator, 121, 201
multi-stage programming, 160
multiobjective chance constrained programming, 197
multiobjective linear programming problem, 4, 74, 244
multiobjective simple recourse optimization, 222
multiobjective stochastic programming problem, 198
multivariate normal distribution, 159
mutually exclusive, 149

N
nonbasis, 40
nondegenerate basic feasible solution, 17
noninferior solution, 123
nonlinear programming, 283
nonnegativity condition, 13
normal distribution, 153
normal random variable, 154, 218
null (empty) set, 149

O
objective function, 12
one-sided goal programming, 89

Index

optimal basic form, 21
optimal canonical form, 21
optimal solution, 17
optimal tableau, 21
optimal value, 17
optimality criterion, 21
over-achievement, 89

P

Pareto optimal solution, 76
Pareto optimality test, 86, 98, 128
pivot element, 22
pivot operation, 23
positive semidefinite, 162
preemptive priority, 89
primal problem, 49
primal variable, 49
primal-dual pair, 49
probability, 149
probability density function, 161
probability mass function, 152
probability model, 168, 174, 210, 265
product fuzzy decision, 114, 121
product operator, 201
production planning problem, 10, 26, 46, 64, 73, 100, 117, 130, 162
profit coefficient, 13
purchase and transportation planning, 233

R

random variable, 152
reduced cost coefficient, 20
redundant, 16
reference membership level, 126, 137, 201, 225, 255, 260
reference point method, 96
relative cost coefficient, 20, 42
revised dual simplex method, 60
revised simplex method, 42
revised simplex tableau, 45
right-hand side constant, 13

S

sample space, 149
satisficing probability level, 165, 198, 210, 268
satisficing solution, 87
scalarization method, 77
sensitivity analysis, 63, 239
sequential quadratic programming method, 162
shadow price, 54
simple recourse problem, 161, 188, 223
simplex criterion, 21
simplex multiplier, 41, 53, 99

simplex tableau, 18
slack variable, 15
standard deviation, 153
standard form, 12
standard normal random variable, 166, 175, 211, 218
standardized normal distribution, 154
standardized normal random variable, 154
standardized random variable, 153
STEP method, 96
stochastic linear programming problem, 164
strong duality theorem, 51
strong law of large numbers, 159
support, 107
supporting hyperplane, 80
surplus variable, 15
symmetric primal–dual pair, 50

T

tangent hyperplane, 99
target value, 175, 210, 268
target variable, 182, 218
total probability formula, 151
trade-off rate, 81, 100, 129, 140, 203
triangular membership function, 111
two-phase method, 31
two-stage model, 222
two-stage programming, 160

U

under achievement, 89
union, 108, 149

V

variance, 153
variance model, 168, 171, 205
variance-covariance matrix, 158, 166, 197
vector-minimization, 75

W

weak duality theorem, 51
weak Pareto optimal solution, 76
weighted ℓ_1 norm, 88
weighted minimax method, 83
weighting coefficient, 78
weighting method, 78
weighting problem, 247

Z

Zimmermann method, 204, 209

MIX
Papier aus verantwortungsvollen Quellen
Paper from responsible sources
FSC® C105338

If you have any concerns about our products,
you can contact us on
ProductSafety@springernature.com

In case Publisher is established outside the EU,
the EU authorized representative is:
**Springer Nature Customer Service Center GmbH
Europaplatz 3, 69115 Heidelberg, Germany**

Printed by Libri Plureos GmbH
in Hamburg, Germany